Lecture Notes in Mathematics 1997

Editors:
J.-M. Morel, Cachan
F. Takens, Groningen
B. Teissier, Paris

Markus Banagl

Intersection Spaces, Spatial Homology Truncation, and String Theory

 Springer

Markus Banagl
University of Heidelberg
Mathematics Institute
Im Neuenheimer Feld 288
69120 Heidelberg
Germany
banagl@mathi.uni-heidelberg.de

ISBN: 978-3-642-12588-1   e-ISBN: 978-3-642-12589-8
DOI: 10.1007/978-3-642-12589-8
Springer Heidelberg Dordrecht London New York

Lecture Notes in Mathematics ISSN print edition: 0075-8434
                              ISSN electronic edition: 1617-9692

Library of Congress Control Number: 2010928327

Mathematics Subject Classification (2000): 55N33; 57P10; 14J17; 81T30; 55P30; 55S36; 14J32; 14J33

Cover design: SPi Publisher Services

Printed on acid-free paper

springer.com

Preface

The primary concern of the work presented here is Poincaré duality for spaces that are not manifolds, but are still put together from manifolds that form the strata of a stratification of the space. Goresky and MacPherson's intersection homology [GM80, GM83], see also [B⁺84, KW06, Ban07], associates to a stratified pseudomanifold X chain complexes $IC_*^{\bar{p}}(X;\mathbb{Q})$ depending on a perversity parameter $\bar{p}$, whose homology $IH_*^{\bar{p}}(X;\mathbb{Q}) = H_*(IC_*^{\bar{p}}(X;\mathbb{Q}))$ satisfies generalized Poincaré duality across complementary perversities when X is closed and oriented. L^2-cohomology [Che80, Che79, Che83] associates to a triangulated pseudomanifold X equipped with a suitable conical Riemannian metric on the top stratum a differential complex $\Omega_{(2)}^*(X)$, the complex of differential L^2-forms ω on the top stratum of X such that $d\omega$ is L^2 as well, whose cohomology $H_{(2)}^*(X) = H^*(\Omega_{(2)}^*(X))$ satisfies Poincaré duality (at least when X has no strata of odd codimension; in more general situations one must choose certain boundary conditions). The linear dual of $IH_*^{\bar{m}}(X;\mathbb{R})$ is isomorphic to $H_{(2)}^*(X)$, by integration. In the present work, we adopt the "spatial philosophy" outlined in the announcement [Ban09], maintaining that a theory of Poincaré duality for stratified spaces benefits from being implemented on the level of spaces, with passage to coarser filters such as chain complexes, homology or homotopy groups occurring as late as possible in the course of the development. Thus we pursue here the following program. To a stratified pseudomanifold X, associate *spaces*

$$I^{\bar{p}}X,$$

the *intersection spaces of* X, such that the *ordinary* homology $\widetilde{H}_*(I^{\bar{p}}X;\mathbb{Q})$ satisfies generalized Poincaré duality when X is closed and oriented. If X has no odd-codimensional strata and $\bar{p}$ is the middle perversity $\bar{p} = \bar{m}$, then we are thus assigning to a singular pseudomanifold a (rational) Poincaré complex. The resulting homology theory $X \rightsquigarrow \widetilde{H}_*(I^{\bar{p}}X)$ is not isomorphic to intersection homology or L^2-cohomology. In fact, it solves a problem in type II string theory related to the existence of massless D-branes, which is neither solved by ordinary homology nor by intersection homology. We show that while $IH_*^{\bar{m}}(X)$ is the correct theory in the realm of type IIA string theory (giving the physically correct counts of massless particles), $\widetilde{H}_*(I^{\bar{m}}X)$ is the correct theory in the realm of type IIB string theory. In other words, the two theories $IH_*^{\bar{m}}(X), \widetilde{H}_*(I^{\bar{m}}X)$ form a mirror pair in the sense of mirror symmetry in algebraic geometry. We will return to these considerations in more detail later in this preface.

The assignment $X \rightsquigarrow I^{\bar{p}}X$ should satisfy the following requirements:

1. $\tilde{H}_*(I^{\bar{p}}X; \mathbb{Q})$ should satisfy generalized Poincaré duality across complementary perversities.
2. $\tilde{H}_*(I^{\bar{p}}X; \mathbb{Q})$ should be a mirror of $IH_*^{\bar{m}}(X; \mathbb{Q})$ in the sense of mirror symmetry.
3. $X \rightsquigarrow I^{\bar{p}}X$ should be as "natural" as possible.
4. X should be modified as little as possible (only near the singularities; the homotopy type away from the singularities should be completely preserved).
5. If X is a finite cell complex, then $I^{\bar{p}}X$ should again be a finite cell complex.
6. $X \rightsquigarrow I^{\bar{p}}X$ should be homotopy-theoretically tractable, so as to facilitate computations.

Note that full naturality in (3) with respect to all continuous maps is too much to expect, since a corresponding property cannot be achieved for intersection homology either. In order to demonstrate (6), we have worked out numerous examples throughout the text, giving concrete intersection spaces for pseudomanifolds ranging from toy examples to complex algebraic threefolds and Calabi–Yau conifolds arising in mathematical physics. In the present monograph, we carry out the above program for pseudomanifolds with isolated singularities as well as, more generally, for two-strata spaces with arbitrary bottom stratum but trivial link bundle. In addition, we make suggestions for how to proceed when there are more than two strata, or when the link bundle is twisted. Future research will have to determine the ultimate domain of pseudomanifolds for which an intersection space is definable. Throughout the general development of the theory, we assume the links of singular strata to be simply connected. In concrete applications, this assumption is frequently unnecessary, see also the paragraph preceding Example 2.15. In the example, we discuss the intersection space of a concrete space whose links are not simply connected. Our construction of intersection spaces is of a homotopy-theoretic nature, resting on technology for spatial homology truncation, which we develop in this book. This technology is completely general, so that it may be of independent interest.

What are the purely mathematical advantages of introducing intersection spaces? Algebraic Topology has developed a vast array of functors defined on spaces, many of which do not factor through chain complexes. For instance, let E_* be any generalized homology theory, defined by a spectrum E, such as K-theory, L-theory, stable homotopy groups, bordism and so on. One may then study the composite assignment

$$X \rightsquigarrow I^{\bar{p}}E_*(X) := E_*(I^{\bar{p}}X).$$

Section 2.7, for example, studies symmetric L-homology, where E_* is given by Ranicki's symmetric L-spectrum $E = \mathbb{L}^\bullet$. We show in Corollary 2.38 that capping with the $\mathbb{L}^\bullet$-homology fundamental class of an n-dimensional oriented compact pseudomanifold X with isolated singularities indeed induces a Poincaré duality isomorphism

$$\tilde{H}^0(I^{\bar{m}}X; \mathbb{L}^\bullet) \otimes \mathbb{Q} \xrightarrow{\cong} \tilde{H}_n(I^{\bar{n}}X; \mathbb{L}^\bullet) \otimes \mathbb{Q}.$$

K-theory is discussed in Section 2.8. A $\bar{p}$-*intersection vector bundle on* X may be defined as an actual vector bundle on $I^{\bar{p}}X$. More generally, given any structure group G, one may define *principal intersection G-bundles over* X as homotopy classes of maps $I^{\bar{p}}X \to BG$. In Example 2.40, we show that there are infinitely many distinct seven-dimensional pseudomanifolds X, whose tangent bundle elements in the KO-theory $\widetilde{KO}(X - \text{Sing})$ of their nonsingular parts do not lift to $\widetilde{KO}(X)$, but do lift to $\widetilde{KO}(I^{\bar{n}}X)$, where $\bar{n}$ is the upper middle perversity. So this framework allows one to formulate the requirement that a pseudomanifold have a $\bar{p}$-intersection tangent bundle, and by varying $\bar{p}$, the severity of this requirement can be adjusted at will. Ultimately, one may want to study the Postnikov tower of $I^{\bar{p}}X$ and view it as a "$\bar{p}$-intersection Postnikov tower" of X.

A further asset of the spatial philosophy is that cochain complexes will automatically come equipped with internal multiplications, making them into differential graded algebras (DGAs). The Goresky–MacPherson intersection chain complexes $IC_*^{\bar{p}}(X)$ are generally not algebras, unless $\bar{p}$ is the zero-perversity, in which case $IC_*^{\bar{0}}(X)$ is essentially the ordinary cochain complex of X. (The Goresky–MacPherson intersection product raises perversities in general.) Similarly, the differential complex $\Omega_{(2)}^*(X)$ of L^2-forms on $X - \text{Sing}$ is not an algebra under wedge product of forms because the product of two L^2-functions need not be L^2 anymore (consider for example $r^{-1/3}$ for small $r > 0$). Using the intersection space framework, the ordinary cochain complex $C^*(I^{\bar{p}}X)$ of $I^{\bar{p}}X$ *is* a DGA, simply by employing the ordinary cup product. For similar reasons, the cohomology of $I^{\bar{p}}X$ is by default endowed with internal cohomology operations, which do not exist for intersection cohomology. These structures, along with Massey triple products and other secondary and higher order operations, remain to be investigated elsewhere. Operations in intersection cohomology that weaken the perversity by a factor of two have been constructed in [Gor84].

In Section 2.6, we construct cap products of the type

$$\widetilde{H}^r(I^{\bar{m}}X) \otimes \widetilde{H}_i(X) \xrightarrow{\cap} \widetilde{H}_{i-r}(I^{\bar{n}}X). \tag{0.1}$$

These products have their applications not only in formulating and proving duality statements, but also in developing various characteristic class formulae, which may lead to extensions of the results of [BCS03, Ban06a]. An $\bar{m}$-intersection vector bundle on X has Chern classes in $H^{\text{even}}(I^{\bar{m}}X)$. Characteristic classes of pseudomanifolds, such as the L-class, generally lie only in $H_*(X; \mathbb{Q})$ and do not lift to intersection homology or to $H_*(I^{\bar{m}}X; \mathbb{Q})$, see for example [GM80, Ban06b]. Consequently, the ordinary cap product $H^r(I^{\bar{m}}X) \otimes H_i(I^{\bar{m}}X) \to H_{i-r}(I^{\bar{m}}X)$ is useless in multiplying the Chern classes of the bundle and the characteristic classes of the pseudomanifold. The above product (0.1) then enables one to carry out such a multiplication. The product (0.1) seems counterintuitive from the point of view of intersection homology because an analogous product

$$IH^r(X) \otimes H_i(X) \dashrightarrow IH_{i-r}(X)$$

on intersection homology cannot exist. The motivational Section 2.6.1 explains why the desired product cannot exist for intersection homology but does exist for intersection space homology. The products themselves are constructed in Section 2.6.3.

Let us briefly indicate how intersection spaces are constructed. We are guided initially by mimicking spatially what intersection homology does algebraically. By Mayer-Vietoris sequences, the overall behavior of intersection homology is primarily controlled by its behavior on cones. If L is a closed n-dimensional manifold, $n > 0$, then

$$IH_r^{\bar{p}}(\overset{\circ}{\text{cone}}(L)) \cong \begin{cases} H_r(L), & r < n - \bar{p}(n+1), \\ 0, & \text{otherwise,} \end{cases}$$

where $\overset{\circ}{\text{cone}}(L)$ denotes the open cone on L and we are using intersection homology built from finite chains. Thus, intersection homology is a process of truncating the homology of a space algebraically above some cut-off degree given by the perversity and the dimension of the space. This is also apparent from Deligne's formula for the intersection chain sheaf. The task at hand is to implement this spatially. Let $\mathbf{C}$ be a category of spaces, that is, a category with a functor $i : \mathbf{C} \to \mathbf{Top}$ to the category $\mathbf{Top}$ of topological spaces and continuous maps. (For instance, $\mathbf{C}$ might be a subcategory of $\mathbf{Top}$ and i the inclusion functor, but it might also be spaces endowed with extra structure with i the forgetful functor, etc.) Let $p : \mathbf{Top} \to \mathbf{HoTop}$ be the natural projection functor to the homotopy category of spaces, sending a continuous map to its homotopy class. Suppose then that we had a functor

$$t_{<k} : \mathbf{C} \longrightarrow \mathbf{HoTop},$$

where k is a positive integer, together with a natural transformation $\text{emb}_k : t_{<k} \to pi$ (think of pi as the "identity functor") such that

$$\text{emb}_k(L)_* : H_r(t_{<k}(L)) \longrightarrow H_r(pi(L))$$

is an isomorphism for $r < k$, while $H_r(t_{<k}(L)) = 0$ for $r \geq k$, for all objects L in $\mathbf{C}$. We refer to such a functor as a *spatial homology truncation functor*. Let X be an n-dimensional closed pseudomanifold with one isolated singular point. Such an X is of the form

$$X = M \cup_{\partial M = L} \text{cone}(L),$$

where L, a closed manifold of dimension $n - 1$, is the link of the singularity, and M, a compact manifold with boundary $\partial M = L$, is the complement of a small open cone-neighborhood of the singularity. Assume that L gives rise to an object L in $\mathbf{C}$. The intersection space $I^{\bar{p}}X$ is defined to be the homotopy cofiber of the composition

$$t_{<k}(L) \xrightarrow{\text{emb}_k(L)} pi(L) = L = \partial M \hookrightarrow M,$$

where $k = n - 1 - \bar{p}(n)$, see Definition 2.10. In other words: we attach the cone on a suitable spatial homology truncation of the link to the exterior of the singularity along the boundary of the exterior. The two extreme cases of this construction arise when $k = 1$ and when k is larger than the dimension of the link. In the former case, $t_{<1}(L)$ is a point (at least when L is path connected) and thus $I^{\bar{p}}X$ is homotopy equivalent to the nonsingular part $X - \text{Sing}$ of X. In the latter case no actual truncation has to be performed, $t_{<k}(L) = L$, $\text{emb}_k(L)$ is the identity map and thus $I^{\bar{p}}X = X$. If there are several isolated singularities, then we perform spatial homology truncation on each of the links. If the singularities are not isolated, a process of fiberwise spatial homology truncation applied to the link bundle has to be used, see Section 2.9. If there are more than two nested strata, then more elaborate homotopy colimit constructions involving iterated truncation techniques can be used.

Theorem 2.12 establishes generalized Poincaré duality for the rational homology of intersection spaces and simultaneously analyzes the relation to intersection homology, both in the isolated singularity case. This relation is of a "reflective" nature (which is also responsible for both theories being mirrors of each other in the context of singular Calabi–Yau threefolds). The requisite abstract language of reflective diagrams is introduced in Section 2.1. Of particular interest here is to understand what happens at the cut-off degree k, which is the middle dimension for the middle perversity. The reflective diagram shows that while $IH_k^{\bar{p}}(X)$ is generally smaller than both $H_k(X - \text{Sing})$ and $H_k(X)$, being a quotient of the former and a subgroup of the latter, $H_k(I^{\bar{p}}X)$, on the other hand, is generally bigger than both $H_k(X - \text{Sing})$ and $H_k(X)$, containing the former as a subgroup and mapping to the latter surjectively. Section 3.9 contains an example of a singular quintic S (a conifold) in $\mathbb{P}^4$ such that $H_3(IS)$ has rank 204, but $IH_3(S)$ has only rank 2. Corollary 2.14 computes the difference of the Euler characteristics of the two theories. As far as Witt groups are concerned, both theories lead to equivalent intersection forms: We prove in Theorem 2.28 that for a pseudomanifold X of dimension $n = 4m$, the symmetric intersection form on $IH_{2m}^{\bar{m}}(X)$ and the symmetric intersection form on $H_{2m}(I^{\bar{m}}X)$ determine the same element in the Witt group of the rationals. In particular, the signature of the two forms are equal. Definition 2.41 contains the construction of $I^{\bar{p}}X$ for a space X with a positive dimensional singular stratum with untwisted link bundle. Theorem 2.47 establishes generalized Poincaré duality in this context.

As our approach relies on the ability to perform spatial homology truncation, Chapter 1 is devoted to a systematic investigation of this problem. The investigation and results are of a general nature and can be read and used independently of any interest in intersection spaces. Throughout the development, we strive to remain firmly on the plane of elementary homotopy theory, using only classical instruments, working unstably, avoiding simplicial or model theoretic language, as such language does not seem to yield any particular advantage here. Our spaces in this chapter will be simply connected CW-complexes because, just as Hilton [Hil65] does, we wish to avail ourselves of the Hurewicz and the Whitehead theorem. Spatial homology truncation *on the object level* has been studied by several researchers: the Eckmann–Hilton dual of the Postnikov decomposition is the homology decomposition (or Moore space decomposition) of a space, see [Hil65, BJCJ59, Moo]. It seems that the problem has not received much attention on the morphism level; see, however, [Bau88] for a tower of categories. Consequently, we focus on aspects of functoriality, and this is where homology truncation turns out to be harder than Postnikov truncation because obstructions surface that do not arise in the Postnikov picture. Given a space X, let $p_n(X) : X \to P_n(X)$ denote a stage-n Postnikov approximation for X. If $f : X \to Y$ is any map, then there exists, uniquely up to homotopy, a map $p_n(f) : P_n(X) \to P_n(Y)$ such that

$$
\begin{array}{ccc}
X & \xrightarrow{\;\;f\;\;} & Y \\
{\scriptstyle p_n(X)}\downarrow & & \downarrow{\scriptstyle p_n(Y)} \\
P_n(X) & \xrightarrow{p_n(f)} & P_n(Y)
\end{array}
$$

homotopy commutes. In the introductory Section 1.1.1 we give an example that shows that this property does not Eckmann–Hilton dualize to spatial homology truncation. Thus a homology truncation functor in this naive sense cannot exist. Our solution proposes to consider spaces endowed with an extra structure. Morphisms should preserve this extra structure; one obtains a category $\mathbf{CW}_{n \supset \partial}$. What is this extra structure? Hilton's homology decomposition really depends on a choice of complement to the group of n-cycles inside of the nth chain group. Such a complement always exists and pairs (space, choice of complement) are the objects of $\mathbf{CW}_{n \supset \partial}$; morphisms are cellular maps that map the complement chosen for the domain to the complement chosen for the codomain. The Compression Theorem 1.32 shows that such morphisms can always be compressed into spatial homology truncations. The upshot at this stage is that we obtain a covariant assignment

$$
t_{<n} : \mathbf{CW}_{n \supset \partial} \longrightarrow \mathbf{HoCW}_{n-1}
$$

of objects and morphisms into the rel $(n-1)$-skeleton homotopy category of CW-complexes together with a natural transformation emb_n from $t_{<n}$ to the identity, such that for every object (K, Y) of $\mathbf{CW}_{n \supset \partial}$, where K is a simply connected CW-complex and Y a complement as discussed above,

$$
\mathrm{emb}_n(K, Y)_* : H_r(t_{<n}(K, Y)) \longrightarrow H_r(K)
$$

is an isomorphism for $r < n$ and $H_r(t_{<n}(K, Y)) = 0$ for $r \geq n$, see the first part of Theorem 1.41. (Note that we do not at this stage claim that $t_{<n}$ is a functor on all of $\mathbf{CW}_{n \supset \partial}$.) This solves the first order problem of the existence of compressions of maps. Immediately, the second order problem of the uniqueness of compressions presents itself. Example 1.9 shows that even when domain and codomain of a map f have unique homological n-truncations and f does have a homological n-truncation, the homotopy class of that truncation may not be uniquely determined by f. The obvious idea of imposing the above requirement of complement-preservation also on homotopies and then just applying the Compression Theorem 1.32 to compress the homotopy into spatial homology truncations does not work. We call a map n-*compression rigid*, if its compression into n-truncations

agrees with f on the $(n-1)$-skeleton and is unique up to rel $(n-1)$-skeleton homotopy, see Definition 1.33 and Proposition 1.34. Example 1.35 exposes a map that is not compression rigid, even though its domain and codomain have unique n-truncations. As an instrument for understanding compression rigidity, we introduce virtual cell groups VC_n of a space, so named because they are homotopy groups which are not themselves cellular chain groups, but they sit naturally between two actual cellular chain groups of certain cylinders. The virtual cell groups come equipped with an endomorphism so that we can formulate the concept of a 1-eigenclass (or eigenclass for short) for elements of VC_n. We show that a map is compression rigid if and only if the homotopies coming from the homotopy commutativity of the transformation square associated to emb_n can be chosen to be eigenclasses in VC_n. For 2-connected spaces, virtual cell groups are computed in Proposition 1.18. An obstruction theory for compression rigidity is set up in Section 1.2. Case studies of compression rigid categories are presented in Section 1.3. The second part of Theorem 1.41 asserts that the covariant assignment $t_{<n}$ is a *functor* on n-compression rigid subcategories of $\mathbf{CW}_{n \supset \partial}$. The dependence of the spatial homology truncation $t_{<n}(K,Y)$ on Y is discussed by Proposition 1.25, Scholium 1.26, Proposition 1.27 and Corollaries 1.30, 1.31. Proposition 1.25 gives a necessary and sufficient condition for $t_{<n}(K,Y)$ and $t_{<n}(K,\overline{Y})$ to be homotopy equivalent rel $(n-1)$-skeleton, where $Y, \overline{Y}$ are two choices of complements. Section 1.4 deals with the truncation of homotopy equivalences, Section 1.5 with the truncation of inclusions, and Section 1.6 with iterated truncation. In Section 1.7, we investigate spatial homology truncation followed by localization at odd primes. Theorem 1.61 establishes that this composite assignment $t_{<n}^{(odd)}$ is a functor on 2-connected spaces. The key ingredients here are the compression rigidity obstruction theory together with Proposition 1.50, which calculates a pertinent homotopy group and shows that it is all 2-torsion.

There are important classes of spaces where no complement Y has to be chosen and the compression rigidity obstructions vanish. We study one such class in detail, namely spaces with vanishing odd-dimensional homology. We refer to this class as the *interleaf category*, **ICW**. It includes for instance simply connected 4-manifolds, smooth compact toric varieties, homogeneous spaces arising as the quotient of a complex simply connected semisimple Lie group by a parabolic subgroup (e.g. flag manifolds, Grassmannians), and smooth Schubert varieties. A truncation functor $t_{<n} : \mathbf{ICW} \to \mathbf{HoCW}$ and cotruncation functor $t_{\geq n} : \mathbf{ICW} \to \mathbf{HoCW}$ are defined. Mostly, but not exclusively, in the context of the interleaf category, we investigate continuity properties of the homology truncation of homeomorphisms. We show in Theorem 1.78 that truncation of cellular self-homeomorphisms of an interleaf space is a continuous H-map into the grouplike topological monoid of self-homotopy equivalences of the homology truncation of the space. In Section 1.11, we discuss fiberwise homology truncation for mapping tori (general simply connected fiber), flat bundles over spaces whose fundamental group G has a $K(G,1)$ of dimension at most 2 (for example flat bundles over closed surfaces other than $\mathbb{R}P^2$; again for general simply connected fiber), and fiber bundles over a sphere of dimension greater than 1, with interleaf fiber.

Since spatial homology truncation of a space L in general requires making a choice of a certain type of subgroup Y in the nth chain group of L in order to obtain an object (L, Y) in $\mathbf{CW}_{n \supset \partial}$, and since the construction of intersection spaces uses this truncation on the links L of singularities, the homotopy type of the intersection space $I^{\bar{p}}X$ may well depend, to some extent, on choices. We show (Theorem 2.18) that the rational homology of $I^{\bar{p}}X$ is well-defined and independent of choices. Furthermore, we give sufficient conditions, in terms of the homology of the links in X and the homology of $X - \text{Sing}$, for the integral homology of $I^{\bar{p}}X$ in the cut-off degree to be independent of choices. Away from the cut-off degree, the integral homology is always independent of choices. The conditions are often satisfied in algebraic geometry for the middle perversity, for instance when X is a complex projective algebraic threefold with isolated hypersurface singularities that are weighted homogeneous and "well-formed," see Theorem 2.24. This class of varieties includes in particular conifolds, to be discussed below. Theorem 2.26 asserts that the homotopy type of $I^{\bar{p}}X$ is well-defined independent of choices when all the links are interleaf spaces.

It was mentioned before that the homology of intersection spaces addresses certain questions in type II string theory – let us expand on this. Our viewpoint is informed by [GSW87, Str95, Hüb97]. In addition to the four dimensions that model space–time, string theory requires six dimensions for a string to vibrate. Due to supersymmetry considerations, these six dimensions must be a Calabi–Yau space, but this still leaves a lot of freedom. It is thus important to have mechanisms to move from one Calabi–Yau space to another. A topologist's take on this might be as follows, disregarding the Calabi–Yau property for a moment. Since any two 6-manifolds are bordant ($\Omega_6^{SO} = 0$) and since, by Morse theory, any bordism is obtained by performing a finite sequence of surgeries, surgery is not an unreasonable vessel to travel between 6-manifolds. Note also that every three-dimensional homology class in a simply connected smooth 6-manifold can be represented, by the Whitney embedding theorem, by an embedded 3-sphere with trivial normal bundle. Physicists' *conifold transition* starts out with a nonsingular Calabi–Yau threefold, passes to a singular variety (the conifold) by a deformation of complex structure, and arrives at a different nonsingular Calabi–Yau threefold by a small resolution of singularities. The deformation collapses embedded 3-spheres to isolated singular points, whose link is $S^3 \times S^2$. The resolution resolves the singular points by replacing each one with a $\mathbb{C}P^1$. As we review in Section 3.6, massless particles in four dimensions should be recorded as classes by good cohomology theories for Calabi–Yau varieties. In type IIA string theory, there are charged twobranes present that wrap around the $\mathbb{C}P^1$ 2-cycles and that become massless when those 2-cycles are collapsed to points by the resolution map, see Section 3.5. We show that intersection homology accounts for all of these massless twobranes and thus is the physically correct homology theory for type IIA string theory. However, in type IIB string theory, there are charged threebranes present that wrap around the 3-spheres and that become massless when those 3-spheres are collapsed to points by the deformation of complex structure. Neither the ordinary homology of the conifold, nor its intersection homology (or L^2-cohomology) accounts for these massless threebranes. In

Proposition 3.6 we prove that the homology of the intersection space of the conifold yields the correct count of these threebranes. From this point of view, the homology of intersection spaces appears to be a physically suitable homology theory in the IIB regime. The theory in particular answers a question posed by [Hüb97] in this regard. Given a Calabi–Yau threefold M, the mirror map associates to it another Calabi–Yau threefold W such that type IIB string theory on $\mathbb{R}^4 \times M$ corresponds to type IIA string theory on $\mathbb{R}^4 \times W$. If M and W are nonsingular, then $b_3(W) = (b_2 + b_4)(M) + 2$ and $b_3(M) = (b_2 + b_4)(W) + 2$ for the Betti numbers of ordinary homology. The preceding discussion suggests that if M and W are singular, $\mathcal{H}_*^{\mathrm{IIA}}$ is a type IIA D-brane-complete homology theory with Poincaré duality, and $\mathcal{H}_*^{\mathrm{IIB}}$ is a type IIB D-brane-complete homology theory with Poincaré duality, then one should expect that

$$
\begin{aligned}
\mathrm{rk}\,\mathcal{H}_3^{\mathrm{IIA}}(M) &= \mathrm{rk}\,\mathcal{H}_2^{\mathrm{IIB}}(W) + \mathrm{rk}\,\mathcal{H}_4^{\mathrm{IIB}}(W) + 2, \\
\mathrm{rk}\,\mathcal{H}_3^{\mathrm{IIA}}(W) &= \mathrm{rk}\,\mathcal{H}_2^{\mathrm{IIB}}(M) + \mathrm{rk}\,\mathcal{H}_4^{\mathrm{IIB}}(M) + 2, \\
\mathrm{rk}\,\mathcal{H}_3^{\mathrm{IIB}}(M) &= \mathrm{rk}\,\mathcal{H}_2^{\mathrm{IIA}}(W) + \mathrm{rk}\,\mathcal{H}_4^{\mathrm{IIA}}(W) + 2, \text{ and} \\
\mathrm{rk}\,\mathcal{H}_3^{\mathrm{IIB}}(W) &= \mathrm{rk}\,\mathcal{H}_2^{\mathrm{IIA}}(M) + \mathrm{rk}\,\mathcal{H}_4^{\mathrm{IIA}}(M) + 2.
\end{aligned}
$$

Corollary 3.14 establishes that this is indeed the case for $\mathcal{H}_*^{\mathrm{IIA}}(-) = IH_*(-)$ and $\mathcal{H}_*^{\mathrm{IIB}}(-) = H_*(I-)$ when M and W are conifolds. Thus $(IH_*(-), H_*(I-))$ is a mirror-pair in this sense. Intersection homology and the homology of intersection spaces reveal themselves as the two sides of one coin.

Prerequisites. In Chapter 1, we assume that the reader is acquainted with the elementary homotopy theory of CW complexes [Whi78, Hil53, Hat02]. In Chapter 2, a rudimentary knowledge of stratification theory, pseudomanifolds, and intersection homology is useful. In addition to the references already mentioned in the beginning of this preface, the reader may wish to consult [GM88, Wei94, Sch03, Pfl01]. A geometric understanding of intersection homology in terms of PL or singular chains is sufficient. Sheaf-theoretic methods are neither used nor required in this book. Regarding Chapter 3, we have made an attempt to collect in Sections 3.1–3.6 all the background material from string theory that we need for our predominantly mathematical arguments in Sections 3.7–3.9. Specific competence in, say, quantum field theory, is not required to read this chapter.

Notation and Conventions. Our convention for the mapping cylinder $Y \cup_f X \times I$ of a map $f : X \to Y$ is that the attaching is carried out at time 1, that is, the points of $X \times \{1\} \subset X \times I$ are attached to Y using f. For products in cohomology and homology, we will use the conventions of Spanier's book [Spa66]. In particular, for an inclusion $i : A \subset X$ of spaces and elements $\xi \in H^p(X)$, $x \in H_n(X, A)$, the formula $\partial_*(\xi \cap x) = i^*\xi \cap \partial_* x$ holds for the connecting homomorphism ∂_* (no sign). For the compatibility between cap- and cross-product, one has the sign

$$
(\xi \times \eta) \cap (x \times y) = (-1)^{p(n-q)}(\xi \cap x) \times (\eta \cap y),
$$

where $\xi \in H^p(X)$, $\eta \in H^q(Y)$, $x \in H_m(X)$, and $y \in H_n(Y)$.

Acknowledgments. We would like to thank physicists Andreas Braun, Arthur Hebecker and Hagen Triendl for several discussions on the string theoretic aspects of this work, mathematicians Dominique Arlettaz, Christian Ausoni and Gérald Gaudens for pointing us to various references on Moore space decompositions, Sylvain Cappell for pointing out future applications of the theory, and Florian Gaisendrees for eliminating some typographical errors. Furthermore, we thank the Deutsche Forschungsgemeinschaft for supporting this research.

Heidelberg,   *Markus Banagl*
February 2010

Contents

Chapter 1
Homotopy Theory

1.1 The Spatial Homology Truncation Machine

1.1.1 Introduction

The Eckmann–Hilton dual of the Postnikov decomposition of a space is the *homology decomposition* (or *Moore space decomposition*) ([Zab76, page 44], [Hil65, BJCJ59, Moo]) of a space. Let us give a brief review of this decomposition, based on dualizing the Postnikov decomposition.

A *Postnikov decomposition* for a simply connected CW-complex X is a commutative diagram

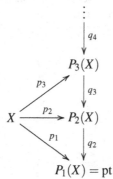

such that $p_{n*} : \pi_r(X) \to \pi_r(P_n(X))$ is an isomorphism for $r \le n$ and $\pi_r(P_n(X)) = 0$ for $r > n$. Let F_n be the homotopy fiber of q_n. Then the exact sequence

$$\pi_{r+1}(P_nX) \xrightarrow{q_{n*}} \pi_{r+1}(P_{n-1}X) \to \pi_r(F_n) \to \pi_r(P_nX) \xrightarrow{q_{n*}} \pi_r(P_{n-1}X)$$

shows that F_n is an Eilenberg–MacLane space $K(\pi_n X, n)$. Constructing $P_{n+1}(X)$ inductively from $P_n(X)$ requires knowing the nth k-invariant, which is a map of the form $k_n : P_n(X) \to Y_n$. The space $P_{n+1}(X)$ is then the homotopy fiber of k_n. Thus there is a homotopy fibration sequence

$$K(\pi_{n+1}X, n+1) \longrightarrow P_{n+1}(X) \xrightarrow{q_{n+1}} P_n(X) \xrightarrow{k_n} Y_n.$$

M. Banagl, *Intersection Spaces, Spatial Homology Truncation, and String Theory*,
Lecture Notes in Mathematics 1997, DOI 10.1007/978-3-642-12589-8_1,
© Springer-Verlag Berlin Heidelberg 2010

This means that $K(\pi_{n+1}X, n+1)$ is homotopy equivalent to the loop space ΩY_n. Consequently,

$$\pi_r(Y_n) \cong \pi_{r-1}(\Omega Y_n) \cong \pi_{r-1}(K(\pi_{n+1}X, n+1)) = \begin{cases} \pi_{n+1}X, & r = n+2, \\ 0, & \text{otherwise,} \end{cases}$$

and we see that Y_n is a $K(\pi_{n+1}X, n+2)$. Thus the nth k-invariant is a map

$$k_n : P_n(X) \longrightarrow K(\pi_{n+1}X, n+2).$$

Note that it induces the zero map on all homotopy groups, but is not necessarily homotopic to the constant map. The original space X is weakly homotopy equivalent to the inverse limit of the $P_n(X)$.

Applying the paradigm of Eckmann–Hilton duality, we arrive at the homology decomposition principle from the Postnikov decomposition principle by changing:

- the direction of all arrows
- π_* to H_*
- loops Ω to suspensions S
- fibrations to cofibrations and fibers to cofibers
- Eilenberg–MacLane spaces $K(G,n)$ to Moore spaces $M(G,n)$
- inverse limits to direct limits

A *homology decomposition* (or *Moore space decomposition*) for a simply connected CW-complex X is a commutative diagram

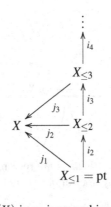

such that $j_{n*} : H_r(X_{\leq n}) \to H_r(X)$ is an isomorphism for $r \leq n$ and $H_r(X_{\leq n}) = 0$ for $r > n$. Let C_n be the homotopy cofiber of i_n. Then the exact sequence

$$H_r(X_{\leq n-1}) \xrightarrow{in_*} H_r(X_{\leq n}) \to H_r(C_n) \to H_{r-1}(X_{\leq n-1}) \xrightarrow{in_*} H_{r-1}(X_{\leq n})$$

shows that C_n is a Moore space $M(H_n X, n)$. Constructing $X_{\leq n+1}$ inductively from $X_{\leq n}$ requires knowing the nth k-invariant, which is a map of the form $k_n : Y_n \to X_{\leq n}$.

The space $X_{\leq n+1}$ is then the homotopy cofiber of k_n. Thus there is a homotopy cofibration sequence

$$Y_n \xrightarrow{k_n} X_{\leq n} \xrightarrow{i_{n+1}} X_{\leq n+1} \longrightarrow M(H_{n+1}X, n+1).$$

This means that $M(H_{n+1}X, n+1)$ is homotopy equivalent to the suspension SY_n. Consequently,

$$\widetilde{H}_r(Y_n) \cong \widetilde{H}_{r+1}(SY_n) \cong \widetilde{H}_{r+1}(M(H_{n+1}X, n+1)) = \begin{cases} H_{n+1}X, & r = n, \\ 0, & \text{otherwise,} \end{cases}$$

and we see that Y_n is an $M(H_{n+1}X, n)$. Thus the nth k-invariant is a map

$$k_n : M(H_{n+1}X, n) \to X_{\leq n}.$$

It induces the zero map on all reduced homology groups, which is a nontrivial statement to make in degree n:

$$k_{n*} : H_n(M(H_{n+1}X, n)) \cong H_{n+1}(X) \longrightarrow H_n(X) \cong H_n(X_{\leq n}).$$

The original space X is homotopy equivalent to the direct limit of the $X_{\leq n}$.

The Eckmann–Hilton duality paradigm, while being a very valuable organizational principle, does have its natural limitations, as we shall now discuss. Postnikov approximations possess rather good functorial properties: Let $p_n(X) : X \to P_n(X)$ be a stage-n Postnikov approximation for X, that is, $p_n(X)_* : \pi_r(X) \to \pi_r(P_n(X))$ is an isomorphism for $r \leq n$ and $\pi_r(P_n(X)) = 0$ for $r > n$. If Z is a space with $\pi_r(Z) = 0$ for $r > n$, then any map $g : X \to Z$ factors up to homotopy uniquely through $P_n(X)$, see [Zab76]. In particular, if $f : X \to Y$ is any map and $p_n(Y) : Y \to P_n(Y)$ is a stage-n Postnikov approximation for Y, then, taking $Z = P_n(Y)$ and $g = p_n(Y) \circ f$, there exists, uniquely up to homotopy, a map $p_n(f) : P_n(X) \to P_n(Y)$ such that

$$\begin{array}{ccc} X & \xrightarrow{\;f\;} & Y \\ \downarrow{\scriptstyle p_n(X)} & & \downarrow{\scriptstyle p_n(Y)} \\ P_n(X) & \xrightarrow{\;p_n(f)\;} & P_n(Y) \end{array}$$

homotopy commutes. One of the starting points for our development of the spatial homology truncation machine presented in this book was the fact that the above functorial property of Postnikov approximations does *not* dualize to homology decompositions. Let us discuss an example based on suggestions of [Zab76] that illustrates this lack of functoriality for Moore space decompositions. Let $X = S^2 \cup_2 e^3$ be a Moore space $M(\mathbb{Z}/2, 2)$ and let $Y = X \vee S^3$. If $X_{\leq 2}$ and $Y_{\leq 2}$ denote stage-2

Moore approximations for X and Y, respectively, then $X_{\leq 2} = X$ and $Y_{\leq 2} = X$. We claim that whatever maps $i : X_{\leq 2} \to X$ and $j : Y_{\leq 2} \to Y$ such that $i_* : H_r(X_{\leq 2}) \to H_r(X)$ and $j_* : H_r(Y_{\leq 2}) \to H_r(Y)$ are isomorphisms for $r \leq 2$ one takes, there is always a map $f : X \to Y$ that cannot be compressed into the stage-2 Moore approximations, i.e. there is no map $f_{\leq 2} : X_{\leq 2} \to Y_{\leq 2}$ such that

$$
\begin{array}{ccc}
X & \xrightarrow{\ f\ } & Y \\
\uparrow{\scriptstyle i} & & \uparrow{\scriptstyle j} \\
X_{\leq 2} & \xrightarrow{\ f_{\leq 2}\ } & Y_{\leq 2}
\end{array}
$$

commutes up to homotopy. We shall employ the universal coefficient exact sequence for homotopy groups with coefficients. If G is an abelian group and $M(G,n)$ a Moore space, then there is a short exact sequence

$$0 \to \text{Ext}(G, \pi_{n+1}Y) \xrightarrow{\ \iota\ } [M(G,n),Y] \xrightarrow{\ \eta\ } \text{Hom}(G, \pi_n Y) \to 0,$$

where Y is any space and $[-,-]$ denotes pointed homotopy classes of maps. The map η is given by taking the induced homomorphism on π_n and using the Hurewicz isomorphism. This universal coefficient sequence is natural in both variables. Hence, the following diagram commutes:

$$
\begin{array}{ccccccccc}
0 & \to & \text{Ext}(\mathbb{Z}/2, \pi_3 Y_{\leq 2}) & \xrightarrow{\iota_{\leq 2}} & [X_{\leq 2}, Y_{\leq 2}] & \xrightarrow{\eta_{\leq 2}} & \text{Hom}(\mathbb{Z}/2, \pi_2 Y_{\leq 2}) & \to & 0 \\
 & & \downarrow{\scriptstyle E_2\pi_3(j)} & & \downarrow{\scriptstyle j_*} & & \downarrow{\scriptstyle \pi_2(j)_*=\text{id}} & & \\
0 & \to & \text{Ext}(\mathbb{Z}/2, \pi_3 Y) & \xrightarrow{\iota} & [X_{\leq 2}, Y] & \xrightarrow{\eta} & \text{Hom}(\mathbb{Z}/2, \pi_2 Y) & \to & 0 \\
 & & \uparrow{\scriptstyle \text{id}=E^Y\pi_2(i)} & & \uparrow{\scriptstyle i^*} & & \uparrow{\scriptstyle \pi_2(i)^*=\text{id}} & & \\
0 & \to & \text{Ext}(\mathbb{Z}/2, \pi_3 Y) & \xrightarrow{\iota} & [X, Y] & \xrightarrow{\eta} & \text{Hom}(\mathbb{Z}/2, \pi_2 Y) & \to & 0
\end{array}
$$

Here we will briefly write $E_2(-) = \text{Ext}(\mathbb{Z}/2, -)$ so that $E_2(G) = G/2G$, and $E^Y(-) = \text{Ext}(-, \pi_3 Y)$. By the Hurewicz theorem, $\pi_2(X) \cong H_2(X) \cong \mathbb{Z}/2$, $\pi_2(Y) \cong H_2(Y) \cong \mathbb{Z}/2$, and $\pi_2(i) : \pi_2(X_{\leq 2}) \to \pi_2(X)$, as well as $\pi_2(j) : \pi_2(Y_{\leq 2}) \to \pi_2(Y)$, are isomorphisms, hence the identity. If a homomorphism $\phi : A \to B$ of abelian groups is onto, then $E_2(\phi) : E_2(A) = A/2A \to B/2B = E_2(B)$ remains onto. By the Hurewicz theorem, $\text{Hur} : \pi_3(Y) \to H_3(Y) = \mathbb{Z}$ is onto. Consequently, the induced map $E_2(\text{Hur}) : E_2(\pi_3 Y) \to E_2(H_3 Y) = E_2(\mathbb{Z}) = \mathbb{Z}/2$ is onto. Let $\xi \in E_2(H_3 Y)$

be the generator. Choose a preimage $x \in E_2(\pi_3 Y)$, $E_2(\text{Hur})(x) = \xi$ and set $[f] = \iota(x) \in [X,Y]$. Suppose there existed a homotopy class $[f_{\leq 2}] \in [X_{\leq 2}, Y_{\leq 2}]$ such that $j_*[f_{\leq 2}] = i^*[f]$. Then

$$\eta_{\leq 2}[f_{\leq 2}] = \pi_2(j)_* \eta_{\leq 2}[f_{\leq 2}] = \eta j_*[f_{\leq 2}] = \eta i^*[f] = \pi_2(i)^* \eta [f] = \pi_2(i)^* \eta \iota(x) = 0.$$

Thus there is an element $\varepsilon \in E_2(\pi_3 Y_{\leq 2})$ such that $\iota_{\leq 2}(\varepsilon) = [f_{\leq 2}]$. From

$$\iota E_2 \pi_3(j)(\varepsilon) = j_* \iota_{\leq 2}(\varepsilon) = j_*[f_{\leq 2}] = i^*[f] = i^* \iota(x) = \iota E^Y \pi_2(i)(x)$$

we conclude that $E_2 \pi_3(j)(\varepsilon) = x$ since ι is injective. By naturality of the Hurewicz map, the square

$$
\begin{array}{ccc}
\pi_3 Y_{\leq 2} & \xrightarrow{\ \pi_3(j)\ } & \pi_3 Y \\
\downarrow{\scriptstyle \text{Hur}} & & \downarrow{\scriptstyle \text{Hur}} \\
0 = H_3 Y_{\leq 2} & \xrightarrow{\ H_3(j)\ } & H_3 Y
\end{array}
$$

commutes and induces a commutative diagram upon application of $E_2(-)$:

$$
\begin{array}{ccc}
E_2(\pi_3 Y_{\leq 2}) & \xrightarrow{\ E_2 \pi_3(j)\ } & E_2(\pi_3 Y) \\
\downarrow{\scriptstyle E_2(\text{Hur})} & & \downarrow{\scriptstyle E_2(\text{Hur})} \\
0 = E_2(H_3 Y_{\leq 2}) & \xrightarrow{\ E_2(H_3(j))\ } & E_2(H_3 Y).
\end{array}
$$

It follows that

$$\xi = E_2(\text{Hur})(x) = E_2(\text{Hur})E_2\pi_3(j)(\varepsilon) = E_2 H_3(j) E_2(\text{Hur})(\varepsilon) = 0,$$

a contradiction. Therefore, no compression $[f_{\leq 2}]$ of $[f]$ exists. We will return to this example at a later point, where an explicit geometric description of the map f will also be given.

From the point of view adopted in this monograph, the lack of functoriality of Moore approximations is due to the wrong choice of morphisms between spaces. The way in which we will approach the problem is to change the categorical setup: Instead of considering CW-complexes and cellular maps between them, we will consider CW-complexes endowed with extra structure and cellular maps that preserve that extra structure. We will show that such morphisms can then be compressed into homology truncations if the latter are constructed correctly. Every CW-complex can indeed be endowed with the requisite extra structure so that this does not limit the class of spaces which the truncation machine can process as an input. (However, there is no way in general to associate the extra structure canonically with every space, although this is possible for certain classes of spaces.) Given a cellular map, it is *not* always possible to adjust the extra structure on the source and on the target of the map so that the map preserves the structures. Thus the category theoretic setup automatically, and in a natural way, singles out those continuous maps that can be compressed into homologically truncated spaces.

Let n be a positive integer.

Definition 1.1. A CW-complex K is called *n-segmented* if it contains a subcomplex $K_{<n} \subset K$ such that

$$H_r(K_{<n}) = 0 \text{ for } r \geq n \tag{1.1}$$

and

$$i_* : H_r(K_{<n}) \xrightarrow{\cong} H_r(K) \text{ for } r < n, \tag{1.2}$$

where i is the inclusion of $K_{<n}$ into K.

Not every n-dimensional complex is n-segmented, but we shall see that every n-dimensional complex K is homotopy equivalent to an n-segmented one, K/n. Let K^r denote the r-skeleton of a CW-complex K.

Lemma 1.2. *Let K be an n-dimensional CW-complex. If its group of n-cycles has a basis of cells, then K is n-segmented.*

Proof. Let $\{z_\beta\}$ be n-cells of K forming a basis for the cycle group $Z_n(K)$. Let $\{y_\alpha\}$ be the rest of the n-cells, generating a subgroup $Y \subset C_n(K)$. Set

$$K_{<n} = K^{n-1} \cup \bigcup_\alpha y_\alpha \subset K.$$

The boundary operator $C_n(K_{<n}) = Y \to C_{n-1}(K_{<n}) = C_{n-1}(K)$ is the restriction of $\partial_n : C_n(K) = Y \oplus Z_n(K) \to C_{n-1}(K)$ to Y, hence injective. Therefore, $H_n(K_{<n}) = 0$. Since the inclusion $K_{<n} \subset K$ induces the identity $Z_{n-1}(K_{<n}) = Z_{n-1}(K)$ and $\operatorname{im} \partial_n| Y = \operatorname{im} \partial_n$, the inclusion induces

$$H_{n-1}(K_{<n}) = \frac{Z_{n-1}(K)}{\operatorname{im}(\partial_n|Y)} = \frac{Z_{n-1}(K)}{\operatorname{im} \partial_n} = H_{n-1}(K).$$

Clearly, $H_r(K_{<n}) = H_r(K)$ for $r \leq n-2$ and $H_r(K_{<n}) = 0$ for $r > n$. $\qquad\square$

If K is any n-dimensional, n-segmented space, then it does not follow automatically that its group of n-cycles $Z_n(K)$ possesses a basis of cells. Nor is the subcomplex $K_{<n}$ unique. As an example, consider the 3-sphere $K = S^3$ with the CW-structure $S^3 = S^2 \cup_1 e_1^3 \cup_1 e_2^3$. This complex is clearly 3-segmented; we may for instance take $K_{<3} = S^2 \cup_1 e_1^3 = D^3$. Neither e_1^3 nor e_2^3 lie in the kernel of the boundary operator, only their difference does. Thus $Z_3(K)$, though nonempty, does not have a basis of cells. The truncation $K_{<3}$ is not unique because the subcomplex $S^2 \cup_1 e_2^3$ would work just as well.

Proposition 1.3. *Let K be an n-dimensional, n-segmented CW-complex and suppose $K_{<n} \subset K$ is a subcomplex with properties (1.1) and (1.2) and such that $(K_{<n})^{n-1} = K^{n-1}$. If the group of n-cycles of K has a basis of cells, then $K_{<n}$ is unique, namely*

$$K_{<n} = K^{n-1} \cup \bigcup_{\alpha} y_{\alpha},$$

where $\{y_\alpha\}$ is the set of n-cells of K that are not cycles.

Proof. Let $\{z_\beta\}$ be n-cells of K forming a basis for the cycle group $Z_n(K)$. Let $\{y_\alpha\}$ be the rest of the n-cells of K. Let $\{e^n_\gamma\}$ be the n-cells of $K_{<n}$. Thus we have

$$K^{n-1} \cup \bigcup_{\gamma} e^n_\gamma = K_{<n} \subset K = K^{n-1} \cup \bigcup_{\alpha} y_\alpha \cup \bigcup_{\beta} z_\beta.$$

The assertion follows once we have established that (1) none of the z_β occur among the e^n_γ, and (2) every y_α appears among the e^n_γ. Suppose (1) were false so that there existed a γ with $e^n_\gamma = z_\beta$ for some β. Since $K_{<n}$ is n-dimensional and $H_n(K_{<n}) = 0$, the cellular boundary operator $\partial^<_n : C_n(K_{<n}) \to C_{n-1}(K_{<n})$ is injective. With $i : C_n(K_{<n}) \hookrightarrow C_n(K)$ the inclusion, we have a commutative diagram

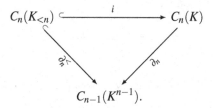

Thus for the above cycle-cell z_β:

$$0 = \partial_n(z_\beta) = \partial_n i(z_\beta) = \partial^<_n(z_\beta) \neq 0,$$

a contradiction. Therefore, $\{e^n_\gamma\}$ must be a subset of $\{y_\alpha\}$.

To establish (2), we observe first that $\mathrm{im}\,\partial^<_n = \mathrm{im}\,\partial_n$: The identity $\partial_n \circ i = \partial^<_n$ shows that $\mathrm{im}\,\partial^<_n \subset \mathrm{im}\,\partial_n$. The inclusion $K_{<n} \subset K$ induces an isomorphism $H_{n-1}(K_{<n}) \xrightarrow{\cong} H_{n-1}(K)$. But the inclusion restricted to $(n-1)$-skeleta is the identity map, whence the identity map induces an isomorphism

$$\frac{Z_{n-1}(K)}{\mathrm{im}\,\partial^<_n} \xrightarrow{\cong} \frac{Z_{n-1}(K)}{\mathrm{im}\,\partial_n}.$$

Now if G is an abelian group and $A \subset B \subset G$ subgroups such that the identity map induces an isomorphism $G/A \xrightarrow{\cong} G/B$, then the injectivity implies $B \subset A$, so that $A = B$. In particular, we conclude for our situation $\mathrm{im}\,\partial^<_n = \mathrm{im}\,\partial_n$. Let $Y \subset C_n(K)$ be the subgroup generated by the cells $\{y_\alpha\}$, giving rise to a decomposition $C_n(K) = Y \oplus Z_n(K)$. The restriction $\partial_n| : Y \to C_{n-1}(K^{n-1})$ is injective and has the same image as ∂_n. Since by (1), $\{e_\gamma^n\} \subset \{y_\alpha\}$, we have $\mathrm{im}(i) \subset Y$. Consequently, there is a restricted diagram

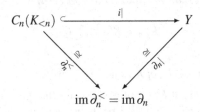

which shows that $i| : C_n(K_{<n}) \xrightarrow{\cong} Y$ is an isomorphism. In particular, every cell $y_\alpha \in Y$ has a preimage in $C_n(K_{<n})$ and that preimage is some n-cell e_γ^n of $K_{<n}$. $\quad\square$

1.1.2 An Example

The example below, due to Peter Hilton, already illustrates all the relevant points and necessary techniques for spatial homology truncation *on the object level*. Let K be the simply connected complex

$$K = S^2 \cup_4 e_1^3 \cup_6 e_2^3.$$

Its homology is

$$H_2(K) = \mathbb{Z}/2, \; H_3(K) = \mathbb{Z}.$$

We claim that K is not 3-segmented. If it were 3-segmented, then there would exist a subcomplex $K_{<3}$ such that

$$H_2(K_{<3}) = \mathbb{Z}/2 \text{ and } H_3(K_{<3}) = 0.$$

The following table shows that no matter which subcomplex we try, each time either the second or third homology is wrong.

$K_{<3}$	$H_2(K_{<3})$	$H_3(K_{<3})$
$*$	0	0
S^2	$\mathbb{Z}$	0
$S^2 \cup_4 e_1^3$	$\mathbb{Z}/4$	0
$S^2 \cup_6 e_2^3$	$\mathbb{Z}/6$	0
K	$\mathbb{Z}/2$	$\mathbb{Z}$

We shall now describe a method to produce a 3-segmented space $K/3$ which is still homotopy equivalent to K. The method is essentially an algebraic change of basis in the third cellular chain group of K. The change of basis is then realized topologically by 3-cell reattachment to yield the desired homotopy equivalence. Let $C_*(K)$ denote the cellular chain complex of K. We equip $C_3(K)$ with the basis $\{e_1^3, e_2^3\}$. The short exact sequence

$$0 \to \ker \partial \longrightarrow C_3(K) \xrightarrow{\ \partial\ } \operatorname{im} \partial \to 0$$

splits since $\operatorname{im} \partial = 2\mathbb{Z} \subset \mathbb{Z}e^2 = C_2(K)$ is free abelian. In fact,

$$s : \operatorname{im} \partial \longrightarrow C_3(K)$$
$$2n \mapsto (-n, n)$$

is an explicit splitting. Set

$$Y_3(K) = \operatorname{im} s = \mathbb{Z}(-1, 1)$$

and

$$Z_3(K) = \ker \partial = \{(n, m) : 2n = -3m\} = \mathbb{Z}(3, -2),$$

so that

$$C_3(K) = Z_3(K) \oplus Y_3(K).$$

This is the change of basis we referred to earlier. The Hurewicz map identifies $C_3(K)$ with $\pi_3(K^3, K^2)$. Under this identification, the element $(3, -2) \in C_3(K)$ corresponds to an element $\zeta \in \pi_3(K^3, K^2)$. Similarly, $(-1, 1)$ corresponds to an $\eta \in \pi_3(K^3, K^2)$. The connecting homomorphism

$$d : \pi_3(K^3, K^2) \longrightarrow \pi_2(K^2) = \pi_2(S^2) \overset{\deg}{\cong} \mathbb{Z}$$

maps a 3-cell e^3, thought of as an element $[\chi(e^3)]$ in $\pi_3(K^3, K^2)$ via its characteristic map $\chi(e^3)$, to the degree of its attaching map. Thus

$$d[\chi(e_1^3)] = 4, \ d[\chi(e_2^3)] = 6,$$

and

$$d\zeta = d(3[\chi(e_1^3)] - 2[\chi(e_2^3)]) = 3d[\chi(e_1^3)] - 2d[\chi(e_2^3)] = 3 \cdot 4 - 2 \cdot 6 = 0,$$

which, of course, confirms that $(3, -2) \in Z_3(K) = \ker \partial$. For the second new basis element we obtain

$$d\eta = d(-[\chi(e_1^3)] + [\chi(e_2^3)]) = -d[\chi(e_1^3)] + d[\chi(e_2^3)] = -4 + 6 = 2.$$

To form $K/3$, take two new 3-cells z and y and attach them to $K^2 = S^2$, using representatives of $d\zeta$ and $d\eta$, respectively, as attaching maps:

$$K/3 := S^2 \cup_{d\zeta=0} z \cup_{d\eta=2} y.$$

Note that the 2-skeleton remains unchanged, $(K/3)^2 = K^2$. Let us describe a rather explicit homotopy equivalence h' (the letter h will be reserved for its homotopy inverse) from K to $K/3$, which realizes the change of basis geometrically. Algebraically, the change of basis on the third cellular chain group is given by the map

$$\theta : \pi_3(K^3, K^2) \longrightarrow \pi_3(K/3, K^2)$$
$$\zeta \longmapsto [\chi(z)]$$
$$\eta \longmapsto [\chi(y)].$$

We observe that the diagram

$$\pi_3(K^3, K^2) \xrightarrow{\theta} \pi_3(K/3, K^2)$$

$$\Big\downarrow{d} \qquad\qquad \Big\downarrow{d}$$

$$\pi_2(K^2) =\!=\!=\!= \pi_2((K/3)^2)$$

commutes, for

$$d\theta(\zeta) = d[\chi(z)] = 0 = d\zeta, \ d\theta(\eta) = d[\chi(y)] = 2 = d\eta.$$

The images of the old basis elements are

$$\theta[\chi(e_1^3)] = \theta(\zeta + 2\eta) = [\chi(z)] + 2[\chi(y)], \ \theta[\chi(e_2^3)] = \theta(\zeta + 3\eta) = [\chi(z)] + 3[\chi(y)].$$
$$(1.3)$$

On the other hand, the $\theta[\chi(e_i^3)]$ are represented by commutative diagrams

$$\partial D^3 \xrightarrow{f_i'|} K^2 = S^2$$

$$\Big\cap \qquad\qquad \Big\cap$$

$$\Big\downarrow \qquad\qquad \Big\downarrow$$

$$D^3 \xrightarrow{\ f_i'\ } K/3$$

Let $g_1 = \chi(e_1^3)|_{\partial e_1^3}$ be the attaching map for e_1^3 in K (a map of degree 4), and let $g_2 = \chi(e_2^3)|_{\partial e_2^3}$ be the attaching map for e_2^3 in K (a map of degree 6). Since $d\theta = d$, the attaching map g_i is homotopic to $f_i'|$. This is confirmed by the degree calculation

$$\deg(f_1'|_{\partial D^3}) = 0 \cdot 1 + 2 \cdot 2 = 4 = \deg(g_1),$$

and

$$\deg(f_2'|_{\partial D^3}) = 0 \cdot 1 + 2 \cdot 3 = 6 = \deg(g_2),$$

using (1.3) and the degrees of the attaching maps for z and y. By the homotopy extension property , there exists, for $i = 1, 2$, a representative $f_i : D^3 \rightarrow K/3$ for $\theta[\chi(e_i^3)]$ that extends g_i. Defining

$$h' : K \longrightarrow K/3$$

by

$$
\begin{aligned}
h'(x) &= x, \quad \text{for } x \in K^2, \\
h'(\chi(e_1^3)(x)) &= f_1(x), \quad \text{for } x \in e_1^3, \\
h'(\chi(e_2^3)(x)) &= f_2(x), \quad \text{for } x \in e_2^3,
\end{aligned}
$$

yields a homotopy equivalence, since h' induces a chain-isomorphism. (It induces θ on the third chain group.)

Let us verify that $K/3$ is 3-segmented. Defining $K_{<3}$ to be the subcomplex

$$K_{<3} = S^2 \cup_2 y \subset K/3,$$

we obtain for the homology:

$$H_1(K_{<3}) = 0, \; H_2(K_{<3}) = \mathbb{Z}/2, \; H_3(K_{<3}) = 0.$$

The desired homological truncation has thus been correctly implemented. In fact, $K/3$ is three-dimensional and its group of 3-cycles has a basis of cells (namely the cell z), so we could have concluded from Lemma 1.2 that $K/3$ is 3-segmented. Moreover, Proposition 1.3 tells us that $K_{<3}$ is unique.

1.1.3 General Spatial Homology Truncation on the Object Level

Functorial spatial homology truncation in the low dimensions $n = 1, 2$ is discussed in Section 1.1.5. In dimensions $n \geq 3$, we shall employ the concept of a homological n-truncation structure. Let $n \geq 3$ be an integer.

Definition 1.4. A (*homological*) *n-truncation structure* is a quadruple $(K, K/n, h, K_{<n})$, where:

1. K is a simply connected CW-complex.
2. K/n is an n-dimensional CW-complex with $(K/n)^{n-1} = K^{n-1}$ and such that the group of n-cycles of K/n has a basis of cells.
3. $h : K/n \rightarrow K^n$ is the identity on K^{n-1} and a cellular homotopy equivalence rel K^{n-1}.
4. $K_{<n} \subset K/n$ is a subcomplex with properties (1.1) and (1.2) with respect to K/n and such that $(K_{<n})^{n-1} = K^{n-1}$.

The first component space K of an n-truncation structure is required to be simply connected because the theory employs the Hurewicz theorem for $n \geq 3$. Since the

$(n-1)$-skeleton of the n-segmentation K/n of K agrees with the $(n-1)$-skeleton of K and $\pi_1(K^{n-1}) = \pi_1(K)$ as $n-1 \geq 2$, it follows that K/n is simply connected as well. The same observation applies to the truncation $K_{<n}$. Note that by Lemma 1.2, K/n is n-segmented, so that $K_{<n}$ does, in fact, exist. Since by Proposition 1.3, $K_{<n}$ is uniquely determined by K/n, it is technically not necessary to include it explicitly as the fourth component into an n-truncation structure. Nevertheless, we find it convenient to do so, as this will automatically fix notation for the n-truncation space. It will also be advantageous when we work with morphisms between n-truncation structures later on. If $(K, K/n, h, K_{<n})$ is an n-truncation structure and $r < n$, then

$$H_r(K_{<n}) \underset{i_*}{\cong} H_r(K/n) \underset{h_*}{\cong} H_r(K^n) \underset{j_*}{\cong} H_r(K),$$

where $i : K_{<n} \subset K/n$ and $j : K^n \subset K$ are the inclusions (while $H_r(K_{<n}) = 0$ for $r \geq n$, of course). Let us recall the following consequence of the homotopy extension property:

Proposition 1.5. *Suppose (X, A) and (Y, A) satisfy the homotopy extension property and $f' : X \to Y$ is a homotopy equivalence with $f'|_A = 1_A$. Then f' is a homotopy equivalence* rel A, *that is, there exists a homotopy inverse f for f' such that $f|_A = 1_A$, $ff' \simeq 1$ rel A and $f'f \simeq 1$ rel A.*

For a proof see [Hat02, Proposition 0.19, page 16].

Proposition 1.6. *Given any integer $n \geq 3$, every simply connected CW-complex K can be completed to an n-truncation structure $(K, K/n, h, K_{<n})$.*

Proof. The proof is based on methods due to Hilton [Hil65], and is suggested by the example in Section 1.1.2. Let $\{e_\gamma^n\}$ be the n-cells of K so that

$$K^n = K^{n-1} \cup \bigcup_\gamma e_\gamma^n.$$

As suggested in the example, we shall carry out a "homotopy-element-to-cell" conversion procedure initiated by an algebraic change of basis in the nth cellular chain group of K. The change of basis is then realized topologically to yield the desired homotopy equivalence. Let $C_*(K)$ denote the cellular chain complex of K. We equip $C_n(K)$ with the basis $\{e_\gamma^n\}$. The short exact sequence

$$0 \to \ker \partial_n \longrightarrow C_n(K) \overset{\partial_n}{\longrightarrow} \operatorname{im} \partial_n \to 0$$

splits since $\operatorname{im} \partial_n \subset C_{n-1}(K)$ is free abelian. Let $s : \operatorname{im} \partial_n \to C_n(K)$ be a splitting. Set $Y = \operatorname{im} s$ and let $Z_n(K) = \ker \partial_n$ be the cycle group so that

$$C_n(K) = Z_n(K) \oplus Y.$$

Since $n \geq 3$, the simple connectivity of K implies the simple connectivity of K^{n-1}. Thus the Hurewicz map identifies $C_n(K)$ with $\pi_n(K^n, K^{n-1})$. Choose elements $\zeta_\beta, \eta_\alpha \in \pi_n(K^n, K^{n-1})$ such that $\{\zeta_\beta\}$ is a basis of $Z_n(K)$ and $\{\eta_\alpha\}$ is a basis of Y. The connecting homomorphism

$$d : \pi_n(K^n, K^{n-1}) \longrightarrow \pi_{n-1}(K^{n-1})$$

maps an n-cell e^n, thought of as an element $[\chi(e^n)]$ in $\pi_n(K^n, K^{n-1})$ (that is, thought of as the homotopy class of its characteristic map), to the class of its attaching map. Let

$$b_\beta : S^{n-1} \longrightarrow K^{n-1}$$

be choices of representatives for the homotopy classes $d\zeta_\beta$ and let

$$a_\alpha : S^{n-1} \longrightarrow K^{n-1}$$

be choices of representatives for the homotopy classes $d\eta_\alpha$.

To form K/n, take new n-cells z_β and y_α and attach them to K^{n-1}, using the attaching maps a_α for the y_α and the b_β for the z_β:

$$K/n := K^{n-1} \cup \bigcup_{a_\alpha} y_\alpha \cup \bigcup_{b_\beta} z_\beta.$$

Let us construct a homotopy equivalence h' from K^n to K/n, which realizes the change of basis geometrically. Algebraically, the change of basis on the nth cellular chain group is given by the isomorphism

$$\begin{aligned} \theta : \pi_n(K^n, K^{n-1}) &\longrightarrow \pi_n(K/n, K^{n-1}) \\ \zeta_\beta &\longmapsto [\chi(z_\beta)] \\ \eta_\alpha &\longmapsto [\chi(y_\alpha)]. \end{aligned}$$

We observe that the diagram

$$\begin{CD} \pi_n(K^n, K^{n-1}) @>\theta>> \pi_n(K/n, K^{n-1}) \\ @VdVV @VVdV \\ \pi_{n-1}(K^{n-1}) @= \pi_{n-1}(K^{n-1}) \end{CD}$$

commutes, for

$$d\theta(\zeta_\beta) = d[\chi(z_\beta)] = [b_\beta] = d\zeta_\beta, \; d\theta(\eta_\alpha) = d[\chi(y_\alpha)] = [a_\alpha] = d\eta_\alpha.$$

The images $\theta[\chi(e_\gamma^n)]$ of the old basis elements are represented by commutative diagrams

$$
\begin{array}{ccc}
\partial D^n & \xrightarrow{\; f_\gamma'| \;} & K^{n-1} \\
\cap \downarrow & & \cap \downarrow \\
D^n & \xrightarrow[\; f_\gamma' \;]{} & K/n
\end{array}
$$

Let $g_\gamma = \chi(e_\gamma^n)| : \partial e_\gamma^n \to K^{n-1}$ be the attaching maps for e_γ^n in K. The map g_γ is homotopic to $f_\gamma'|$ because

$$[g_\gamma] = d[\chi(e_\gamma^n)] = d\theta[\chi(e_\gamma^n)] = d[f_\gamma'] = [f_\gamma'|].$$

By the homotopy extension property, there exists, for every γ, a representative $f_\gamma : D^n \to K/n$ for $\theta[\chi(e_\gamma^n)]$ that extends g_γ. Defining

$$h' : K^n \longrightarrow K/n$$

by

$$h'(x) \quad = \quad x, \text{ for } x \in K^{n-1},$$

$$h'(\chi(e_\gamma^n)(x)) = f_\gamma(x), \text{ for } x \in e_\gamma^n,$$

yields a map, since $\chi(e_\gamma^n)|_{\partial e_\gamma^n} = g_\gamma = f_\gamma|_{\partial e_\gamma^n}$. It is a homotopy equivalence, since it induces a chain-isomorphism. (It induces θ on the nth chain group.) Note that K/n is simply connected since its $(n-1)$-skeleton is K^{n-1} and $\pi_1(K^{n-1}) \to \pi_1(K/n)$ as well as $\pi_1(K^{n-1}) \to \pi_1(K)$ are isomorphisms as $n-1 \geq 2$.

Let us verify that the cycle group $Z_n(K/n)$ possesses a basis of cells. The commutativity of the diagram

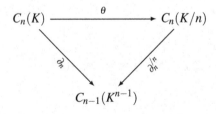

can be established in various ways. It follows, for instance, from the commutativity of the diagram

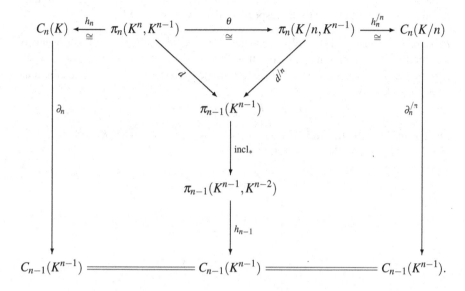

Here $h_n, h_n^{/n}$ and h_{n-1} are Hurewicz homomorphisms. Since $n \geq 3$, h_n and $h_n^{/n}$ are isomorphisms, and if $n \geq 4$, then h_{n-1} is an isomorphism as well. If $n = 3$, then $C_2(K^2)$ cannot in general be identified with $\pi_2(K^2, K^1)$. For example, if $\pi_1(K^1)$ contains two noncommuting elements then $\pi_2(K^2, K^1)$ will not be abelian because the homomorphism $\pi_2(K^2, K^1) \to \pi_1(K^1)$ is surjective as K^2 is simply connected. Alternatively, one can argue that θ is part of a chain map induced by the continuous map h', and that map is the identity on K^{n-1}. If $\zeta \in Z_n(K)$, then $\partial_n^{/n} \theta \zeta = \partial_n \zeta = 0$, so $\theta \zeta \in Z_n(K/n)$ and $\theta Z_n(K) \subset Z_n(K/n)$. Conversely, for $z \in Z_n(K/n)$, let $\zeta = \theta^{-1}(z)$. Then

$$\partial_n \zeta = \partial_n^{/n} \theta \zeta = \partial_n^{/n}(z) = 0$$

so that $\zeta \in Z_n(K)$ and $z = \theta \zeta \in \theta Z_n(K)$. Therefore, $\theta Z_n(K) = Z_n(K/n)$ and there is a restriction $\theta| : Z_n(K) \overset{\cong}{\longrightarrow} Z_n(K/n)$. This restriction sends the basis $\{\zeta_\beta\}$ of $Z_n(K)$ to $\{\theta(\zeta_\beta)\}$, which must thus be a basis of $Z_n(K/n)$. Now $\theta(\zeta_\beta) = z_\beta$ and the z_β are n-cells of K/n. Hence, $Z_n(K/n)$ has a basis of cells.

As noted before, Lemma 1.2 implies that K/n is n-segmented and by Proposition 1.3, the required subcomplex $K_{<n}$ of K/n is uniquely determined. Explicitly,

$$K_{<n} = K^{n-1} \cup \bigcup_{a_\alpha} y_\alpha.$$

Finally, being CW pairs, (K^n, K^{n-1}) and $(K/n, K^{n-1})$ satisfy the homotopy extension property. Applying Proposition 1.5 to $h' : (K^n, K^{n-1}) \to (K/n, K^{n-1})$, which is indeed the identity on K^{n-1}, we get a homotopy inverse $h : K/n \to K^n$ such that hh' and $h'h$ are homotopic to the respective identity maps rel K^{n-1}.   □

Remark 1.7. Since K is simply connected, one may up to homotopy equivalence assume that its 1-skeleton is a 0-cell. It follows then that for $n = 3$, we may always assume that in the 3-segmentation of a space, the cycle-cells z_β are wedged on, that is, $K/3$ has the form

$$K/3 = K^2 \cup \bigcup y_\alpha \vee \bigvee z_\beta.$$

Indeed, in this situation K^2 is a wedge of 2-spheres and $\pi_2(K^2, K^1) \cong \pi_2(K^2) \cong H_2(K^2) \cong C_2(K)$, so the factorization

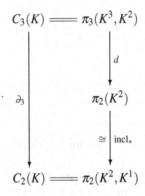

shows that already $d\zeta_\beta = 0$. Thus for the representatives b_β we could take constant maps. This remark applies to higher n as well if K is such that

$$\mathrm{im}(d : \pi_n(K^n, K^{n-1}) \to \pi_{n-1}(K^{n-1})) \cap \ker(\pi_{n-1}(K^{n-1}) \to \pi_{n-1}(K^{n-1}, K^{n-2})) = 0.$$

Example 1.8. Suppose K is such that the boundary map on the nth cellular chain group of K vanishes. Then $C_n(K) = Z_n(K)$ and $Y = 0$. Thus in this situation, the complementary space Y is unique and no choice has to be made. We have

$$K/n = K^{n-1} \cup \bigcup_\beta z_\beta,$$

since there are no cells y_α. It follows, as expected, that

$$K_{<n} = K^{n-1},$$

the $(n-1)$-skeleton of K. If K in fact has only even-dimensional cells, then all boundary maps in the cellular chain complex vanish and hence $K_{<n} = K^{n-1}$ for any n. We shall return to this scenario in Section 1.9 on the interleaf category.

1.1.4 Virtual Cell Groups and Eigenclasses

In order to obtain functoriality for spatial homology truncation on suitable cellular maps $f : K \to L$, one must deal successfully with roughly two major issues: First, the map f must be compressible into a truncated map $f_{<n} : K_{<n} \to L_{<n}$. Second, if $g : L \to P$ is another compressible map with truncation $g_{<n} : L_{<n} \to P_{<n}$, then gf ought to be compressible with $(gf)_{<n}$ homotopic to $g_{<n} \circ f_{<n}$. The second issue is harder and involves certain homotopy groups $VC_n(\Lambda)$ associated to a homological n-truncation structure Λ.

Example 1.9. Let us exhibit an example of a map $f : K \to L$, where K and L are simply connected 4-segmented CW-complexes ($K = K/4$, $L = L/4$) with unique 4-truncation subcomplexes $K_{<4} \subset K$, $L_{<4} \subset L$, such that there are two nonhomotopic maps $f_{<4}, f'_{<4} : K_{<4} \to L_{<4}$ with

$$
\begin{array}{ccc}
K & \xleftarrow{\;i_K\;} & K_{<4} \\
f \downarrow & & \downarrow\downarrow{\scriptstyle f'_{<4}\;f_{<4}} \\
L & \xleftarrow{\;i_L\;} & L_{<4}
\end{array}
\qquad (1.4)
$$

homotopy commutative. The example thus demonstrates that in general, contrary to Postnikov truncation, the diagram (1.4) may not uniquely determine the homological compression of a map f. Let $K = S^3$ and let L be the suspension of three-dimensional real projective space $\mathbb{R}P^3$. Clearly, $K_{<4} = S^3 = K$ is the unique subcomplex that truncates the homology of K above degree 3. The space L has the cell structure

$$
L = S^2 \cup_2 e^3 \cup_b e^4
$$

and its homology is

$$
H_0(L) \cong \mathbb{Z},\; H_1(L) = 0,\; H_2(L) \cong \mathbb{Z}/2,\; H_3(L) = 0,\; H_4(L) \cong \mathbb{Z}.
$$

The cycle group $Z_4(L) = C_4(L) = \mathbb{Z}e^4$ has a basis of cells. Hence L is 4-segmented by Lemma 1.2. Necessarily, $Y_4(L) = 0$. The 4-truncation is $L_{<4} = L^3 = S^2 \cup_2 e^3$, unique by Proposition 1.3. The interesting feature is that while the attaching map $b : S^3 = \partial e^4 \to L_{<4}$ is sufficiently trivial to produce a trivial cellular chain boundary map $C_4(L) \to C_3(L)$, one can show that nevertheless $[b] \neq 0 \in \pi_3(L_{<4}) \cong \mathbb{Z}/4$. Since the 4-cell in L is attached by b, we have

$$
i_{L*}[b] = 0,\; i_{L*} : \pi_3(L_{<4}) \to \pi_3(L).
$$

Set

$$
f_{<4} = b : K_{<4} = S^3 \longrightarrow L_{<4},
$$

let $f'_{<4}$ be the constant map $K_{<4} \to L_{<4}$, and let $f : K \to L$ be the composition

$$K = S^3 \xrightarrow{\;b\;} L_{<4} \xhookrightarrow{\;i_L\;} L.$$

By definition,

$$
\begin{array}{ccc}
K & \xLongequal{\;i_K\;} & K_{<4} \\
{\scriptstyle f}\downarrow & & \downarrow{\scriptstyle f_{<4}} \\
L & \xleftarrow{\;i_L\;} & L_{<4}
\end{array}
$$

commutes. The square

$$
\begin{array}{ccc}
K & \xLongequal{\;i_K\;} & K_{<4} \\
{\scriptstyle f}\downarrow & & \downarrow{\scriptstyle f'_{<4}} \\
L & \xleftarrow{\;i_L\;} & L_{<4}
\end{array}
$$

homotopy commutes because

$$f = i_L b \simeq \mathrm{const} = i_L f'_{<4}.$$

Lastly, $f_{<4}$ and $f'_{<4}$ are not homotopic, for $[f_{<4}] = [b] \neq 0 = [f'_{<4}]$ in $\pi_3(L_{<4})$. This finishes the construction of the example. Since the maps $f_{<4}$ and $f'_{<4}$ do not agree on the 3-skeleton, this example is not rel 3-skeleton. A much deeper example is Example 1.35, which shows that even when one requires the homological n-truncation of a map f to agree with f on the nose on the $(n-1)$-skeleton, and requires all homotopies to be rel $(n-1)$-skeleton, the truncation may not be unique.

Definition 1.10. Let $n \geq 3$ be an integer. The *nth virtual cell group* $VC_n(\Lambda)$ of an n-truncation structure $\Lambda = (L, L/n, h, L_{<n})$ is the homotopy group

$$VC_n(\Lambda) = \pi_{n+1}(L/n \times I, L_{<n} \times \partial I \cup L^{n-1} \times I).$$

If an n-truncation structure Λ has been fixed for a space L then we shall also write $VC_n(L)$.

The choice of terminology arises from the fact that $VC_n(L)$ naturally sits between two actual cellular chain groups: The inclusion of pairs

$$(L_{<n} \times I, L_{<n} \times \partial I \cup L^{n-1} \times I) \subset (L/n \times I, L_{<n} \times \partial I \cup L^{n-1} \times I)$$

induces a map ϕ,

$$C_{n+1}(L_{<n} \times I) \cong \pi_{n+1}(L_{<n} \times I, L_{<n} \times \partial I \cup L^{n-1} \times I) \longrightarrow VC_n(L),$$

where the first isomorphism derives from the fact that $L_{<n} \times \partial I \cup L^{n-1} \times I$ is precisely the n-skeleton of $L_{<n} \times I$: Since L is simply connected, L and $L_{<n}$ have the same $(n-1)$-skeleton, and $n-1 \geq 2$, we have $L_{<n}$ simply connected. Thus the cylinder $L_{<n} \times I$ and its n-skeleton are simply connected, again using $n \geq 3$. Therefore, the Hurewicz map is an isomorphism. Similar remarks apply to the cylinder $L/n \times I$. The inclusion of the pairs

$$(L/n \times I, L_{<n} \times \partial I \cup L^{n-1} \times I) \subset (L/n \times I, L/n \times \partial I \cup L^{n-1} \times I)$$

induces a map ψ,

$$VC_n(L) \longrightarrow \pi_{n+1}(L/n \times I, L/n \times \partial I \cup L^{n-1} \times I) \cong C_{n+1}(L/n \times I).$$

The virtual cell group of L comes equipped with an important endomorphism $E_L \in \text{End}(VC_n(L))$. To construct it, we observe first that

$$\psi\phi : C_{n+1}(L_{<n} \times I) \longrightarrow C_{n+1}(L/n \times I)$$

is the canonical inclusion

$$\bigoplus_\alpha \mathbb{Z}[\chi(y_\alpha \times I)] \hookrightarrow \bigoplus_\alpha \mathbb{Z}[\chi(y_\alpha \times I)] \oplus \bigoplus_\beta \mathbb{Z}[\chi(z_\beta \times I)],$$

where the y_α are the n-cells of $L_{<n}$ and the z_β are the rest of the n-cells in L/n, constituting a basis of the cycle group $Z_n(L/n)$. (Note that thus ϕ is injective.) Let

$$p : C_{n+1}(L/n \times I) \longrightarrow C_{n+1}(L_{<n} \times I)$$

be the projection

$$p(\sum_\alpha \lambda_\alpha[\chi(y_\alpha \times I)] + \sum_\beta \mu_\beta[\chi(z_\beta \times I)]) = \sum_\alpha \lambda_\alpha[\chi(y_\alpha \times I)],$$

and set
$$E_L = \phi \circ p \circ \psi : VC_n(L) \longrightarrow VC_n(L).$$

Definition 1.11. An element $x \in VC_n(L)$ is called an *eigenclass* if $x \in \ker(E_L - 1)$.

In other words, x is an eigenclass iff $x = \sum_\alpha \lambda_\alpha \phi[\chi(y_\alpha \times I)]$, where $\psi(x)$ dictates the coefficients λ_α.

Lemma 1.12. *If $x \in VC_n(L)$ is an eigenclass, then x is not torsion.*

Proof. Suppose $kx = 0$, $k \in \mathbb{Z}$, $k > 1$. From $x = \phi p \psi(x)$, it follows that $\phi(k \cdot p\psi(x)) = kx = 0$. By the injectivity of ϕ, $k \cdot p\psi(x) = 0$, so that $p\psi(x) \in C_{n+1}(L_{<n} \times I)$ is torsion if not zero. But $C_{n+1}(L_{<n} \times I)$ is free abelian, so $p\psi(x) = 0$. This implies $x = \phi p\psi(x) = \phi(0) = 0$. $\square$

Example 1.13. Let us work out the case $L = \mathbb{C}P^2$, complex projective space with the usual CW-structure, and $n = 4$. This space is already 4-segmented, so that $L/4 = \mathbb{C}P^2$. The single 4-cell is a cycle. Therefore, there are no cells y_α and $L_{<4} = L^3 = L^2 = S^2$. The virtual cell group $VC_4(\mathbb{C}P^2)$ is nontrivial, in fact, contains an infinite cyclic subgroup. To see this, we note that

$$VC_4(\mathbb{C}P^2) = \pi_5(\mathbb{C}P^2 \times I, S^2 \times I) \cong \pi_5(\mathbb{C}P^2, S^2)$$

and consider the exact homotopy group sequence of the pair $(\mathbb{C}P^2, S^2)$:

$$\pi_5(S^2) \longrightarrow \pi_5(\mathbb{C}P^2) \longrightarrow \pi_5(\mathbb{C}P^2, S^2).$$

Using the fiber bundle $S^1 \to S^5 \to \mathbb{C}P^2$, we find $\pi_5(\mathbb{C}P^2) \cong \pi_5(S^5) \cong \mathbb{Z}$. Since $\pi_5(S^2) \cong \mathbb{Z}/2$, the left map is the zero map. Consequently, the right-hand map injects an infinite cyclic subgroup into $\pi_5(\mathbb{C}P^2, S^2)$.

As for the maps ϕ and ψ, we have

$$0 = C_5(S^2 \times I) \xrightarrow{\phi=0} VC_4(\mathbb{C}P^2) \xrightarrow{\psi} C_5(\mathbb{C}P^2 \times I) \cong \mathbb{Z}.$$

It follows that the endomorphism $E_{\mathbb{C}P^2}$ is zero and none of the nontrivial elements of $VC_4(\mathbb{C}P^2)$ are eigenclasses.

Example 1.14. We work out $L = S^3 \cup_2 e^4$, that is, the Moore space obtained by attaching a 4-cell to the 3-sphere by a map of degree 2, again for $n = 4$. Here the 4-cell is not a cycle. Thus there are no cells z_β, e^4 is a cell y_α and there are no other y_α. We conclude that $L_{<4} = L/4 = L$. To analyze $VC_4(L)$ and the action of E_L, we consider the diagram

$$
\begin{array}{ccccc}
C_5(L_{<4} \times I) & \xhookrightarrow{\phi} & VC_4(L) & \xrightarrow{\psi} & C_5(L/4 \times I) \\
\big\| & & & & \big\| \\
\mathbb{Z}[\chi(e^4 \times I)] & \xrightarrow{\psi\phi=\mathrm{id}} & & & \mathbb{Z}[\chi(e^4 \times I)]
\end{array}
$$

Let $A = L/n \times \partial I \cup L^{n-1} \times I$ and $B = L_{<n} \times \partial I \cup L^{n-1} \times I$. The map ψ fits into the exact homotopy sequence of the triple $(L/n \times I, A, B)$:

$$\pi_{n+1}(A, B) \longrightarrow VC_n(L) \xrightarrow{\psi} C_{n+1}(L/n \times I) \longrightarrow \pi_n(A, B).$$

In the present situation, $A = B$, so $\pi_5(A, B) = \pi_4(A, B) = 0$ and thus ψ is an isomorphism. In particular, $VC_4(L) \cong \mathbb{Z}$. If $\phi\psi(1) = m$, then

$$\psi(1) = \psi\phi\psi(1) = \psi(m).$$

Therefore, $m = 1$ and so $\phi\psi$ is the identity. The projection p is the identity as well. Thus the endomorphism $E_L = \phi p \psi = \phi \psi = 1$ is the identity. It follows that *every* element of $VC_4(L)$ is an eigenclass.

The concept of eigenclasses leads to the concept of an eigenhomotopy. If $H : K \times I \to L$ is a homotopy, then we may regard it as a map $H' : K \times I \to L \times I$ by setting $H'(k,t) = (H(k,t),t)$. (Caution: If H is cellular, then H' need not be cellular.) For any cell e in a CW-complex, $\chi(e)$ denotes its characteristic map.

Definition 1.15. Let K be a simply connected n-dimensional CW-complex and $\Lambda = (L, L/n, h, L_{<n})$ an n-truncation structure. Let $H : K \times I \to L/n$ be a cellular homotopy rel K^{n-1} such that $H(K \times \partial I) \subset L_{<n} \subset L/n$. The homotopy H is called an *eigenhomotopy* if $H'_*[\chi(y \times I)]$ is an eigenclass in $VC_n(\Lambda)$ for every n-cell y of K. Here H'_* is the induced map on homotopy groups,

$$H'_* : C_{n+1}(K \times I) \cong \pi_{n+1}(K \times I, K \times \partial I \cup K^{n-1} \times I) \longrightarrow VC_n(\Lambda).$$

Note that H' does in fact map $K \times \partial I \cup K^{n-1} \times I$ into the subcomplex $L_{<n} \times \partial I \cup L^{n-1} \times I$ because $H'(K \times \partial I) \subset L_{<n} \times \partial I$ and for $k \in K^{n-1}$ and $t \in I$ we have $H(k,t) = H(k,0)$ as H is rel K^{n-1} and

$$H(k,0) \in H(K^{n-1} \times \partial I) \subset H((K \times I)^{n-1}) \subset L^{n-1}$$

(since H is cellular), whence

$$H'(k,t) = (H(k,0),t) \in L^{n-1} \times I.$$

Eigenhomotopies will be used later on (Definition 1.33) in defining compression rigid maps.

Example 1.16. Suppose K is a space whose cellular boundary operator in degree 4 vanishes. For instance, K might not have any 3-cells. In this situation, every homotopy $H : K_{<4} \times I \to \mathbb{CP}^2/4 = \mathbb{CP}^2$ with $H(K_{<4} \times \partial I) \subset \mathbb{CP}^2_{<4} = S^2$ is an eigenhomotopy, even though $VC_4(\mathbb{CP}^2)$ has no nontrivial eigenclasses according to Example 1.13. The reason is that by Example 1.8, $K_{<4} = K^3$ and thus $C_5(K_{<4} \times I) = C_5(K^3 \times I) = 0$. Therefore, H'_* is the zero map.

Example 1.17. Every cellular homotopy $H : K \times I \to S^3 \cup_2 e^4$ which is rel K^3 is an eigenhomotopy for $n = 4$, as follows from Example 1.14.

Proposition 1.18. *Let $n \geq 3$ be an integer and $\Lambda = (L, L/n, h, L_{<n})$ an n-truncation structure such that L/n has finitely many n-cells. Let G be the abelian group*

$$G = \mathbb{Z}^{c-b} \oplus (\mathbb{Z}/2)^b,$$

where $b = b_n(L^n)$ is the nth Betti number of L^n and c is the number of n-cells of L/n. Then

(1) $VC_n(\Lambda)$ maps onto G.

(2) If $H_2(L^{n-1}) = 0$, then $VC_n(\Lambda) \cong G$.

The free abelian part $\mathbb{Z}^{c-b}$ in G corresponds to the cells of type y_α in L/n, the torsion part $(\mathbb{Z}/2)^b$ in G corresponds to the cells of type z_β in L/n.

Proof. Since L/n has finitely many n-cells, we can write

$$L/n = L^{n-1} \cup y_1 \cup \cdots \cup y_{c-a} \cup z_1 \cup \cdots \cup z_a,$$

where $\{z_1, \ldots, z_a\}$ is a basis for $Z_n(L/n)$ and $y_1, \ldots, y_{c-a}$ are the n-cells of $L_{<n} \subset L/n$. As L/n is n-dimensional, we have $H_n(L/n) = Z_n(L/n)$. The homotopy equivalence h induces an isomorphism $H_n(L/n) \cong H_n(L^n)$. It follows that $a = b$. We shall use the following consequence of the homotopy excision theorem; see [Hat02, page 364, Proposition 4.28]: If a CW-pair (X, A) is r-connected and A is s-connected, with $r, s \geq 0$, then the map $\pi_i(X, A) \to \pi_i(X/A)$ induced by the quotient map $X \to X/A$ is an isomorphism for $i \leq r + s$ and a surjection for $i = r + s + 1$. A CW-pair (X, A) is r-connected if all the cells in $X - A$ have dimension greater than r. The complement $(L/n \times I) - (L_{<n} \times \partial I \cup L^{n-1} \times I)$ contains cells of dimension $n+1$, namely the $y_j \times (0, 1)$ and the $z_i \times (0, 1)$, as well as cells of dimension n, namely the $z_i \times \{0\}$ and $z_i \times \{1\}$. Thus the CW-pair $(L/n \times I, L_{<n} \times \partial I \cup L^{n-1} \times I)$ is $r = (n-1)$-connected. The subspace $P = L_{<n} \times \partial I \cup L^{n-1} \times I$ is $s = 1$-connected, being the n-skeleton of the simply connected space $L_{<n} \times I$ ($n \geq 3$). Thus, as $n + 1 \leq r + s + 1 = n + 1$,

$$VC_n(\Lambda) = \pi_{n+1}(L/n \times I, P) \longrightarrow \pi_{n+1}((L/n \times I)/P)$$

is surjective. We shall show that $\pi_{n+1}((L/n \times I)/P) \cong G$. Let us investigate the homotopy type of the quotient space

$$\frac{L/n \times I}{P} = \frac{L^{n-1} \times I \cup \bigcup_{j=1}^{c-b} y_j \times I \cup \bigcup_{i=1}^{b} z_i \times I}{L^{n-1} \times I \cup \bigcup_{j=1}^{c-b} y_j \times \partial I}.$$

The boundary of an $(n+1)$-cell $y_j \times I$ is attached to $L^{n-1} \times I \cup y_j \times \partial I$, which is being collapsed to a point. Thus every $y_j \times I$ becomes an $(n+1)$-sphere S_j^{n+1} in the quotient. The boundary of an n-cell $z_i \times \{t\}$, $t \in \{0, 1\}$, is attached to $L^{n-1} \times \{t\}$, which is being collapsed to a point. Thus every $z_i \times \{t\}$ becomes an n-sphere $S_i^n \times \{t\}$ in the quotient. The boundary of an $(n+1)$-cell $z_i \times I$ is attached to $L^{n-1} \times I \cup z_i \times \partial I$, but, as we have seen, $z_i \times \partial I$ is *not* collapsed to a point, rather to spheres $S_i^n \times \partial I$. Consequently, every $z_i \times I$ becomes $(S_i^n \times I)/(* \times I)$ in the quotient, where $*$ is the base point in the sphere. The space $(S_i^n \times I)/(* \times I)$ is homotopy equivalent to S_i^n, since $* \times I$ is contractible, so $(S_i^n \times I)/(* \times I) \simeq S_i^n \times I \simeq S_i^n$. Therefore,

$$\frac{L/n \times I}{P} \simeq \bigvee_{j=1}^{c-b} S_j^{n+1} \vee \bigvee_{i=1}^{b} S_i^n,$$

and we need to show that

$$\pi_{n+1}(\bigvee S_j^{n+1} \vee \bigvee S_i^n) \cong G.$$

In order to do so, we use the natural decomposition

$$\pi_{n+1}(X \vee Y) \cong \pi_{n+1}(X) \oplus \pi_{n+1}(Y) \oplus \pi_{n+2}(X \times Y, X \vee Y)$$

together with the fact that for a p-connected X and a q-connected Y, $\pi_{n+2}(X \times Y, X \vee Y)$ vanishes when $n+2 \leq p+q+1$. Let $X = \bigvee_j S_j^{n+1}$, a $p = n$-connected space, and $Y = \bigvee_i S_i^n$ a $q = (n-1)$-connected space. As $n+2 \leq p+q+1 = 2n$ (recall $n \geq 3$), we have $\pi_{n+2}(X \times Y, X \vee Y) = 0$ and

$$\pi_{n+1}(\bigvee S_j^{n+1} \vee \bigvee S_i^n) \cong \pi_{n+1}(\bigvee S_j^{n+1}) \oplus \pi_{n+1}(\bigvee S_i^n).$$

By the Hurewicz theorem,

$$\pi_{n+1}(\bigvee S_j^{n+1}) \cong H_{n+1}(\bigvee_{j=1}^{c-b} S_j^{n+1}) \cong \mathbb{Z}^{c-b}.$$

For the n-spheres, we have the formula

$$\pi_{n+1}(\bigvee S_i^n) = \bigoplus_{i=1}^{b} \pi_{n+1}(S_i^n),$$

since $\pi_{n+2}(S_1^n \times (S_2^n \vee \cdots \vee S_b^n), \bigvee_i S_i^n) = 0$, as follows from the $(n-1)$-connectivity of S_1^n and $S_2^n \vee \cdots \vee S_b^n$, observing that $n+2 \leq 2(n-1)+1$ (again using $n \geq 3$), together with an induction on b. As $n \geq 3$, we have

$$\pi_{n+1}(S_i^n) = \mathbb{Z}/2.$$

This establishes statement (1). To prove statement (2), we assume $H_2(L^{n-1}) = 0$. The homeomorphism

$$\frac{P}{L^{n-1} \times I} \cong \bigvee_{j=1}^{c-b} (S_j^n \times \{0\} \vee S_j^n \times \{1\})$$

implies

$$H_2(P, L^{n-1} \times I) \cong H_2(P/(L^{n-1} \times I)) = 0.$$

From the exact sequence

$$0 = H_2(L^{n-1}) \cong H_2(L^{n-1} \times I) \longrightarrow H_2(P) \longrightarrow H_2(P, L^{n-1} \times I) = 0$$

of the pair $(P, L^{n-1} \times I)$ we conclude that $H_2(P) = 0$. Since P is simply connected, it follows from the Hurewicz theorem that P is $s = 2$-connected. Thus, as $n + 1 \leq r + s = (n-1) + 2$,

$$VC_n(\Lambda) = \pi_{n+1}(L/n \times I, P) \longrightarrow \pi_{n+1}((L/n \times I)/P)$$

is an isomorphism.   □

If $H_2(L^{n-1})$ is not zero in Proposition 1.18 then $VC_n(\Lambda)$ need not be isomorphic to G. Consider as an example the space $L = \mathbb{C}P^2$ with its standard 4-truncation structure $\Lambda = (\mathbb{C}P^2, \mathbb{C}P^2, \mathrm{id}, S^2)$. Note that $H_2(L^{n-1}) = H_2(S^2) \cong \mathbb{Z} \neq 0$. The fourth Betti number $b = b_4(\mathbb{C}P^2) = 1$ and the number of 4-cells of $L/4 = \mathbb{C}P^2$ is $c = 1$. Thus $G = \mathbb{Z}/2$, and Proposition 1.18 asserts that $VC_4(\mathbb{C}P^2)$ maps onto $\mathbb{Z}/2$. However, according to Example 1.13, $VC_4(\mathbb{C}P^2)$ contains $\mathbb{Z}$. Thus $VC_4(\mathbb{C}P^2) \not\cong G$.

Example 1.19. In Example 1.14, we have seen that $VC_4(\Lambda) \cong \mathbb{Z}$ for the 4-truncation structure $\Lambda = (S^3 \cup_2 e^4, S^3 \cup_2 e^4, \mathrm{id}, S^3 \cup_2 e^4)$. This is confirmed by Proposition 1.18: As $b = b_4(S^3 \cup_2 e^4) = 0$ and $c = 1$, we have $G = \mathbb{Z}$. Since $H_2(L^{n-1}) = H_2(L^3) = H_2(S^3) = 0$, the proposition implies $VC_4(\Lambda) \cong G = \mathbb{Z}$.

1.1.5 Functoriality in Low Dimensions

Let **CW** be the category of CW-complexes and cellular maps, let $\mathbf{CW}^0$ be the full subcategory of path connected CW-complexes and let $\mathbf{CW}^1$ be the full subcategory of simply connected CW-complexes. Let **HoCW** denote the category of CW-complexes and homotopy classes of cellular maps. Let $\mathbf{HoCW}_n$ denote the category of CW-complexes and rel n-skeleton homotopy classes of cellular maps.

Dimension $n = 1$: It is straightforward to define a covariant truncation functor

$$t_{<n} = t_{<1} : \mathbf{CW}^0 \longrightarrow \mathbf{HoCW}$$

together with a natural transformation

$$\mathrm{emb}_1 : t_{<1} \longrightarrow t_{<\infty},$$

where $t_{<\infty} : \mathbf{CW}^0 \to \mathbf{HoCW}$ is the natural "inclusion-followed-by-quotient" functor given by $t_{<\infty}(K) = K$ for objects K and $t_{<\infty}(f) = [f]$ for morphisms f, such that for all objects K, $\mathrm{emb}_{1*} : H_0(t_{<1}K) \to H_0(t_{<\infty}K)$ is an isomorphism and $H_r(t_{<1}K) = 0$ for $r \geq 1$. The details are as follows: For a path connected CW-complex K, set $t_{<1}(K) = k^0$, where k^0 is a 0-cell of K. Let $\mathrm{emb}_1(K) : t_{<1}(K) = k^0 \to t_{<\infty}(K) = K$ be the inclusion of k^0 in K. Then emb_{1*} is an isomorphism on H_0 as K is path connected. Clearly $H_r(t_{<1}K) = 0$ for $r \geq 1$. Let $f : K \to L$ be a cellular map between objects of $\mathbf{CW}^0$. The morphism $t_{<1}(f) : t_{<1}(K) = k^0 \to l^0 = t_{<1}(L)$ is the homotopy class of the unique map from a point to a point. In particular, $t_{<1}(\mathrm{id}_K) = [\mathrm{id}_{k^0}]$ and for a cellular map $g : L \to P$ we have $t_{<1}(gf) = t_{<1}(g) \circ t_{<1}(f)$, so that $t_{<1}$ is indeed a functor. To show that emb_1 is a natural transformation, we need to see that

$$
\begin{array}{ccc}
t_{<1}(K) & \xrightarrow{\mathrm{emb}_1(K)} & t_{<\infty}(K) \\
{\scriptstyle t_{<1}(f)}\Big\downarrow & & \Big\downarrow{\scriptstyle t_{<\infty}(f)} \\
t_{<1}(L) & \xrightarrow{\mathrm{emb}_1(L)} & t_{<\infty}(L),
\end{array}
$$

that is

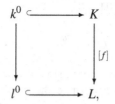

commutes in **HoCW**. This is where we need the functor $t_{<1}$ to have values only in **HoCW**, not in **CW**, because the square need certainly not commute in **CW**. (The points k^0 and l^0 do not know anything about f, so l^0 need not be the image of k^0 under f.) Since L is path connected, there is a path $\omega : I \to L$ from $l^0 = \omega(0)$ to $f(k^0) = \omega(1)$. Then $H : \{k^0\} \times I \to L$, $H(k^0, t) = \omega(t)$, defines a homotopy from $k^0 \to l^0 \hookrightarrow L$ to $k^0 \hookrightarrow K \xrightarrow{f} L$.

Dimension $n = 2$: We will define a covariant truncation functor

$$
t_{<n} = t_{<2} : \mathbf{CW}^1 \longrightarrow \mathbf{HoCW}
$$

together with a natural transformation

$$
\mathrm{emb}_2 : t_{<2} \longrightarrow t_{<\infty},
$$

where $t_{<\infty} : \mathbf{CW}^1 \to \mathbf{HoCW}$ is as above (only restricted to simply connected spaces), such that for all objects K, $\mathrm{emb}_{2*} : H_r(t_{<2}K) \to H_r(t_{<\infty}K)$ is an isomorphism for $r = 0, 1$, and $H_r(t_{<2}K) = 0$ for $r \geq 2$. For a simply connected CW-complex K, set $t_{<2}(K) = k^0$, where k^0 is a 0-cell of K. Let $\mathrm{emb}_2(K) : t_{<2}(K) = k^0 \to t_{<\infty}(K) = K$ be the inclusion as in the case $n = 1$. It follows that emb_{2*} is an isomorphism both on H_0 as K is path connected and on H_1 as $H_1(k^0) = 0 = H_1(K)$, while trivially $H_r(t_{<2}K) = 0$ for $r \geq 2$. On a cellular map f, $t_{<2}(f)$ is defined as in the case $n = 1$. As in the case $n = 1$, this yields a functor and emb_2 is a natural transformation.

1.1.6 Functoriality in Dimensions $n \geq 3$

Let $n \geq 3$ be an integer.

Definition 1.20. A *morphism*

$$(K, K/n, h_K, K_{<n}) \longrightarrow (L, L/n, h_L, L_{<n})$$

of homological n-truncation structures is a commutative diagram

$$
\begin{array}{ccccccc}
K & \xleftarrow{\;j_K\;} & K^n & \xleftarrow{\;h_K\;} & K/n & \xleftarrow{\;i_K\;} & K_{<n} \\
\downarrow{\scriptstyle f} & & \downarrow{\scriptstyle f|} & & \downarrow{\scriptstyle f/n} & & \downarrow{\scriptstyle f_{<n}} \\
L & \xleftarrow{\;j_L\;} & L^n & \xleftarrow{\;h_L\;} & L/n & \xleftarrow{\;i_L\;} & L_{<n}
\end{array}
$$

in **CW**. The composition of two morphisms of n-truncation structures is defined in the obvious way. Let $\mathbf{CW}_{\supset <n}$ denote the resulting category of n-truncation structures.

Commutativity on the nose is rarely achieved in practice. More important is thus the associated rel $(n-1)$-skeleton homotopy category $\mathbf{HoCW}_{\supset <n}$ whose objects are n-truncation structures as before, but whose morphisms are now commutative diagrams

$$
\begin{array}{ccccccc}
K & \xleftarrow{\;[j_K]\;} & K^n & \xleftarrow{\;[h_K]\;} & K/n & \xleftarrow{\;[i_K]\;} & K_{<n} \\
\downarrow & & \downarrow & & \downarrow & & \downarrow \\
L & \xleftarrow{\;[j_L]\;} & L^n & \xleftarrow{\;[h_L]\;} & L/n & \xleftarrow{\;[i_L]\;} & L_{<n}
\end{array}
$$

in $\mathbf{HoCW}_{n-1}$, where $[-]$ denotes the rel $(n-1)$-skeleton homotopy class of a cellular map. Thus a morphism $F : (K, K/n, h_K, K_{<n}) \to (L, L/n, h_L, L_{<n})$ in $\mathbf{HoCW}_{\supset <n}$ is a quadruple $F = ([f], [f^n], [f/n], [f_{<n}])$ represented by a diagram

$$
\begin{array}{ccccccc}
K & \xleftarrow{\;j_K\;} & K^n & \xleftarrow{\;h_K\;} & K/n & \xleftarrow{\;i_K\;} & K_{<n} \\
\downarrow{\scriptstyle f} & & \downarrow{\scriptstyle f^n} & & \downarrow{\scriptstyle f/n} & & \downarrow{\scriptstyle f_{<n}} \\
L & \xleftarrow{\;j_L\;} & L^n & \xleftarrow{\;h_L\;} & L/n & \xleftarrow{\;i_L\;} & L_{<n}
\end{array}
$$

with $f j_K \simeq j_L f^n$ rel K^{n-1}, $h_L(f/n) \simeq f^n h_K$ rel K^{n-1}, and $(f/n)i_K \simeq i_L f_{<n}$ rel K^{n-1}. (The map f^n is not required to be the restriction of f to K^n.) Two morphisms

$([f],[f^n],[f/n],[f_{<n}])$ and $([g],[g^n],[g/n],[g_{<n}])$ are equal iff $f \simeq g$ rel K^{n-1}, $f^n \simeq g^n$ rel K^{n-1}, $f/n \simeq g/n$ rel K^{n-1}, and $f_{<n} \simeq g_{<n}$ rel K^{n-1}. Note that it is necessary to record the four components of the quadruple $([f],[f^n],[f/n],[f_{<n}])$, since not even $[f^n]$, for example, is determined by $[f]$: Consider the n-truncation structures $(K,K/n,h_K,K_{<n}) = (S^n = e^0 \cup e^n, S^n, \mathrm{id}_{S^n}, e^0)$ and $(L,L/n,h_L,L_{<n}) = (e^0 \cup e^n \cup_2 e^{n+1}, S^n, \mathrm{id}_{S^n}, e^0)$. Let $f : K \to L$ be the map $S^n \xrightarrow{2} S^n \hookrightarrow L$ and let $g : K \to L$ be the constant map to e^0. Then $f \simeq g$ rel $K^{n-1} = e^0$, but $f^n = f|_{K^n} : K^n = S^n \xrightarrow{2} S^n = L^n$ is not homotopic to $g^n = g|_{L^n} = \mathrm{const}_{e_0} : K^n \to L^n$. However, $[f/n]$ is determined uniquely by $[f^n]$: Let $h'_L : L^n \to L/n$ be a rel L^{n-1} homotopy inverse for h_L. Then the requirement $[h_L] \circ [f/n] = [f^n] \circ [h_K]$ implies the formula

$$[f/n] = [h'_L] \circ [f^n] \circ [h_K].$$

This formula determines $[f/n]$, since if $h''_L : L^n \to L/n$ is another rel L^{n-1} homotopy inverse for h_L, then $[h'_L] = [h'_L] \circ [h_L] \circ [h''_L] = [h''_L]$.

Lemma 1.21. *A morphism*

$$F = ([f],[f^n],[f/n],[f_{<n}]) : (K,K/n,h_K,K_{<n}) \to (L,L/n,h_L,L_{<n})$$

in $\mathbf{HoCW}_{\supset <n}$ *is an isomorphism if, and only if,* $f, f^n, f/n$ *and* $f_{<n}$ *are homotopy equivalences rel* K^{n-1}.

Proof. Suppose there exists a morphism $G : (L,L/n,h_L,L_{<n}) \to (K,K/n,h_K,K_{<n})$ such that $G \circ F = \mathrm{id}$ and $F \circ G = \mathrm{id}$ in $\mathbf{HoCW}_{\supset <n}$. With $G = ([g],[g^n],[g/n],[g_{<n}])$,

$$([\mathrm{id}_K],[\mathrm{id}_{K^n}],[\mathrm{id}_{K/n}],[\mathrm{id}_{K_{<n}}]) = \mathrm{id} = G \circ F$$
$$= ([g \circ f],[g^n \circ f^n],[(g/n) \circ (f/n)],[g_{<n} \circ f_{<n}])$$

implies $g \circ f \simeq \mathrm{id}_K$ rel K^{n-1}, $g^n \circ f^n \simeq \mathrm{id}_{K^n}$ rel K^{n-1}, $g/n \circ f/n \simeq \mathrm{id}_{K/n}$ rel K^{n-1}, and $g_{<n} \circ f_{<n} \simeq \mathrm{id}_{K_{<n}}$ rel K^{n-1}. Similarly, homotopies $f \circ g \simeq \mathrm{id}_L$ rel L^{n-1}, etc. are obtained from $F \circ G = \mathrm{id}$.

Conversely, assume that $f, f^n, f/n$ and $f_{<n}$ are homotopy equivalences rel K^{n-1}. Let $g, g^n, g/n$ and $g_{<n}$ be homotopy inverses rel $(n-1)$-skeleta for $f, f^n, f/n$ and $f_{<n}$, respectively, and set $G = ([g],[g^n],[g/n],[g_{<n}])$. Then G is indeed a morphism in $\mathbf{HoCW}_{\supset <n}$, for in the diagram

$$
\begin{array}{ccccccc}
L & \xleftarrow{\ j_L\ } & L^n & \xleftarrow{\ h_L\ } & L/n & \xleftarrow{\ i_L\ } & L_{<n} \\
\downarrow g & & \downarrow g^n & & \downarrow g/n & & \downarrow g_{<n} \\
K & \xleftarrow{\ j_K\ } & K^n & \xleftarrow{\ h_K\ } & K/n & \xleftarrow{\ i_K\ } & K_{<n}
\end{array}
$$

we have homotopy commutativity, rel L^{n-1}, in all three squares: Since $j_L f^n \simeq f j_K$ rel K^{n-1}, we have $g j_L f^n g^n \simeq g f j_K g^n$ rel L^{n-1} and so $g j_L \simeq j_K g^n$ rel L^{n-1}. Since

$f^n h_K \simeq h_L(f/n)$ rel K^{n-1}, we have $g^n f^n h_K(g/n) \simeq g^n h_L(f/n)(g/n)$ rel L^{n-1} and so $h_K(g/n) \simeq g^n h_L$ rel L^{n-1}. Finally, since $(f/n)i_K \simeq i_L f_{<n}$ rel K^{n-1}, we have $(g/n)(f/n)i_K g_{<n} \simeq (g/n)i_L f_{<n} g_{<n}$ rel L^{n-1} and so $i_K g_{<n} \simeq (g/n)i_L$ rel L^{n-1}. Clearly, G is an inverse for F in $\mathbf{HoCW}_{\supset <n}$ □

Definition 1.22. The category $\mathbf{CW}_{n\supset\partial}$ of *n-boundary-split CW-complexes* consists of the following objects and morphisms: Objects are pairs (K,Y), where K is a simply connected CW-complex and $Y \subset C_n(K)$ is a subgroup of the nth cellular chain group of K that arises as the image $Y = s(\text{im}\,\partial)$ of some splitting $s : \text{im}\,\partial \to C_n(K)$ of the boundary map $\partial : C_n(K) \to \text{im}\,\partial(\subset C_{n-1}(K))$. (Given K, such a splitting always exists, since $\text{im}\,\partial$ is free abelian.) A morphism $(K,Y_K) \to (L,Y_L)$ is a cellular map $f : K \to L$ such that $f_*(Y_K) \subset Y_L$. The composition of morphisms is defined, since for a second morphism $(L,Y_L) \to (P,Y_P)$, given by a cellular map $g : L \to P$ with $g_*(Y_L) \subset Y_P$, we have $(g \circ f)_*(Y_K) = g_*(f_*(Y_K)) \subset g_*(Y_L) \subset Y_P$.

Example 1.23. This example expands on the theme of Example 1.8. Suppose K is homotopy equivalent to a space L whose nth cellular boundary map is zero. Let $f : K \to L$ be a homotopy equivalence with homotopy inverse $g : L \to K$. Further, choose a homotopy H from gf to the identity. Then H induces a canonical choice Y_K so that $(K,Y_K) \in \mathbf{CW}_{n\supset\partial}$: We have an induced diagram

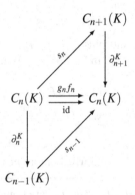

where g_n, f_n are the chain maps induced by g, f, respectively, and $\{s_n\}$ is the chain homotopy induced by H. Applying ∂_n^K to the equation

$$\partial_{n+1}^K s_n + s_{n-1}\partial_n^K = \text{id} - g_n f_n,$$

we obtain

$$\partial_n^K s_{n-1}\partial_n^K = \partial_n^K$$

because $\partial_n^K \partial_{n+1}^K = 0$ and $\partial_n^K g_n = g_{n-1}\partial_n^L = 0$. Thus

$$s = s_{n-1}| : \text{im}\,\partial_n^K \longrightarrow C_n(K)$$

is a splitting for ∂_n^K on its image, giving $Y_K = s_{n-1}|(\text{im}\,\partial_n^K)$.

We shall construct a covariant assignment

$$\tau_{<n} : \mathbf{CW}_{n \supset \partial} \longrightarrow \mathbf{HoCW}_{\supset <n}$$

of objects and morphisms. We will see later that the assignment is a functor on subcategories of $\mathbf{CW}_{n \supset \partial}$ whose morphisms have n-compression rigid image under $\tau_{<n}$ (see Definition 1.33). Let (K, Y_K) be an object of $\mathbf{CW}_{n \supset \partial}$. By Proposition 1.6, (K, Y_K) can be completed to an n-truncation structure $(K, K/n, h_K, K_{<n})$ in $\mathbf{CW}_{\supset <n}$ such that $h_{K*} i_{K*} C_n(K_{<n}) = Y_K$, where $i_{K*} : C_n(K_{<n}) \to C_n(K/n)$ is the monomorphism induced by the inclusion $i_K : K_{<n} \hookrightarrow K/n$. Choose such a completion and set

$$\boxed{\tau_{<n}(K, Y_K) = (K, K/n, h_K, K_{<n}).}$$

We will see in Scholium 1.26 below that the rel $(n-1)$-skeleton homotopy type of $K_{<n}$ does not depend on the choice of n-truncation structure completion of (K, Y_K). If the n-skeleton of K already has a cell-basis for its n-cycle group (which implies that it is n-segmented, Lemma 1.2) and Y_K is the canonical subgroup, that is, generated by those n-cells that are not cycles, then we will assume that we have chosen

$$\tau_{<n}(K, Y_K) = (K, K^n, \mathrm{id}_{K^n}, K_{<n}),$$

i.e. $K/n = K^n$ and $h_K = \mathrm{id}_{K^n}$. In this case $K_{<n}$ is uniquely determined by K, Proposition 1.3. However, even if K^n has a cell-basis for its n-cycle group, the subspace Y_K is not unique: The complex $K^3 = (S^2 \cup_2 e^3) \vee S^3$ is 3-segmented, $C_3(K^3) = \mathbb{Z}e^3 \oplus \mathbb{Z}S^3$, $Z_3(K^3) = \mathbb{Z}S^3$, $\mathrm{im}\, \partial_3 = 2\mathbb{Z}S^2$. Any $m \in \mathbb{Z}$ defines a splitting $s : \mathrm{im}\, \partial_3 \to C_3(K^3)$ by $s(2S^2) = e^3 + mS^3$. Thus the possible choices for Y_K are parametrized by m, $Y_K(m) = \mathbb{Z}(e^3 + mS^3) \subset C_3(K^3)$.

Remark 1.24. Knowing that h_K is a homotopy equivalence which restricts to the identity on the $(n-1)$-skeleton implies that the chain map h_{K*} induced by h_K on the cellular chain complexes is in fact a chain isomorphism, not just a chain equivalence. This can be seen as follows: Let K^{n-1} be an $(n-1)$-dimensional CW-complex and let $\{\xi_\alpha^n\}, \{\eta_\alpha^n\}$ be two collections of n-cells indexed by the same set $\{\alpha\}$. Let $X^n = K^{n-1} \cup \bigcup \xi_\alpha^n$ and $Y^n = K^{n-1} \cup \bigcup \eta_\alpha^n$ be n-dimensional CW-complexes obtained from K^{n-1} by attaching the cells ξ_α^n and η_α^n, respectively. Suppose $f : X \to Y$ is a cellular homotopy equivalence which is the identity on K^{n-1}. Then $f_* : C_r(X) \to C_r(Y)$ is the identity for $r < n$ and the zero map between zero groups for $r > n$. So in order to show that f_* is a chain isomorphism, it remains to show this in degree $r = n$. The map of pairs $f : (X, K^{n-1}) \to (Y, K^{n-1})$ induces a commutative ladder on homology exact sequences,

$$
\begin{array}{ccccccccc}
H_n(K^{n-1}) & \longrightarrow & H_n(X) & \longrightarrow & H_n(X, K^{n-1}) & \longrightarrow & H_{n-1}(K^{n-1}) & \longrightarrow & H_{n-1}(X) \\
\downarrow{\scriptstyle f_*=\mathrm{id}} & & {\scriptstyle\cong}\downarrow{\scriptstyle f_*} & & \downarrow{\scriptstyle f_*} & & \downarrow{\scriptstyle f_*=\mathrm{id}} & & {\scriptstyle\cong}\downarrow{\scriptstyle f_*} \\
H_n(K^{n-1}) & \longrightarrow & H_n(Y) & \longrightarrow & H_n(Y, K^{n-1}) & \longrightarrow & H_{n-1}(K^{n-1}) & \longrightarrow & H_{n-1}(Y).
\end{array}
$$

By the 5-Lemma,

$$C_n(X) = H_n(X, K^{n-1}) \xrightarrow{f_*} H_n(Y, K^{n-1}) = C_n(Y)$$

is an isomorphism.

Given a fixed space K, let us proceed to investigate the homotopy theoretic dependence of the truncated space $K_{<n}$, where $\tau_{<n}(K,Y) = (K, K/n, h_K, K_{<n})$, on different choices of Y.

Proposition 1.25. *Let $(K,Y), (K,\overline{Y})$ be two completions of a simply connected CW-complex K to objects in $\mathbf{CW}_{n \supset \partial}$. Let $(K, K/n, h_K, K_{<n}) = \tau_{<n}(K,Y)$ and $(K, \overline{K}/n, \overline{h}_K, \overline{K}_{<n}) = \tau_{<n}(K,\overline{Y})$. Then $K_{<n}$ and $\overline{K}_{<n}$ are cellularly homotopy equivalent rel $(n-1)$-skeleton if and only if $d(Y) = d(\overline{Y})$, where $d : \pi_n(K^n, K^{n-1}) \to \pi_{n-1}(K^{n-1})$ is the boundary homomorphism.*

Proof. Let $f : K_{<n} \to \overline{K}_{<n}$ be a cellular homotopy equivalence rel K^{n-1}. The induced chain map $f_* : C_n(K_{<n}) \to C_n(\overline{K}_{<n})$ in degree n is an isomorphism by Remark 1.24. In particular

$$f_* C_n(K_{<n}) = C_n(\overline{K}_{<n}). \tag{1.5}$$

By the naturality of both the Hurewicz isomorphism and the homotopy boundary homomorphism, the square

$$
\begin{array}{ccc}
C_n(K_{<n}) & \xrightarrow[\cong]{f_*} & C_n(\overline{K}_{<n}) \\
\| & & \| \\
\pi_n(K_{<n}, K^{n-1}) & & \pi_n(\overline{K}_{<n}, K^{n-1}) \\
\downarrow{\scriptstyle d_{<n}} & & \downarrow{\scriptstyle \overline{d}_{<n}} \\
\pi_{n-1}(K^{n-1}) & \xrightarrow{f_* = \mathrm{id}} & \pi_{n-1}(K^{n-1})
\end{array}
$$

commutes, so that

$$\overline{d}_{<n} f_* = d_{<n}. \tag{1.6}$$

By the construction of $\tau_{<n}$, we have

$$h_{K*} i_* C_n(K_{<n}) = Y \tag{1.7}$$

and

$$\overline{h}_{K*} \overline{i}_* C_n(\overline{K}_{<n}) = \overline{Y}, \tag{1.8}$$

where $i : K_{<n} \hookrightarrow K/n$ and $\overline{\imath} : \overline{K}_{<n} \hookrightarrow \overline{K}/n$ are the subspace inclusions. The commutative diagram

$$
\begin{array}{ccccc}
C_n(K_{<n}) & \xrightarrow{\ i_* \ } & C_n(K/n) & \xrightarrow{\ h_{K*} \ } & C_n(K^n) \\
\| & & \| & & \| \\
\pi_n(K_{<n}, K^{n-1}) & & \pi_n(K/n, K^{n-1}) & & \pi_n(K^n, K^{n-1}) \\
\downarrow{\scriptstyle d_{<n}} & & \downarrow{\scriptstyle d/n} & & \downarrow{\scriptstyle d} \\
\pi_{n-1}(K^{n-1}) & \xrightarrow{\ i_*=\mathrm{id}\ } & \pi_{n-1}(K^{n-1}) & \xrightarrow{\ h_{K*}=\mathrm{id}\ } & \pi_{n-1}(K^{n-1})
\end{array}
$$

shows that

$$d \circ h_{K*} \circ i_* = d_{<n}. \tag{1.9}$$

Similarly,

$$d \circ \overline{h}_{K*} \circ \overline{\imath}_* = \overline{d}_{<n}, \tag{1.10}$$

where $\overline{d}_{<n} : \pi_n(\overline{K}_{<n}, K^{n-1}) \to \pi_{n-1}(K^{n-1})$. We conclude

$$
\begin{aligned}
d(Y) &= d h_{K*} i_*(C_n K_{<n}) && \text{by (1.7)} \\
&= d_{<n}(C_n K_{<n}) && \text{by (1.9)} \\
&= \overline{d}_{<n} f_*(C_n K_{<n}) && \text{by (1.6)} \\
&= \overline{d}_{<n}(C_n \overline{K}_{<n}) && \text{by (1.5)} \\
&= d \overline{h}_{K*} \overline{\imath}_*(C_n \overline{K}_{<n}) && \text{by (1.10)} \\
&= d(\overline{Y}) && \text{by (1.8)}.
\end{aligned}
$$

Conversely, assume $d(Y) = d(\overline{Y})$. In the first step, we will construct an isomorphism $\theta : C_n(K_{<n}) \to C_n(\overline{K}_{<n})$ such that

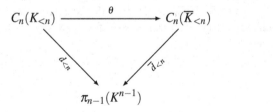

commutes. In the second step, we will realize θ by a continuous map. We claim that $d_{<n}$ and $\overline{d}_{<n}$ are injective. Indeed, the chain boundary

$$\partial_n : C_n(K_{<n}) \longrightarrow C_{n-1}(K_{<n})$$

is injective since

$$\ker \partial_n = H_n(K_{<n}) = 0,$$

and factors as

$$
\begin{array}{ccc}
C_n(K_{<n}) & \xrightarrow{\;d_{<n}\;} & \pi_{n-1}(K^{n-1}) \\
\cap \downarrow \scriptstyle{\partial_n} & & \downarrow \scriptstyle{\text{incl}_*} \\
C_{n-1}(K_{<n}) & \xleftarrow{\;\text{Hur}\;} & \pi_{n-1}(K^{n-1}, K^{n-2}),
\end{array}
$$

which implies that $d_{<n}$ is injective. The same argument applied to the chain boundary operator of $\overline{K}_{<n}$ yields the injectivity of $\overline{d}_{<n}$. Equations $(1.7) - (1.10)$ above still hold in the present context (as they do not involve the homotopy equivalence f). Thus

$$d_{<n}(C_n K_{<n}) = d(Y) = d(\overline{Y}) = \overline{d}_{<n}(C_n \overline{K}_{<n}).$$

Since $\overline{d}_{<n}$ is an isomorphism onto its image, there is an inverse

$$\overline{d}_{<n}^{-1} : \overline{d}_{<n}(C_n \overline{K}_{<n}) \xrightarrow{\;\cong\;} C_n \overline{K}_{<n}.$$

We define θ to be the composition

$$
\begin{array}{ccc}
C_n(K_{<n}) & \xrightarrow[\cong]{\;d_{<n}\;} & d_{<n}(C_n K_{<n}) = \overline{d}_{<n}(C_n \overline{K}_{<n}) \\
& \searrow \scriptstyle{\theta} & \downarrow \scriptstyle{\cong}\; \overline{d}_{<n}^{-1} \\
& & C_n(\overline{K}_{<n}).
\end{array}
$$

In order to realize θ topologically, we proceed as in the proof of Proposition 1.6. Let $\{y_\alpha\}$ be the n-cells of $K_{<n}$ and let $\chi(y_\alpha) : y_\alpha \to K_{<n}$ be their characteristic maps. Let $a_\alpha = \chi(y_\alpha)| : \partial y_\alpha \to K^{n-1}$ be the corresponding attaching maps. The homotopy classes $\{[\chi(y_\alpha)]\}$ form a basis for $\pi_n(K_{<n}, K^{n-1}) = C_n(K_{<n})$. Choose representatives

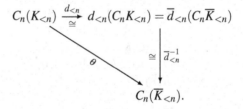

for the images $\theta[\chi(y_\alpha)] \in \pi_n(\overline{K}_{<n}, K^{n-1})$. The attaching map a_α is homotopic to $f'_\alpha|$ because

$$[a_\alpha] = d_{<n}[\chi(y_\alpha)] = \overline{d}_{<n}\theta[\chi(y_\alpha)] = \overline{d}_{<n}[f'_\alpha] = [f'_\alpha|].$$

By the homotopy extension property, there exists, for every α, a representative $f_\alpha : y_\alpha \to \overline{K}_{<n}$ for $\theta[\chi(y_\alpha)]$ that extends a_α, $f_\alpha|_{\partial y_\alpha} = a_\alpha$. Defining

$$f : K_{<n} \longrightarrow \overline{K}_{<n}$$

by

$$\begin{aligned} f(x) &= x, \text{ for } x \in K^{n-1}, \\ f(\chi(y_\alpha)(x)) &= f_\alpha(x), \text{ for } x \in y_\alpha, \end{aligned}$$

yields a map, since $\chi(y_\alpha)|_{\partial y_\alpha} = a_\alpha = f_\alpha|_{\partial y_\alpha}$. It is a homotopy equivalence, since it induces a chain-isomorphism. (It induces θ on the nth chain group.) It is moreover a homotopy equivalence rel K^{n-1} by Proposition 1.5. □

Taking $Y = \overline{Y}$ in the preceding proof, we obtain in particular:

Scholium 1.26. *If $(K, K/n, h, K_{<n})$ and $(K, \overline{K}/n, \overline{h}, \overline{K}_{<n})$ are two n-truncation structure completions of an object (K, Y) in $\mathbf{CW}_{n \supset \partial}$ such that*

$$h_* i_{K*} C_n(K_{<n}) = Y = \overline{h}_* i_{\overline{K}*} C_n(\overline{K}_{<n})$$

then $K_{<n}$ and $\overline{K}_{<n}$ are homotopy equivalent rel K^{n-1}.

Thus, up to rel $(n-1)$-skeleton homotopy equivalence, the definition of $\tau_{<n}(K, Y_K)$ given above is independent of choices. Some applications of Proposition 1.25 follow.

Proposition 1.27. *In the following statement, assume $K^1 = \text{pt}$ when $n = 3$. If the skeletal inclusion $K^{n-2} \subset K^{n-1}$ induces the zero map*

$$\pi_{n-1}(K^{n-2}) \xrightarrow{0} \pi_{n-1}(K^{n-1})$$

then the rel $(n-1)$-skeleton homotopy type of $K_{<n}$ is independent of the choice of Y, where $(K, K/n, h_K, K_{<n}) = \tau_{<n}(K, Y)$.

Proof. The exact sequence

$$\pi_{n-1}(K^{n-2}) \xrightarrow{0} \pi_{n-1}(K^{n-1}) \xrightarrow{\text{incl}_*} \pi_{n-1}(K^{n-1}, K^{n-2}) = C_{n-1}(K)$$

shows that incl_* is injective so that the restriction

$$\text{incl}_* : \pi_{n-1}(K^{n-1}) \xrightarrow{\cong} \text{im incl}_*$$

is an isomorphism. We have thus the following factorization of the homotopy boundary homomorphism d:

$$
\begin{array}{ccc}
C_n(K) & \xrightarrow{\quad d \quad} & \pi_{n-1}(K^{n-1}) \\
\Big\downarrow {\partial_n} & & \Big\uparrow {\cong} \;\; \mathrm{incl}_*^{-1} \\
\mathrm{im}\,\partial_n = \mathrm{im}(\mathrm{incl}_* \circ d) & \lhook\joinrel\longrightarrow & \mathrm{im}\,\mathrm{incl}_*
\end{array}
$$

If $(K,Y),(K,\overline{Y}) \in Ob\,\mathbf{CW}_{n \supset \partial}$, then ∂_n by definition maps both Y and $\overline{Y}$ onto $\mathrm{im}\,\partial_n$. Hence,

$$d(Y) = \mathrm{incl}_*^{-1}\partial_n(Y) = \mathrm{incl}_*^{-1}\,\mathrm{im}\,\partial_n = \mathrm{incl}_*^{-1}\partial_n(\overline{Y}) = d(\overline{Y}).$$

By Proposition 1.25, $K_{<n} \simeq \overline{K}_{<n}$ rel K^{n-1}, where $(K,\overline{K}/n,\overline{h}_K,\overline{K}_{<n}) = \tau_{<n}(K,\overline{Y})$.

$\square$

Examples 1.28. Let p be an odd prime and q a positive integer:

1. Suppose the $(n-2)$-skeleton of K has the form $K^{n-2} = S^{n-3} \cup e^{n-2}$, where e^{n-2} is attached to S^{n-3} by a map of degree p^q. Then the assumption of Proposition 1.27 is satisfied as $\pi_{n-1}(S^{n-3} \cup e^{n-2}) = 0$, see [Hil53].
2. Suppose the $(n-1)$-skeleton of K has the form $K^{n-1} = S^{n-2} \cup e^{n-1}$, where e^{n-1} is attached to S^{n-2} by a map of degree p^q. Then the assumption of Proposition 1.27 is satisfied as $\pi_{n-1}(K^{n-1}) = 0$.
3. $(n \geq 6.)$ Suppose the $(n-1)$-skeleton of K has the form $K^{n-1} = S^{n-3} \cup e^{n-1}$, where e^{n-1} is attached to S^{n-3} by an essential map. Then $\pi_{n-1}(K^{n-2}) = \pi_{n-1}(S^{n-3}) = \mathbb{Z}/2$ (since $n-3 \geq 3$) and $\pi_{n-1}(K^{n-1}) = \mathbb{Z}$, [Hil53]. Thus the map

$$\pi_{n-1}(K^{n-2}) = \mathbb{Z}/2 \longrightarrow \mathbb{Z} = \pi_{n-1}(K^{n-1})$$

is trivial.

Let us recall the definition of a J_m-complex due to J. H. C. Whitehead [Whi49].

Definition 1.29. A CW-complex K is a J_m-*complex*, if the skeletal inclusions induce zero maps $\pi_r(K^{r-1}) \to \pi_r(K^r)$ for all $r = 2,\ldots,m$.

The space $S^3 \cup_3 e^4$, for example, is a J_5-complex. If K is a simply connected J_m-complex, then the Hurewicz map $\pi_r(K) \to H_r(K)$ is an isomorphism for $r \leq m$ (and a surjection in degree $r = m+1$). We obtain the following corollary to Proposition 1.27:

Corollary 1.30. *If K is a J_{n-1}-complex, then the rel $(n-1)$-skeleton homotopy type of $K_{<n}$ is independent of the choice of Y, where $(K,K/n,h_K,K_{<n}) = \tau_{<n}(K,Y)$.*

For the value $n = 3$, the proposition implies:

Corollary 1.31. *For $n = 3$ and $K^1 = \mathrm{pt}$, the rel 2-skeleton homotopy type of $K_{<3}$ is independent of the choice of Y, where $(K,K/3,h_K,K_{<3}) = \tau_{<3}(K,Y)$.*

Proof. For $n = 3$, K^{n-2} is a point and so $\pi_2(K^1) = 0$. The conclusion follows from Proposition 1.27. ☐

In order to define $\tau_{<n}$ on morphisms, we prove the existence of morphism completions:

Theorem 1.32 (Compression Theorem). *Any morphism* $f : (K, Y_K) \to (L, Y_L)$ *in* $\mathbf{CW}_{n \supset \partial}$ *can be completed to a morphism* $\tau_{<n}(K, Y_K) \to \tau_{<n}(L, Y_L)$ *in* $\mathbf{HoCW}_{\supset <n}$.

Proof. The map $f : K \to L$ is cellular and $f_*(Y_K) \subset Y_L$. With

$$\tau_{<n}(K, Y_K) = (K, K/n, h_K, K_{<n}), \quad \tau_{<n}(L, Y_L) = (L, L/n, h_L, L_{<n}),$$

our task is to complete the diagram

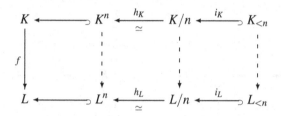

by filling in the three dotted arrows in such a way that all three squares commute up to homotopy rel K^{n-1}. Since f is cellular, it restricts to a map between the n-skeleta. This defines $f^n = f| : K^n \to L^n$. Choose a cellular homotopy inverse $h'_L : L^n \to L/n$ for h_L such that h'_L is the identity on L^{n-1} and $h_L h'_L \simeq \mathrm{id}$ rel L^{n-1}, $h'_L h_L \simeq \mathrm{id}$ rel L^{n-1}. Set

$$f/n = h'_L \circ f| \circ h_K : K/n \longrightarrow L/n.$$

Then the middle square commutes up to homotopy rel K^{n-1}. It remains to be shown that the map $(f/n)i_K : K_{<n} \to L/n$ can be deformed into the subcomplex $L_{<n}$ rel K^{n-1}. By definition,

$$K/n = K^{n-1} \cup \bigcup_\alpha y_\alpha \cup \bigcup_\beta z_\beta, \quad K_{<n} = K^{n-1} \cup \bigcup_\alpha y_\alpha,$$

$$L/n = L^{n-1} \cup \bigcup_\gamma y'_\gamma \cup \bigcup_\delta z'_\delta, \quad L_{<n} = L^{n-1} \cup \bigcup_\gamma y'_\gamma,$$

where the z_β are n-cells constituting a basis for the cycle group $Z_n(K/n)$, the y_α are the remaining n-cells of K/n, the z'_δ constitute a basis for $Z_n(L/n)$ and the y'_γ are the remaining n-cells of L/n. The various characteristic maps form bases for the homotopy groups rel $(n-1)$-skeleton:

$$\pi_n(K/n, K^{n-1}) = \bigoplus_\alpha \mathbb{Z}[\chi(y_\alpha)] \oplus \bigoplus_\beta \mathbb{Z}[\chi(z_\beta)],$$

$$\pi_n(L/n, L^{n-1}) = \bigoplus_\gamma \mathbb{Z}[\chi(y'_\gamma)] \oplus \bigoplus_\delta \mathbb{Z}[\chi(z'_\delta)].$$

Set

$$\zeta_\beta = h_{K*}[\chi(z_\beta)], \ \eta_\alpha = h_{K*}[\chi(y_\alpha)],$$
$$\zeta_\delta' = h_{L*}[\chi(z_\delta')], \ \eta_\gamma' = h_{L*}[\chi(y_\gamma')].$$

Since h_{K*} and h_{L*} are chain maps, the elements ζ_β and ζ_δ' are cycles, i.e. $\zeta_\beta \in Z_n(K)$ and $\zeta_\delta' \in Z_n(L)$. By definition of $\tau_{<n}(K, Y_K)$, the η_α lie in Y_K. The η_γ' lie in Y_L. As both h_{K*} and h_{L*} are isomorphisms, $\{\eta_\alpha\}$ is a basis for Y_K, $\{\zeta_\beta\}$ is a basis for $Z_n(K)$, $\{\eta_\gamma'\}$ is a basis for Y_L and $\{\zeta_\delta'\}$ is a basis for $Z_n(L)$. The situation is summarized in the following commutative diagram.

$$
\begin{array}{ccccc}
& \begin{matrix} \zeta_\beta \leftarrow [\chi(z_\beta)] \\ \eta_\alpha \leftarrow [\chi(y_\alpha)] \end{matrix} & & & \\
\underset{\{\eta_\alpha\} \quad \{\zeta_\beta\}}{Y_K \oplus Z_n(K)} & \longleftarrow & \bigoplus_\alpha \mathbb{Z}[\chi(y_\alpha)] \oplus \bigoplus_\beta \mathbb{Z}[\chi(z_\beta)] & \hookleftarrow & \bigoplus_\alpha \mathbb{Z}[\chi(y_\alpha)] \\
\Vert & & \Vert & & \Vert \\
\pi_n(K^n, K^{n-1}) & \xleftarrow[h_{K*}]{\cong} & \pi_n(K/n, K^{n-1}) & \xleftarrow[i_{K*}]{} & \pi_n(K_{<n}, K^{n-1}) \\
f|_* \downarrow & & \downarrow (f/n)_* & & \Vert \\
\pi_n(L^n, L^{n-1}) & \xleftarrow[h_{L*}]{\cong} & \pi_n(L/n, L^{n-1}) & \xleftarrow[i_{L*}]{} & \pi_n(L_{<n}, L^{n-1}) \\
\Vert & & \Vert & & \Vert \\
\underset{\{\eta_\gamma'\} \quad \{\zeta_\delta'\}}{Y_L \oplus Z_n(L)} & \longleftarrow & \bigoplus_\gamma \mathbb{Z}[\chi(y_\gamma')] \oplus \bigoplus_\delta \mathbb{Z}[\chi(z_\delta')] & \hookleftarrow & \bigoplus_\gamma \mathbb{Z}[\chi(y_\gamma')] \\
& \begin{matrix} \zeta_\delta' \leftarrow [\chi(z_\delta')] \\ \eta_\gamma' \leftarrow [\chi(y_\gamma')] \end{matrix} & & &
\end{array}
$$

(We have $h_{L*} \circ (f/n)_* = f|_* \circ h_{K*}$ because $h_L \circ f/n \simeq h_L \circ h_L' \circ f| \circ h_K \simeq f| \circ h_K$ by a homotopy rel K^{n-1}.) The commutative square

$$
\begin{array}{ccc}
\partial y_\alpha & \xrightarrow{\chi(y_\alpha)|} & K^{n-1} \\
\downarrow & & \downarrow \\
y_\alpha & \xrightarrow{\chi(y_\alpha)} & K_{<n}
\end{array}
$$

represents the element $[\chi(y_\alpha)] \in \pi_n(K_{<n}, K^{n-1})$, and

$$[\chi(y_\alpha)|_{\partial y_\alpha}] = d/^n[\chi(y_\alpha)] = d/^n h_{K*}(\eta_\alpha) = d\eta_\alpha$$

holds. Since $f_*(Y_K) \subset Y_L$, we can write $f|_*(\eta_\alpha) = \sum_\gamma \lambda_\gamma \eta'_\gamma$ for some integers λ_γ. Thus,

$$
\begin{aligned}
(f/n)_* i_{K*} [\chi(y_\alpha)] &= h'_{L*} f|_* h_{K*} i_{K*} [\chi(y_\alpha)] = h'_{L*} f|_*(\eta_\alpha) \\
&= h'_{L*}(\sum_\gamma \lambda_\gamma \eta'_\gamma) = \sum_\gamma \lambda_\gamma h'_{L*}(\eta'_\gamma) \\
&= \sum_\gamma \lambda_\gamma h'_{L*} h_{L*} [\chi(y'_\gamma)] = \sum_\gamma \lambda_\gamma [\chi(y'_\gamma)] = \sum_\gamma \lambda_\gamma i_{L*} [\chi(y'_\gamma)] \\
&= i_{L*}(\sum_\gamma \lambda_\gamma [\chi(y'_\gamma)]),
\end{aligned}
$$

whence $(f/n)_* i_{K*} [\chi(y_\alpha)]$ is in the image of i_{L*}. Hence, by exactness of the sequence

$$
\pi_n(L_{<n}, L^{n-1}) \xrightarrow{i_{L*}} \pi_n(L/n, L^{n-1}) \xrightarrow{j_*} \pi_n(L/n, L_{<n})
$$

associated to the triple $(L/n, L_{<n}, L^{n-1})$,

$$
j_*(f/n)_* i_{K*} [\chi(y_\alpha)] = 0 \in \pi_n(L/n, L_{<n}).
$$

This element is explicitly represented by the composition

This means that the composition

$$
y_\alpha \xrightarrow{\chi(y_\alpha)} K_{<n} \overset{i_K}{\hookrightarrow} K/n \xrightarrow{f/n} L/n
$$

is homotopic, rel ∂y_α, to a map into $L_{<n}$ (see [Bre93], Theorem 5.8 in Chapter VII, p. 448). Consequently there exist homotopies

$$
H^\alpha : y_\alpha \times I \longrightarrow L/n
$$

such that

    (i) $H^\alpha(-,0) = (f/n) \circ i_K \circ \chi(y_\alpha)$,
    (ii) $H^\alpha(y_\alpha \times \{1\}) \subset L_{<n}$,
    (iii) $H^\alpha(x,t) = (f/n \circ \chi(y_\alpha)|)(x)$, for all $x \in \partial y_\alpha,\ t \in I$.

In order to assemble these homotopies to a homotopy

$$
H : K_{<n} \times I \longrightarrow L/n
$$

rel K^{n-1} such that

$$H(-,0) = (f/n)i_K, \quad H(K_{<n} \times \{1\}) \subset L_{<n},$$

set

$$H(x,t) = (i_L j(f/n)|)(x)$$

for $x \in K^{n-1}$ and

$$H(\chi(y_\alpha)(x),t) = H^\alpha(x,t)$$

for $x \in y_\alpha$. Then H is indeed a map because for $x \in \partial y_\alpha$,

$$H^\alpha(x,t) = (f/n \circ \chi(y_\alpha)|)(x) = (i_L \circ j \circ f/n \circ \chi(y_\alpha)|)(x),$$

by (*iii*) above. In other words, H is the unique map determined by the universal property of the pushout:

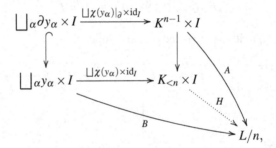

where $A(x,t) = (i_L j(f/n)|)(x)$ for $(x,t) \in K^{n-1} \times I$ and $B(x,t) = H^\alpha(x,t)$ for $x \in y_\alpha$, $t \in I$, observing that for $x \in \partial y_\alpha$, $t \in I$,

$$\begin{aligned}
A(\chi(y_\alpha)(x),t) &= i_L j(f/n)\chi(y_\alpha)(x) \\
&= (f/n)\chi(y_\alpha)(x) \\
&= H^\alpha(x,t) \text{ by } (iii) \\
&= B(x,t).
\end{aligned}$$

Defining

$$f_{<n} = H(-,1),$$

we obtain the desired morphism, represented by

$$
\begin{array}{ccccccc}
K & \longleftarrow & K^n & \xleftarrow{\ h_K\ }_{\simeq} & K/n & \xleftarrow{\ i_K\ } & K_{<n} \\
\Big\downarrow{\scriptstyle f} & & \Big\downarrow{\scriptstyle f|} & & \Big\downarrow{\scriptstyle f/n} & & \Big\downarrow{\scriptstyle f_{<n}} \\
L & \longleftarrow & L^n & \xleftarrow{\ h_L\ }_{\simeq} & L/n & \xleftarrow{\ i_L\ } & L_{<n}.
\end{array}
$$

$\square$

At this point, it is instructive to return to the example discussed in the introduction 1.1.1. There we constructed a (homotopy class of a) map $f : K \to L$ with $K = S^2 \cup_2 e^3$ a Moore space $M(\mathbb{Z}/2, 2)$ and $L = K \vee S^3$ that could not be compressed to a map $f_{<3} : K_{<3} \to L_{<3}$. In light of Theorem 1.32, this must mean that f cannot be promoted to a morphism $f : (K, Y_K) \to (L, Y_L)$ in $\mathbf{CW}_{3 \supset \partial}$, no matter which Y_K and Y_L one takes. Let us prove directly that this is indeed the case, by giving an explicit geometric description of f. The cofibration sequence

$$S^2 \xrightarrow{i=2} S^2 \longrightarrow K = \text{cone}(i) \xrightarrow{\iota} S^3 \xrightarrow{\Sigma i=2} S^3,$$

where ι collapses the 2-skeleton S^2 of K to a point, induces an exact sequence

$$\pi_3(L) \xrightarrow{\Sigma i=2} \pi_3(L) \xrightarrow{\iota} [K, L]$$

and the cokernel of Σi is $\text{Ext}(\mathbb{Z}/2, \pi_3 L)$. Let $g : S^3 \hookrightarrow K \vee S^3 = L$ be the inclusion which is the identity onto the second wedge summand. Then the composition

$$K \xrightarrow{\iota} S^3 \xrightarrow{g} L$$

is homotopic to f. To see this, we only have to verify that $E_2(\text{Hur})[g] = \xi$, where $E_2(-) = \text{Ext}(\mathbb{Z}/2, -)$, Hur : $\pi_3(L) \to H_3(L) = \mathbb{Z}$ is the Hurewicz map so that $E_2(\text{Hur}) : E_2(\pi_3 L) \to E_2(H_3 L) = \mathbb{Z}/2$, and $\xi \in E_2(H_3 L)$ is the generator. Let $[S^3] \in H_3(L)$ denote the preferred generator of $H_3(L)$. Then ξ is the residue class of $[S^3]$ modulo 2. The map $E_2(\text{Hur})$ sends the residue class of $[g]$ in $\pi_3(L)/2\pi_3(L)$ to the residue class of $g_*[S^3]$, $[S^3] \in H_3(S_3)$ the fundamental class, in $H_3(L)/2H_3(L)$. Since g is the identity on the second wedge summand, we have indeed $g_*[S^3] = [S^3]$. Given this geometric description of f, its action on chains is easily obtained: $C_3(K) = \mathbb{Z}e_K^3$, where e_K^3 is the 3-cell of K and $C_3(L) = \mathbb{Z}e_L^3 \oplus \mathbb{Z}[S^3]$, where e_L^3 is the 3-cell in L contained in $K \subset L$, and where we wrote $[S^3]$ for the other 3-cell of L, contained in the 3-sphere in L. Then $f_* : C_3(K) \to C_3(L)$ is given by

$$f_*(e_K^3) = g_* \iota_*(e_K^3) = g_*[S^3] = [S^3].$$

The boundary operator $\partial_3^K : C_3(K) \to C_2(K) = \mathbb{Z}e^2$ is multiplication by 2. Thus $Z_3(K) = \ker \partial_3^K = 0$ and $Y_K = C_3(K)$ is uniquely determined. For $\partial_3^L : C_3(L) \to C_2(L)$ we have $\partial_3^L(e_L^3) = 2e^2$ and $\partial_3^L[S^3] = 0$. Hence $Z_3(L) = \ker \partial_3^L = \mathbb{Z}[S^3]$ and in the decomposition $C_3(L) = Z_3(L) \oplus Y_L$, Y_L is any subgroup of the form $\mathbb{Z}(e_L^3 + m[S^3])$ with $m \in \mathbb{Z}$. We conclude that since

$$f_*(Y_K) = f_*C_3(K) = \mathbb{Z}[S^3] = Z_3(L),$$

there is no admissible Y_L such that $f_*(Y_K) \subset Y_L$ and f does not give rise to a morphism in $\mathbf{CW}_{3 \supset \partial}$.

Definition 1.33. Let $(K, K/n, h_K, K_{<n})$ and $(L, L/n, h_L, L_{<n})$ be n-truncation structures. A morphism $([f], [f^n], [f/n], [f_{<n}]) : (K, K/n, h_K, K_{<n}) \to (L, L/n, h_L, L_{<n})$ in $\mathbf{HoCW}_{\supset <n}$ is called n-compression rigid if for any two cellular maps $g_1, g_2 : K_{<n} \to L_{<n}$ such that

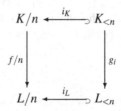

homotopy commutes rel K^{n-1} for $i = 1, 2$, the homotopy $H : K_{<n} \times I \to L/n$ between $i_L g_1$ and $i_L g_2$ can be chosen to be an eigenhomotopy (still rel K^{n-1}).

The property of n-compression rigidity is indeed a well-defined property of a morphism in $\mathbf{HoCW}_{\supset <n}$, for it does not depend on the choice of representative: Suppose that $([f], [f^n], [f/n], [f_{<n}]) = ([g], [g^n], [g/n], [g_{<n}])$ and this morphism is n-compression rigid with respect to f/n. Given $g_1, g_2 : K_{<n} \to L_{<n}$ with $i_L g_1 \simeq (g/n)i_K \simeq i_L g_2$ rel K^{n-1}, we use $f/n \simeq g/n$ rel K^{n-1}, and therefore $(f/n)i_K \simeq (g/n)i_K$ rel K^{n-1}, to obtain homotopies $i_L g_1 \simeq (f/n)i_K \simeq i_L g_2$ rel K^{n-1}. By n-compression rigidity with respect to f/n, the homotopy between $i_L g_1$ and $i_L g_2$ can be chosen to be an eigenhomotopy. Hence, the morphism is n-compression rigid with respect to g/n.

On the other hand, compression rigidity is not expected to be a property of $[f]$ alone because $[f] = [g]$ does not imply $[f/n] = [g/n]$, as noted before.

An obstruction theory for deciding compression rigidity in practice is provided in Section 1.2.

Morphisms $f : (K, Y_K) \to (L, Y_L)$ in $\mathbf{CW}_{n \supset \partial}$ are required to satisfy $f_*(Y_K) \subset Y_L$. This ensures that f can be pushed down to a map $f_{<n} : K_{<n} \to L_{<n}$ between n-truncations. If one wants any two such maps $K_{<n} \to L_{<n}$, both truncating f, to be homotopic, which is necessary to obtain functoriality, then one needs an additional condition – a higher order analog of the previous condition – to ensure that *homotopies* can be pushed down to the truncated spaces. Unfortunately, it turns out to be subtler than just requiring "$H_*(Y_{K \times I}) \subset Y_{L \times I}$" and then applying Theorem 1.32 in degree $n + 1$ to H instead of f. The difficulty is related to the fact that the n-skeleton of a cylinder $K \times I$, where K is an n-dimensional complex, is not $K^{n-1} \times I$, but $K^{n-1} \times I \cup K^n \times \partial I$. Rather, the eigenhomotopy property is precisely the condition needed. The following proposition shows that two truncation versions of a map are homotopic if, and only if, the map being truncated is compression rigid.

Proposition 1.34. *Let* $(K, K/n, h_K, K_{<n})$ *and* $(L, L/n, h_L, L_{<n})$ *be* n-truncation *structures and* $F = ([f], [f^n], [f/n], [f_{<n}]) : (K, K/n, h_K, K_{<n}) \to (L, L/n, h_L, L_{<n})$ *a morphism in* $\mathbf{HoCW}_{\supset <n}$. *Then any two cellular maps* $g_1, g_2 : K_{<n} \to L_{<n}$ *such that*

$$K/n \xleftarrow{\quad i_K \quad} K_{<n}$$

(diagram: $K/n \xleftarrow{i_K} K_{<n}$ top row, vertical maps f/n on left and g_i on right, bottom row $L/n \xleftarrow{i_L} L_{<n}$)

homotopy commutes rel K^{n-1} for $i = 1, 2$ are homotopic rel K^{n-1} if, and only if, F is n-compression rigid.

Proof. Assume that F is n-compression rigid. We have $i_L g_1 \simeq (f/n) i_K \simeq i_L g_2$ rel K^{n-1}. By n-compression rigidity, the homotopy $H : K_{<n} \times I \to L/n$ between $i_L g_1$ and $i_L g_2$ can be taken to be an eigenhomotopy rel K^{n-1}. Define $H' : K_{<n} \times I \to L/n \times I$ by $H'(k,t) = (H(k,t),t)$. By cellularity, g_1 sends K^{n-1} to L^{n-1}. Thus H' restricts to a map

$$H'|_{K^{n-1} \times I} = g_1|_{K^{n-1}} \times \mathrm{id}_I = g_2|_{K^{n-1}} \times \mathrm{id}_I : K^{n-1} \times I \longrightarrow L^{n-1} \times I.$$

Furthermore, $H'(K_{<n} \times \partial I) \subset L_{<n} \times \partial I$ via $g_1 \cup g_2$. Hence, setting

$$A = K_{<n} \times I, \ A_0 = K_{<n} \times \partial I \cup K^{n-1} \times I,$$
$$B' = L/n \times I, \ B = L_{<n} \times I, \ B_0 = L_{<n} \times \partial I \cup L^{n-1} \times I,$$

we have a map of pairs

$$H' : (A, A_0) \longrightarrow (B', B_0).$$

Let $y = y_\alpha^n$ be an n-cell of $K_{<n}$ with characteristic map $\chi(y) : y \to K_{<n}$ and attaching map $\chi(y)|_{\partial y} : \partial y \to K^{n-1}$. The characteristic map $\chi(y \times I)$ of the $(n+1)$-cell $y \times I$ of $K_{<n} \times I$ is then

(diagram: $y \times I \xrightarrow{\chi(y \times I) = \chi(y) \times \mathrm{id}_I} K_{<n} \times I$ with vertical inclusions, bottom row $\partial(y \times I) = (\partial y) \times I \cup y \times \partial I \xrightarrow{\chi(y)|\times \mathrm{id}_I \cup \chi(y) \times \mathrm{id}_{\partial I}} K^{n-1} \times I \cup K_{<n} \times \partial I$)

and represents an element $[\chi(y \times I)] \in \pi_{n+1}(A, A_0)$. Applying the induced map $H'_* : \pi_{n+1}(A, A_0) \to \pi_{n+1}(B', B_0)$, we obtain an eigenclass $x_\alpha = H'_*[\chi(y \times I)] \in \pi_{n+1}(B', B_0) = VC_n(L)$. Thus

$$x_\alpha = E_L(x_\alpha) = \phi p \psi(x_\alpha).$$

The long exact homotopy sequence of the triple (B', B, B_0) yields the exact sequence

$$C_{n+1}(L_{<n} \times I) \xrightarrow{\phi} VC_n(L) \xrightarrow{\varepsilon} \pi_{n+1}(B', B).$$

Since x_α is in the image of ϕ, we have $\varepsilon(x_\alpha) = 0$. This means that the composition

$$y \times I \xrightarrow{\chi(y \times I)} K_{<n} \times I \xrightarrow{H'} L/n \times I$$

is homotopic, rel $\partial(y \times I)$, to a map $H_\alpha^<$ into $L_{<n} \times I$. This map $H_\alpha^< : y \times I \to L_{<n} \times I$ is equal to $H'| \circ (\chi(y)| \times \mathrm{id}_I)$ when restricted to $(\partial y) \times I$ and is equal to $(g_1 \cup g_2) \circ (\chi(y) \times \mathrm{id}_{\partial I})$ when restricted to $y \times \partial I$. Let us assemble these $H_\alpha^<$ to a homotopy $H^< : K_{<n} \times I \to L_{<n}$. For $x \in K^{n-1}$, set

$$H^<(x, t) = g_1(x) = g_2(x).$$

For $x \in y_\alpha^n$, set

$$H^<(\chi(y_\alpha^n)(x), t) = \pi_1 H_\alpha^<(x, t),$$

where $\pi_1 : L_{<n} \times I \to L_{<n}$ is the first-factor projection. Then $H^<$ is indeed a map because for $x \in \partial y_\alpha$,

$$\begin{aligned} H^<(\chi(y_\alpha)|_{\partial y_\alpha}(x), t) &= \pi_1 H_\alpha^<(x, t) = \pi_1 \circ H' \circ \chi(y_\alpha \times I)(x, t) \\ &= \pi_1 H'(\chi(y_\alpha)(x), t) = H(\chi(y_\alpha)(x), t) \\ &= g_1(\chi(y_\alpha)(x)). \end{aligned}$$

In other words, $H^<$ is the unique map determined by the universal property of the pushout:

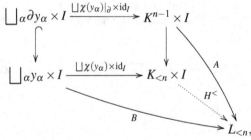

where $A(x, t) = g_1(x)$ for $(x, t) \in K^{n-1} \times I$ and $B(x, t) = \pi_1 H_\alpha^<(x, t)$ for $x \in y_\alpha, t \in I$, observing that for $x \in \partial y_\alpha, t \in I$,

$$\begin{aligned} A(\chi(y_\alpha)(x), t) &= g_1(\chi(y_\alpha)(x)) \\ &= H(\chi(y_\alpha)(x), t) \text{ (since } \chi(y_\alpha)(x) \in K^{n-1}) \\ &= \pi_1 H'(\chi(y_\alpha)(x), t) \\ &= \pi_1 H_\alpha^<(x, t) \\ &= B(x, t). \end{aligned}$$

For $t = 0$ we have $H^<(x,0) = g_1(x)$ when $x \in K^{n-1}$ and $H^<(\chi(y_\alpha)(x),0) = \pi_1 H^<_\alpha(x,0) = g_1(\chi(y_\alpha)(x))$ when $x \in y_\alpha$. Therefore, $H^<(-,0) = g_1$, and similarly $H^<(-,1) = g_2$. The map $H^<$ is the desired homotopy rel K^{n-1} between g_1 and g_2.

Let us now prove the converse direction. We assume that whenever g_1 and g_2 are cellular maps such that $i_L g_1 \simeq (f/n) i_K \simeq i_L g_2$ rel K^{n-1} then in fact $g_1 \simeq g_2$ rel K^{n-1}. We have to show that F is n-compression rigid. Let g_1, g_2 be maps as above and let $H : K_{<n} \times I \to L_{<n}$ be a homotopy rel K^{n-1} between g_1 and g_2. The associated map $H' : K_{<n} \times I \to L_{<n} \times I$ is a map of pairs $H' : (A, A_0) \to (B, B_0)$ which induces on homotopy groups a homomorphism $H'_\# : C_{n+1}(K_{<n} \times I) \to C_{n+1}(L_{<n} \times I)$. Regarding H' as a map $(A, A_0) \to (B', B_0)$, it induces a homomorphism $H'_* : C_{n+1}(K_{<n} \times I) \to VC_n(L)$ such that $H'_* = \phi H'_\#$. Let

$$j : \bigoplus_\alpha \mathbb{Z}[\chi(y_\alpha \times I)] \hookrightarrow \bigoplus_\alpha \mathbb{Z}[\chi(y_\alpha \times I)] \oplus \bigoplus_\beta \mathbb{Z}[\chi(z_\beta \times I)]$$

be the canonical inclusion so that $pj = \text{id}$ and $\psi\phi = j$. We will show that $x = H'_*[\chi(y_\alpha \times I)]$ is an eigenclass. Let us calculate the action of the endomorphism E_L on x:

$$\begin{aligned}
E_L(x) &= \phi p \psi(x) \\
&= \phi p \psi \phi H'_\#[\chi(y_\alpha \times I)] \\
&= \phi p j H'_\#[\chi(y_\alpha \times I)] \\
&= \phi H'_\#[\chi(y_\alpha \times I)] \\
&= H'_*[\chi(y_\alpha \times I)] \\
&= x.
\end{aligned}$$

Hence x is an eigenclass as claimed. $\qquad\qquad\square$

Example 1.35. We exhibit an example of a map $f : K \to L$, where K and L are simply connected 5-segmented CW-complexes ($K = K/5$, $L = L/5$) with unique 5-truncation subcomplexes $K_{<5} \subset K$, $L_{<5} \subset L$, such that there are two nonhomotopic maps $g_1, g_2 : K_{<5} \to L_{<5}$, which are equal on the 4-skeleton K^4 of K and such that

$$\begin{array}{ccc}
K & \xleftarrow{\ i_K\ } & K_{<5} \\
{\scriptstyle f}\downarrow & & \downarrow{\scriptstyle g_i} \\
L & \xleftarrow{\ i_L\ } & L_{<5}
\end{array}$$

homotopy commutes rel K^4, $i = 1,2$. This, then, furnishes an example of a map that is not compression rigid. Let

$$K = S^4 \cup_4 e^5, \quad L = S^3 \cup_2 e^4 \cup e^5,$$

where the 5-cell in L is attached to S^3 by an essential map $\partial e^5 \to S^3$. The complex K is a Moore space $M(\mathbb{Z}/4, 4)$ and the 4-skeleton $S^3 \cup_2 e^4$ of L is a Moore space $M(\mathbb{Z}/2, 3)$. The cycle group $Z_5(K)$ is zero and $Y_5(K) = C_5(K) = \mathbb{Z}e^5$ is unique. The space K is 5-segmented with 5-truncation $K_{<5} = K$, unique by Proposition 1.3. The cycle group $Z_5(L) = C_5(L) = \mathbb{Z}e^5$ has a basis of cells. Hence L is 5-segmented by Lemma 1.2. Necessarily, $Y_5(L) = 0$. The 5-truncation is $L_{<5} = L^4 = S^3 \cup_2 e^4$, unique by Proposition 1.3. By classical homotopy theoretic arguments,

$$\pi_5(S^3 \cup_2 e^4) \cong \mathbb{Z}/4$$

and

$$\pi_5(S^3 \cup_2 e^4 \cup e^5) \cong \mathbb{Z}/2 \oplus \mathbb{Z}.$$

Since L is 2-connected, we may apply Proposition 1.50 to obtain $\pi_6(L, L_{<5}) \cong \mathbb{Z}/2$, using $H_5(L) \cong \mathbb{Z}$. The exact sequence of the pair,

$$\pi_6(L, L_{<5}) \longrightarrow \pi_5(L_{<5}) \xrightarrow{i_{L*}} \pi_5(L),$$

then shows that the kernel of i_{L*} is either zero or isomorphic to $\mathbb{Z}/2$. Since every homomorphism $\mathbb{Z}/4 \to \mathbb{Z}/2 \oplus \mathbb{Z}$ has a nontrivial kernel, $\ker i_{L*}$ is isomorphic to $\mathbb{Z}/2$. Write $\mathbb{Z}/4 = \{0, 1, 2, 3\}$. The only subgroup of $\mathbb{Z}/4$ isomorphic to $\mathbb{Z}/2$ is $\{0, 2\}$. We deduce that $\ker i_{L*} = \{0, 2\} \subset \pi_5(L_{<5})$. Let $h : S^5 \to L_{<5}$ be a map representing $2 = [h]$. Let $\mathrm{coll} : K = S^4 \cup_4 e^5 \to S^5$ be the map that collapses S^4 to a point, which then becomes the base point s_0 of S^5. The Puppe cofibration sequence

$$S^4 \xrightarrow{4} S^4 \longrightarrow \mathrm{cone}(4) = K \xrightarrow{\mathrm{coll}} S^5 = S(S^4) \xrightarrow{4} S^5 = S(S^4)$$

induces the exact rows of the commutative diagram

$$
\begin{array}{ccccc}
\pi_5(L_{<5}) & \xrightarrow{4=0} & \pi_5(L_{<5}) & \xhookrightarrow{\mathrm{coll}_*} & [K, L_{<5}] \\
\downarrow{\scriptstyle i_{L*}} & & \downarrow{\scriptstyle i_{L*}} & & \downarrow{\scriptstyle i_{L*}} \\
\pi_5(L) & \xrightarrow{4} & \pi_5(L) & \xrightarrow{\mathrm{coll}_*} & [K, L].
\end{array}
$$

Since the element $2 = [h] \in \pi_5(L_{<5})$ is not divisible by 4 (none of the nontrivial elements of $\pi_5(L_{<5})$ are), it is by exactness not in the kernel of coll_*. Thus

$$[h \circ \mathrm{coll}] = \mathrm{coll}_*[h] \neq 0 \in [K, L_{<5}].$$

As $i_{L*}[h] = 0 \in \pi_5(L)$, there exists a base point preserving homotopy $H : S^5 \times I \to L$ from $H_0 = i_L h$ to the constant map H_1, which sends every point to the base point l_0 of $L_{<5} \subset L$. Thus $H(s_0, t) = l_0$ for all $t \in I$. Define a homotopy $G : K_{<5} \times I \to L$ by

$$G(x, t) = H(\mathrm{coll}(x), t), \quad x \in K_{<5}, \ t \in I.$$

It is a homotopy from

$$G(x,0) = H(\text{coll}(x),0) = i_L h \text{coll}(x)$$

to the constant map

$$G(x,1) = H(\text{coll}(x),1) = l_0.$$

It is rel K^4, as for $x \in K^4 = S^4$,

$$G(x,t) = H(\text{coll}(x),t) = H(s_0,t) = l_0$$

for all $t \in I$. Let $g_1 : K_{<5} \to L_{<5}$ be the composition

$$K_{<5} = K \xrightarrow{\text{coll}} S^5 \xrightarrow{h} L_{<5}$$

and let $f : K \to L$ be the composition

$$K = K_{<5} \xrightarrow{g_1} L_{<5} \xrightarrow{i_L} L.$$

By construction,

$$\begin{array}{ccc}
K & \xrightarrow{i_K} & K_{<5} \\
{\scriptstyle f}\downarrow & & \downarrow{\scriptstyle g_1} \\
L & \xleftarrow{i_L} & L_{<5}
\end{array}$$

commutes. Taking $g_2 : K_{<5} \to L_{<5}$ to be the constant map to l_0, the square

$$\begin{array}{ccc}
K & \xrightarrow{i_K} & K_{<5} \\
{\scriptstyle f}\downarrow & & \downarrow{\scriptstyle g_2} \\
L & \xleftarrow{i_L} & L_{<5}
\end{array}$$

homotopy commutes rel K^4, as via the rel K^4 homotopy G,

$$f = i_L h \text{coll} \underset{G}{\simeq} \text{const}_{l_0} = i_L g_2.$$

Thus g_1 and g_2 are both valid homological 5-truncations of f, agreeing with f on the 4-skeleton. However, g_1 and g_2 are not homotopic, since

$$[g_1] = [h \circ \text{coll}] \neq 0 = [g_2] \in [K_{<5}, L_{<5}].$$

Proposition 1.36 (Homotopy Invariance of Compression Rigidity). *Let*

$$(K, K/n, h_K, K_{<n}) \xrightarrow[U]{\cong} (K', K'/n, h_{K'}, K'_{<n})$$

$$F \downarrow \qquad\qquad\qquad \downarrow F'$$

$$(L, L/n, h_L, L_{<n}) \xrightarrow[V]{\cong} (L', L'/n, h_{L'}, L'_{<n})$$

be a commutative square in $\mathbf{HoCW}_{\supset <n}$, *with* U, V *isomorphisms. If* F *is n-compression rigid, then* F' *is n-compression rigid.*

Proof. The morphism F has the form $F = ([f], [f^n], [f/n], [f_{<n}])$, and F' has the form $F' = ([f'], [f'^m], [f'/n], [f'_{<n}])$. Suppose $g'_i : K'_{<n} \to L'_{<n}$, $i = 1, 2$, are two cellular maps such that the squares

$$\begin{array}{ccc} K'/n & \xleftarrow{\ i_{K'}\ } & K'_{<n} \\ {\scriptstyle f'/n}\downarrow & & \downarrow{\scriptstyle g'_i} \\ L'/n & \xleftarrow{\ i_{L'}\ } & L'_{<n} \end{array} \qquad\qquad (1.11)$$

commute up to homotopy rel $(K')^{n-1}$. We have to show that $g'_1 \simeq g'_2$ rel $(K')^{n-1}$. The morphism U has components $U = ([u], [u^n], [u/n], [u_{<n}])$, and V has components $V = ([v], [v^n], [v/n], [v_{<n}])$. The cellular maps $u, u^n, u/n, u_{<n}$ are all homotopy equivalences rel K^{n-1} and the cellular maps $v, v^n, v/n, v_{<n}$ are all homotopy equivalences rel L^{n-1}, see Lemma 1.21. Let $u'_{<n}, v'_{<n}, v'/n$ be rel $(n-1)$-skeleta homotopy inverses for $u_{<n}, v_{<n}, v/n$, respectively. Set

$$g_i = v'_{<n} g'_i u_{<n} : K_{<n} \longrightarrow L_{<n}, \ i = 1, 2.$$

Since V is a morphism and $[v'/n], [v'_{<n}]$ are the third and fourth component of the inverse V^{-1} (see Lemma 1.21), the diagram

$$\begin{array}{ccc} L'/n & \xleftarrow{\ i_{L'}\ } & L'_{<n} \\ {\scriptstyle v'/n}\downarrow{\scriptstyle \simeq} & & {\scriptstyle \simeq}\downarrow{\scriptstyle v'_{<n}} \\ L/n & \xleftarrow{\ i_L\ } & L_{<n} \end{array} \qquad\qquad (1.12)$$

homotopy commutes rel $(L')^{n-1}$. Since U is a morphism, the diagram

$$
\begin{array}{ccc}
K/n & \xleftarrow{\quad i_K \quad} & K_{<n} \\
{\scriptstyle u/n}\Big\downarrow {\scriptstyle\simeq} & & {\scriptstyle\simeq}\Big\downarrow {\scriptstyle u_{<n}} \\
K'/n & \xleftarrow{\quad i_{K'} \quad} & K'_{<n}
\end{array}
\qquad (1.13)
$$

homotopy commutes rel K^{n-1}. From $F'U = VF$, we get a rel K^{n-1} homotopy commutative diagram

$$
\begin{array}{ccc}
K/n & \xrightarrow{\;\;\simeq\;\;} & K'/n \\
 & {\scriptstyle u/n} & \\
{\scriptstyle f/n}\Big\downarrow & & \Big\downarrow{\scriptstyle f'/n} \\
L/n & \xrightarrow[{\;\;v/n\;\;}]{\;\;\simeq\;\;} & L'/n
\end{array}
$$

which implies

$$
(v'/n)(f'/n)(u/n) \simeq (v'/n)(v/n)(f/n) \simeq f/n \qquad (1.14)
$$

rel K^{n-1}. Therefore,

$$
\begin{aligned}
i_L g_i &= i_L v'_{<n} g'_i u_{<n} \\
&\simeq (v'/n) i_{L'} g'_i u_{<n} && \text{(by (1.12))} \\
&\simeq (v'/n)(f'/n) i_{K'} u_{<n} && \text{(by (1.11))} \\
&\simeq (v'/n)(f'/n)(u/n) i_K && \text{(by (1.13))} \\
&\simeq (f/n) i_K && \text{(by (1.14))},
\end{aligned}
$$

$i = 1, 2$, rel K^{n-1}. Since F is n-compression rigid, Proposition 1.34 implies $g_1 \simeq g_2$ rel K^{n-1}, i.e. $v'_{<n} g'_1 u_{<n} \simeq v'_{<n} g'_2 u_{<n}$ rel K^{n-1}. Hence,

$$
g'_1 \simeq v_{<n} v'_{<n} g'_1 u_{<n} u'_{<n} \simeq v_{<n} v'_{<n} g'_2 u_{<n} u'_{<n} \simeq g'_2
$$

rel $(K')^{n-1}$, whence F' is n-compression rigid by Proposition 1.34. $\qquad\square$

Corollary 1.37 (Inversion Invariance of Compression Rigidity). *Let*

$$
F : (K, K/n, h_K, K_{<n}) \longrightarrow (L, L/n, h_L, L_{<n})
$$

be an isomorphism in **HoCW**$_{\supset <n}$. *If F is n-compression rigid, then F^{-1} is n-compression rigid as well.*

Proof. In Proposition 1.36, take $U = F$, $V = F^{-1}$ and $F' = F^{-1}$. □

Let $f : (K, Y_K) \to (L, Y_L)$ be a morphism in $\mathbf{CW}_{n \supset \partial}$. If f is the identity, set

$$\tau_{<n}(f) = \mathrm{id}_{\tau_{<n}(K,Y_K)} = ([\mathrm{id}_K], [\mathrm{id}_{K^n}], [\mathrm{id}_{K/n}], [\mathrm{id}_{K_{<n}}]),$$

where $\tau_{<n}(K, Y_K) = (K, K/n, h_K, K_{<n})$. If not, proceed as follows: By Theorem 1.32, f can be completed to a morphism

$$([f], [f^n], [f/n], [f_{<n}]) : \tau_{<n}(K, Y_K) \longrightarrow \tau_{<n}(L, Y_L)$$

in $\mathbf{HoCW}_{\supset <n}$ such that

1. $f^n = f|_{K^n}$ and
2. $f/n = h'_L \circ f^n \circ h_K$, where $h'_L : L^n \to L/n$ is a homotopy inverse rel L^{n-1} for h_L.

Choose such a completion and set

$$\boxed{\tau_{<n}(f) = ([f], [f^n], [f/n], [f_{<n}]).}$$

The truncation $\tau_{<n}$ is now defined on objects and morphisms. For a morphism $F = ([f], [f^n], [f/n], [f_{<n}])$ in $\mathbf{HoCW}_{\supset <n}$, we shall also write F^n for the second component $[f^n]$ of F, F/n for the third component $[f/n]$ of F, and $F_{<n}$ for the fourth component $[f_{<n}]$ of F.

Lemma 1.38. *If $f : (K, Y_K) \to (L, Y_L)$ and $g : (L, Y_L) \to (P, Y_P)$ are morphisms in* $\mathbf{CW}_{n \supset \partial}$, *then*

$$\tau_{<n}(g)/n \circ \tau_{<n}(f)/n = \tau_{<n}(g \circ f)/n.$$

Proof. Let $(K, K/n, h_K, K_{<n}) = \tau_{<n}(K, Y_K)$, $(L, L/n, h_L, L_{<n}) = \tau_{<n}(L, Y_L)$, and $(P, P/n, h_P, P_{<n}) = \tau_{<n}(P, Y_P)$. Let h'_L, h'_P be homotopy inverses rel $(n-1)$-skeleta for h_L, h_P, respectively. Set $h = gf$. By definition of $\tau_{<n}$ on morphisms, we have

$$\tau_{<n}(f) = ([f], [f|_{K^n}], [h'_L \circ f|_{K^n} \circ h_K], [f_{<n}]),$$

$$\tau_{<n}(g) = ([g], [g|_{L^n}], [h'_P \circ g|_{L^n} \circ h_L], [g_{<n}]),$$

and

$$\tau_{<n}(h) = ([h], [h|_{K^n}], [h''_P \circ h|_{K^n} \circ h_K], [h_{<n}]),$$

where h''_P is some homotopy inverse rel P^{n-1} for h_P. The maps h'_P and h''_P need not be equal, but they are homotopic rel P^{n-1}, so that $[h'_P] = [h''_P]$. The assertion is established by the following calculation on rel $(n-1)$-skeleta homotopy classes:

$$\begin{aligned}
\tau_{<n}(g)/n \circ \tau_{<n}(f)/n &= [h'_P \circ g|_{L^n} \circ h_L] \circ [h'_L \circ f|_{K^n} \circ h_K] \\
&= [h'_P \circ g|_{L^n}] \circ [h_L \circ h'_L] \circ [f|_{K^n} \circ h_K] \\
&= [h'_P \circ g|_{L^n} \circ f|_{K^n} \circ h_K] \\
&= [h'_P] \circ [(gf)|_{K^n} \circ h_K] \\
&= [h''_P] \circ [h|_{K^n} \circ h_K] \\
&= \tau_{<n}(h)/n.
\end{aligned}$$

□

Theorem 1.39. *Let $f : (K, Y_K) \to (L, Y_L)$ and $g : (L, Y_L) \to (P, Y_P)$ be morphisms in* $\mathbf{CW}_{n \supset \partial}$ *such that $\tau_{<n}(g \circ f)$ is n-compression rigid. Then*

$$\tau_{<n}(g \circ f) = \tau_{<n}(g) \circ \tau_{<n}(f)$$

in $\mathbf{HoCW}_{\supset <n}$.

Proof. Set $h = gf$. If

$$\tau_{<n}(f) = ([f], [f^n], [f/n], [f_{<n}]), \quad \tau_{<n}(g) = ([g], [g^n], [g/n], [g_{<n}])$$

and

$$\tau_{<n}(h) = ([h], [h^n], [h/n], [h_{<n}])$$

then

$$\tau_{<n}(g) \circ \tau_{<n}(f) = ([g] \circ [f], [g^n] \circ [f^n], [g/n] \circ [f/n], [g_{<n}] \circ [f_{<n}]),$$

and thus

$$\tau_{<n}(g \circ f) = \tau_{<n}(g) \circ \tau_{<n}(f)$$

iff

$(1)\ [g] \circ [f] = [h], \qquad (2)\ [g^n] \circ [f^n] = [h^n],$

$(3)\ [g/n] \circ [f/n] = [h/n], (4)\ [g_{<n}] \circ [f_{<n}] = [h_{<n}].$

Equality holds in (1) by definition, and follows in (2) from

$$h^n = (gf)|_{K^n} = g|_{L^n} \circ f|_{K^n} = g^n \circ f^n.$$

Equality in (3) holds by Lemma 1.38. Using the two homotopy commutative diagrams

$$
\begin{array}{ccc}
K/n & \xleftarrow{\ i_K\ } & K_{<n} \\
{\scriptstyle f/n}\downarrow & & \downarrow{\scriptstyle f_{<n}} \\
L/n & \xleftarrow{\ i_L\ } & L_{<n}
\end{array}
\qquad
\begin{array}{ccc}
L/n & \xleftarrow{\ i_L\ } & L_{<n} \\
{\scriptstyle g/n}\downarrow & & \downarrow{\scriptstyle g_{<n}} \\
P/n & \xleftarrow{\ i_P\ } & P_{<n}
\end{array}
$$

where both homotopies may be assumed to be rel $(n-1)$-skeleta, we obtain

$$(h/n)i_K \simeq (g/n)(f/n)i_K \simeq (g/n)i_L f_{<n} \simeq i_P g_{<n} f_{<n}$$

rel K^{n-1}, where the first homotopy comes from (3). Also,

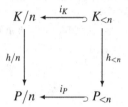

commutes up to homotopy rel K^{n-1}, whence

$$i_P h_{<n} \simeq (h/n) i_K \simeq i_P g_{<n} f_{<n}$$

rel K^{n-1}. By Proposition 1.34, $h_{<n} \simeq g_{<n} f_{<n}$ rel K^{n-1}, since $([h], [h^n], [h/n], [h_{<n}])$ is n-compression rigid. This establishes equality (4).   □

Let us call a subcategory $\mathbf{C} \subset \mathbf{CW}_{n \supset \partial}$ *(n-)compression rigid*, if the image under $\tau_{<n}$ of every morphism in $\mathbf{C}$ is n-compression rigid. We have seen in Proposition 1.34 that the truncation $f_{<n}$ is homotopy-theoretically well-defined precisely for n-compression rigid morphisms.

Corollary 1.40. *Let $\mathbf{C} \subset \mathbf{CW}_{n \supset \partial}$ be any compression rigid subcategory. Then the assignment $\tau_{<n}$ is a covariant functor $\tau_{<n} : \mathbf{C} \longrightarrow \mathbf{HoCW}_{\supset <n}$.*

Recall that $\mathbf{HoCW}_{n-1}$ denotes the category whose objects are CW-complexes and whose morphisms are rel $(n-1)$-skeleton homotopy classes of cellular maps. Let

$$P_4 : \mathbf{HoCW}_{\supset <n} \longrightarrow \mathbf{HoCW}_{n-1}$$

be the functor given by projection to the fourth component, that is, for an object $(K, K/n, h, K_{<n})$ in $\mathbf{HoCW}_{\supset <n}$, $P_4(K, K/n, h, K_{<n}) = K_{<n}$ and for a morphism $([f], [f^n], [f/n], [f_{<n}])$ in $\mathbf{HoCW}_{\supset <n}$, $P_4([f], [f^n], [f/n], [f_{<n}]) = [f_{<n}]$. Let

$$t_{<\infty} : \mathbf{CW}_{n \supset \partial} \longrightarrow \mathbf{HoCW}_{n-1}$$

be the natural projection functor, that is, $t_{<\infty}(K, Y_K) = K$ for an object (K, Y_K) in $\mathbf{CW}_{n \supset \partial}$, and $t_{<\infty}(f) = [f]$ for a morphism $f : (K, Y_K) \to (L, Y_L)$ in $\mathbf{CW}_{n \supset \partial}$. Define a covariant assignment of objects and morphisms

$$\boxed{t_{<n} = P_4 \circ \tau_{<n} : \mathbf{CW}_{n \supset \partial} \longrightarrow \mathbf{HoCW}_{n-1}}.$$

By Corollary 1.40, $t_{<n}$ is a functor on all n-compression rigid subcategories of $\mathbf{CW}_{n \supset \partial}$. The assignment $t_{<n}$ comes with a natural transformation

$$\mathrm{emb}_n : t_{<n} \longrightarrow t_{<\infty},$$

which we shall now describe. Let (K,Y) be an object of $\mathbf{CW}_{n\supset\partial}$. Applying $\tau_{<n}$, we obtain an n-truncation structure $\tau_{<n}(K,Y) = (K,K/n,h,K_{<n})$. Let

$$\mathrm{emb}_n(K,Y) : t_{<n}(K,Y) = K_{<n} \longrightarrow K = t_{<\infty}(K,Y)$$

be the rel K^{n-1} homotopy class of the composition

$$K_{<n} \hookrightarrow K/n \xrightarrow{h} K^n \hookrightarrow K.$$

This is a natural transformation: Given a morphism $f : (K,Y_K) \to (L,Y_L)$ in $\mathbf{CW}_{n\supset\partial}$, we apply $\tau_{<n}$ to obtain $\tau_{<n}(f) = ([f],[f^n],[f/n],[f_{<n}])$ so that $t_{<n}(f) = [f_{<n}]$. Then the required commutativity in $\mathbf{HoCW}_{n-1}$ of the square

$$
\begin{array}{ccc}
t_{<n}(K,Y_K) & \xrightarrow{\mathrm{emb}_n(K,Y_K)} & t_{<\infty}(K,Y_K) \\
\downarrow{\scriptstyle t_{<n}(f)} & & \downarrow{\scriptstyle t_{<\infty}(f)} \\
t_{<n}(L,Y_L) & \xrightarrow{\mathrm{emb}_n(L,Y_L)} & t_{<\infty}(L,Y_L)
\end{array}
$$

follows from the commutativity in $\mathbf{HoCW}_{n-1}$ of the diagram

$$
\begin{array}{ccccccc}
K_{<n} & \xrightarrow{[i_K]} & K/n & \xrightarrow{[h_K]} & K^n & \xrightarrow{[j_K]} & K \\
\downarrow{\scriptstyle t_{<n}(f)=[f_{<n}]} & & \downarrow{\scriptstyle [f/n]} & & \downarrow{\scriptstyle [f^n]} & & \downarrow{\scriptstyle [f]=t_{<\infty}(f)} \\
L_{<n} & \xrightarrow{[i_L]} & L/n & \xrightarrow{[h_L]} & L^n & \xrightarrow{[j_L]} & L
\end{array}
$$

where $\tau_{<n}(K,Y_K) = (K,K/n,h_K,K_{<n})$ and $\tau_{<n}(L,Y_L) = (L,L/n,h_L,L_{<n})$. We have proved:

Theorem 1.41. *Let $n \geq 3$ be an integer. There is a covariant assignment $t_{<n}$: $\mathbf{CW}_{n\supset\partial} \longrightarrow \mathbf{HoCW}_{n-1}$ of objects and morphisms together with a natural transformation $\mathrm{emb}_n : t_{<n} \to t_{<\infty}$ such that for an object (K,Y) of $\mathbf{CW}_{n\supset\partial}$, one has $H_r(t_{<n}(K,Y)) = 0$ for $r \geq n$, and*

$$\mathrm{emb}_n(K,Y)_* : H_r(t_{<n}(K,Y)) \xrightarrow{\cong} H_r(K)$$

is an isomorphism for $r < n$. The assignment $t_{<n}$ is a functor on all n-compression rigid subcategories of $\mathbf{CW}_{n\supset\partial}$.

For the degrees $n < 3$, the functor $t_{<n}$ has been constructed in Section 1.1.5.

Remark 1.42 (Effect on Cohomology). If $r > n$, then

$$H^r(t_{<n}(K,Y)) \cong \mathrm{Hom}(H_r(t_{<n}(K,Y)),\mathbb{Z}) \oplus \mathrm{Ext}(H_{r-1}(t_{<n}(K,Y)),\mathbb{Z}) = 0.$$

For the borderline case $r = n$,

$$H^n(t_{<n}(K,Y)) \cong \mathrm{Hom}(H_n(t_{<n}(K,Y)),\mathbb{Z}) \oplus \mathrm{Ext}(H_{n-1}(t_{<n}(K,Y)),\mathbb{Z})$$
$$\cong \mathrm{Ext}(H_{n-1}(K),\mathbb{Z})$$

(this is the torsion subgroup of $H_{n-1}(K)$ if $H_{n-1}(K)$ is finitely generated), while for $r < n$,

$$H^r(t_{<n}(K,Y)) \cong \mathrm{Hom}(H_r(t_{<n}(K,Y)),\mathbb{Z}) \oplus \mathrm{Ext}(H_{r-1}(t_{<n}(K,Y)),\mathbb{Z})$$
$$\cong \mathrm{Hom}(H_r(K),\mathbb{Z}) \oplus \mathrm{Ext}(H_{r-1}(K),\mathbb{Z})$$
$$\cong H^r(K).$$

Thus, $t_{<n}(K,Y)$ is only up to degree-$(n-1)$-torsion a spatial cohomology truncation. In particular, over the rationals, $t_{<n}(K,Y)$ *is* a valid spatial cohomology truncation.

1.2 Compression Rigidity Obstruction Theory

The Compression Theorem 1.32 asserts that every cellular map f that preserves chosen direct sum complements of the n-cycle groups, that is, every morphism in the category $\mathbf{CW}_{n\supset\partial}$ of n-boundary-split CW-complexes, possesses a homological truncation $t_{<n}(f)$. We have also seen that f does not in general determine the homotopy class $t_{<n}(f)$ uniquely, not even when the domain and codomain of f are n-segmented with unique n-truncating subcomplexes. We called f n-compression rigid if it determines a unique homotopy class $t_{<n}(f)$. Compression rigidity was defined in terms of eigenhomotopies in Definition 1.33, and then characterized as being equivalent to the above uniqueness property in Proposition 1.34. On compression rigid categories, spatial homology truncation is a functor (Theorem 1.41). It is in practice not always easy to decide directly from Definition 1.33 or Proposition 1.34, whether a given map is compression rigid. The present section addresses this by systematically identifying obstruction cocycles. A characterization of the notion of compression rigidity in terms of obstruction cocycles is provided by Theorem 1.44. Regarding the question as to when a given homotopy H can be compressed into an n-truncation, we shall see in Proposition 1.48 that for a homotopy $H : K_{<n} \times I \to L/n$, the obstruction cocycle lies in $C^{n+1}(K_{<n} \times I; \pi_{n+1}(L/n, L_{<n}))$. The homotopy group $\pi_{n+1}(L/n, L_{<n})$ thus plays a key role and is studied in Proposition 1.50. Some simple sufficient conditions for compression rigidity are deduced from the general obstruction theory.

1.2.1 Existence of Compressed Homotopies

In order to fix notation, let us begin by recalling some basic obstruction theory.

Lemma 1.43. *Let X and Y be CW-complexes with X of dimension n and Y n-simple (i.e. $\pi_1(Y)$ acts trivially on $\pi_n(Y)$, for example Y simply connected). Let $g_1, g_2 : X \to Y$ be two maps such that $g_1|_{X^{n-1}} = g_2|_{X^{n-1}}$. Then g_1 and g_2 are homotopic rel X^{n-1} if, and only if, a single obstruction cocycle*

$$\omega(g_1, g_2) \in C^{n+1}(X \times I; \pi_n(Y))$$

vanishes. The obstruction cocycle is natural, that is, if $f : Y \to Y'$ is a map into an n-simple CW-complex Y', then

$$f_* \omega(g_1, g_2) = \omega(fg_1, fg_2) \in C^{n+1}(X \times I; \pi_n(Y')),$$

where

$$f_* : C^{n+1}(X \times I; \pi_n(Y)) \longrightarrow C^{n+1}(X \times I; \pi_n(Y'))$$

composes a cochain with the induced map $f_* : \pi_n(Y) \to \pi_n(Y')$.

Proof. The n-skeleton of $Z = X \times I$ is given by $Z^n = X \times \partial I \cup X^{n-1} \times I \subset Z$. Set

$$g = (g_1 \times \{0\} \cup g_2 \times \{1\}) \cup (g_1|_{X^{n-1}} \times \mathrm{id}_I) : Z^n \longrightarrow Y.$$

Let e^{n+1} be an $(n+1)$-cell in Z with attaching map

$$\chi(e^{n+1})| : S^n = \partial e^{n+1} \longrightarrow Z^n.$$

Composing with g defines a map

$$S^n \xrightarrow{\chi(e^{n+1})|} Z^n \xrightarrow{g} Y.$$

Define

$$\omega(g_1, g_2)(e^{n+1}) = [g \circ \chi(e^{n+1})|] \in \pi_n(Y).$$

(Since Y is n-simple, any map of an oriented n-sphere into Y represents a well-defined element of $\pi_n(Y)$.) Then the core theorem of obstruction theory asserts that g extends to a map $Z = Z^{n+1} \to Y$ if, and only if, $\omega(g_1, g_2) = 0$.

For a map $f : Y \to Y'$, we have

$$\begin{aligned}
f_* \omega(g_1, g_2)(e^{n+1}) &= f_*[g \circ \chi(e^{n+1})|] \\
&= [f \circ g \circ \chi(e^{n+1})|] \\
&= \omega(fg_1, fg_2)(e^{n+1}) \in \pi_n(Y')
\end{aligned}$$

because

$$fg = ((fg_1) \times \{0\} \cup (fg_2) \times \{1\}) \cup ((fg_1)|_{X^{n-1}} \times \mathrm{id}_I).$$

□

Theorem 1.44. *Let (K, Y_K) and (L, Y_L) be objects of $\mathbf{CW}_{n \supset \partial}$ with $\tau_{<n}(K, Y_K) = (K, K/n, h_K, K_{<n})$ and $\tau_{<n}(L, Y_L) = (L, L/n, h_L, L_{<n})$. Let $i_L : L_{<n} \hookrightarrow L/n$ denote the subcomplex inclusion. A morphism $([f], [f^n], [f/n], [f_{<n}]) : (K, K/n, h_K, K_{<n}) \to (L, L/n, h_L, L_{<n})$ in $\mathbf{HoCW}_{\supset <n}$ is n-compression rigid if, and only if, the following statement holds: For every $f'_{<n} : K_{<n} \to L_{<n}$ such that*

$$i_{L*}\omega(f_{<n}, f'_{<n}) = 0 \in C^{n+1}(K_{<n} \times I; \pi_n(L/n))$$

one actually has

$$\omega(f_{<n}, f'_{<n}) = 0 \in C^{n+1}(K_{<n} \times I; \pi_n(L_{<n})).$$

Proof. In order to prove the only if-direction, suppose that $([f], [f^n], [f/n], [f_{<n}])$ is n-compression rigid. Let $f'_{<n} : K_{<n} \to L_{<n}$ be a map such that

$$i_{L*}\omega(f_{<n}, f'_{<n}) = 0 \in C^{n+1}(K_{<n} \times I; \pi_n(L/n)).$$

By Lemma 1.43,

$$i_{L*}\omega(f_{<n}, f'_{<n}) = \omega(i_L f_{<n}, i_L f'_{<n}),$$

and the latter cocycle is the obstruction for finding a homotopy rel K^{n-1} between $i_L f_{<n}$ and $i_L f'_{<n}$. As this cocycle vanishes, there is a homotopy $i_L f_{<n} \simeq i_L f'_{<n}$ rel K^{n-1}. Since

$$
\begin{CD}
K/n @<i_K<< K_{<n} \\
@Vf/nVV @VVf_{<n}V \\
L/n @<i_L<< L_{<n}
\end{CD}
$$

homotopy commutes rel K^{n-1}, we also have a homotopy commutative diagram

$$
\begin{CD}
K/n @<i_K<< K_{<n} \\
@Vf/nVV @VVf'_{<n}V \\
L/n @<i_L<< L_{<n}
\end{CD}
$$

rel K^{n-1}. Thus, by Proposition 1.34, $f_{<n} \simeq f'_{<n}$ rel K^{n-1}. Hence the obstruction $\omega(f_{<n}, f'_{<n})$ vanishes.

To prove the if-direction, assume that $i_{L*}\omega(f_{<n}, f'_{<n}) = 0$ implies $\omega(f_{<n}, f'_{<n}) = 0$ for all $f'_{<n}$. By Proposition 1.34, n-compression rigidity of $([f], [f^n], [f/n], [f_{<n}])$ follows once we have shown that whenever $f'_{<n}, f''_{<n}$ are such that $i_L f'_{<n} \simeq (f/n)i_K$ rel K^{n-1} and $i_L f''_{<n} \simeq (f/n)i_K$ rel K^{n-1}, one can conclude $f'_{<n} \simeq f''_{<n}$ rel K^{n-1}. If $i_L f'_{<n} \simeq (f/n)i_K \simeq i_L f''_{<n}$ rel K^{n-1} then $(f/n)i_K \simeq i_L f_{<n}$ rel K^{n-1} implies $\omega(i_L f_{<n}, i_L f'_{<n}) = 0$ and $\omega(i_L f_{<n}, i_L f''_{<n}) = 0$. Thus $i_{L*}\omega(f_{<n}, f'_{<n}) = 0$ and $i_{L*}\omega(f_{<n}, f''_{<n}) = 0$, which implies $\omega(f_{<n}, f'_{<n}) = 0$ and $\omega(f_{<n}, f''_{<n}) = 0$. Consequently, there exist homotopies $f'_{<n} \simeq f_{<n} \simeq f''_{<n}$ rel K^{n-1}. $\qquad\square$

Corollary 1.45. *A morphism*

$$([f], [f^n], [f/n], [f_{<n}]) : (K, K/n, h_K, K_{<n}) \rightarrow (L, L/n, h_L, L_{<n})$$

in **HoCW**$_{\supset <n}$ *is n-compression rigid if $i_{L*} : \pi_n(L_{<n}) \rightarrow \pi_n(L/n)$ is injective.*

Proof. If $i_{L*} : \pi_n(L_{<n}) \rightarrow \pi_n(L/n)$ is injective then

$$\mathrm{Hom}(C_{n+1}(K_{<n} \times I), i_{L*}) : C^{n+1}(K_{<n} \times I; \pi_n(L_{<n})) \longrightarrow C^{n+1}(K_{<n} \times I; \pi_n(L/n))$$

is injective as well. $\qquad\square$

1.2.2 Compression of a Given Homotopy

Let $n \geq 3$ be an integer. In investigating the n-compression rigidity of a morphism

$$([f], [f^n], [f/n], [f_{<n}]) : (K, K/n, h_K, K_{<n}) \rightarrow (L, L/n, h_L, L_{<n})$$

it may sometimes be useful to know whether a *particular* homotopy can be compressed into the truncated spaces. We will here determine the obstructions to deforming, rel $K_{<n} \times \partial I \cup K^{n-1} \times I$, a given rel K^{n-1} homotopy $H : K_{<n} \times I \rightarrow L/n$ from $i_L g_1$ to $i_L g_2$ to a homotopy $K_{<n} \times I \rightarrow L_{<n}$. The resulting homotopy would then be rel K^{n-1} and from g_1 to g_2.

We begin by turning the inclusion $L_{<n} \hookrightarrow L/n$ into a fibration, that is, we choose a homotopy equivalence $\lambda : L_{<n} \xrightarrow{\simeq} \mathcal{L}_{<n}$ and a fibration $p : \mathcal{L}_{<n} \rightarrow L/n$ such that

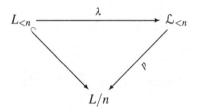

commutes. We may take λ to be an inclusion such that $\mathcal{L}_{<n}$ deformation retracts onto $L_{<n}$. In particular, there is a homotopy inverse λ' for λ such that $\lambda'\lambda = \mathrm{id}_{L_{<n}}$, see [Whi78], Theorem I.7.30. Let F denote the fiber of p and let

$$g_0 = (g_1 \times \{0\} \cup g_2 \times \{1\}) \cup (g_1|_{K^{n-1}} p_1) : K_{<n} \times \partial I \cup K^{n-1} \times I \longrightarrow \mathcal{L}_{<n},$$

where $p_1 : K^{n-1} \times I \to K^{n-1}$ is the first factor projection. We need to solve the relative lifting problem

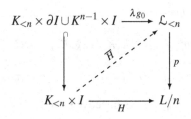

For if a solution $\overline{H}$ exists, then $H_{<n} = \lambda' \circ \overline{H}$ satisfies

$$H_{<n}(k,0) = g_1(k), \ H_{<n}(k,1) = g_2(k) \text{ for } k \in K_{<n}$$

because

$$H_{<n}(k,0) = \lambda'\overline{H}(k,0) = \lambda'\lambda g_0(k,0) = g_0(k,0) = g_1(k)$$

(similarly for $t = 1$) and $H_{<n}(k,t) = g_1(k)$ for $k \in K^{n-1}$ and all $t \in I$, since for $k \in K^{n-1}$,

$$H_{<n}(k,t) = \lambda'\lambda g_0(k,t) = g_0(k,t) = g_1(k)$$

for all t. Thus $H_{<n}$ is the sought compression of H.

Lemma 1.46. *The homotopy fiber F of $i : L_{<n} \hookrightarrow L/n$ is $(n-2)$-connected. It is not $(n-1)$-connected unless i is the identity.*

Proof. The CW pair $(L/n, L_{<n})$ is $(n-1)$-connected and the subcomplex $L_{<n}$ is 1-connected. Thus the quotient map induces an isomorphism

$$\pi_j(L/n, L_{<n}) \cong \pi_j((L/n)/L_{<n}) \cong \pi_j(\bigvee_\beta S_\beta^n)$$

for $j \leq (n-1)+1 = n$. For $0 < j < n$, $\pi_j(\bigvee_\beta S_\beta^n) \cong H_j(\bigvee_\beta S_\beta^n) = 0$ by the Hurewicz theorem. Therefore,

$$\pi_k(F) \cong \pi_{k+1}(L/n, L_{<n}) = 0$$

when $k \leq n-2$. For $k = n-1$,

$$\pi_{n-1}(F) \cong \pi_n(L/n, L_{<n}) \cong \pi_n(\bigvee_\beta S_\beta^n) \cong H_n(\bigvee_\beta S_\beta^n) \neq 0$$

unless there are no cells z_β, in which case i is the identity. $\square$

Lemma 1.47. *The group* $G = H^{k+1}(K_{<n} \times I, K_{<n} \times \partial I \cup K^{n-1} \times I; \pi_k F)$ *vanishes unless* $k = n$. *For* $k = n$, $G \cong C^{n+1}(K_{<n} \times I; \pi_n F)$.

Proof. The complex $A = K_{<n} \times \partial I \cup K^{n-1} \times I$ is the n-skeleton $(K_{<n} \times I)^n$ of $K_{<n} \times I = (K_{<n} \times I)^{n+1}$. By the universal coefficient theorem,

$$G \cong \mathrm{Hom}(H_{k+1}(K_{<n} \times I, A), \pi_k F) \oplus \mathrm{Ext}(H_k(K_{<n} \times I, A), \pi_k F).$$

The group $H_j(K_{<n} \times I, A)$ is zero for $j \neq n+1$ and isomorphic to the cellular chain group $C_{n+1}(K_{<n} \times I)$ for $j = n+1$. Thus $G = 0$ for $k \notin \{n, n+1\}$. For $k = n+1$, $G \cong \mathrm{Ext}(C_{n+1}(K_{<n} \times I), \pi_{n+1} F) = 0$, since $C_{n+1}(K_{<n} \times I)$ is free abelian. For $k = n$, $G \cong \mathrm{Hom}(C_{n+1}(K_{<n} \times I), \pi_n F) = C^{n+1}(K_{<n} \times I; \pi_n F)$. $\quad\square$

To solve the relative lifting problem, we consider the Moore-Postnikov tower of principal fibrations of the map p:

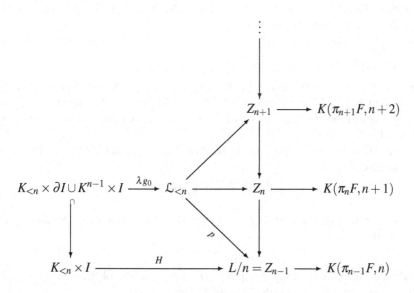

By Lemma 1.46, the Moore-Postnikov factorization begins with Z_{n-1}. The composition across the bottom of the diagram gives a primary obstruction

$$\omega_{n-1} \in H^n(K_{<n} \times I, K_{<n} \times \partial I \cup K^{n-1} \times I; \pi_{n-1} F).$$

According to Lemma 1.47, this group is zero and the primary obstruction vanishes, so that a lift of H to Z_n exists. The obstruction to lifting further to Z_{n+1} is a class

$$\omega_n \in H^{n+1}(K_{<n} \times I, K_{<n} \times \partial I \cup K^{n-1} \times I; \pi_n F).$$

This cohomology group is nonzero by Lemma 1.47 and Proposition 1.50 below, unless $L_{<n} \hookrightarrow L/n$ is the identity or $K_{<n}$ has no n-cells, i.e. $K^{n-1} \hookrightarrow K_{<n}$ is the identity. If $\omega_n = 0$, then the rest of the obstructions are classes

$$\omega_k \in H^{k+1}(K_{<n} \times I, K_{<n} \times \partial I \cup K^{n-1} \times I; \pi_k F),$$

$k > n$. But these all vanish by Lemma 1.47. Observing that $\pi_n(F) \cong \pi_{n+1}(L/n, L_{<n})$, we have shown:

Proposition 1.48. *The homotopy $H : K_{<n} \times I \to L/n$ can be compressed into $L_{<n}$ rel K^{n-1} if, and only if, a single obstruction*

$$\omega_n(H) \in C^{n+1}(K_{<n} \times I; \pi_{n+1}(L/n, L_{<n}))$$

vanishes.

Corollary 1.49. *A morphism $F : (K, K/n, h_K, K_{<n}) \to (L, L/n, h_L, L_{<n})$ in $\mathrm{HoCW}_{\supset<n}$ is n-compression rigid, if*

(1) $\partial_n = 0 : C_n(K) \longrightarrow C_{n-1}(K)$, or
(2) $\partial_n : C_n(L) \longrightarrow C_{n-1}(L)$ is injective.

Proof. (1) Since $\partial_n = 0$, we have $C_n(K) = Z_n(K)$ and $Y_K = 0$. Thus K/n has no cells y_α and $K_{<n} = K^{n-1}$. Consequently, $K_{<n} \times I$ has no $(n+1)$-cells, $C_{n+1}(K_{<n} \times I) = 0$ and $\omega_n(H) = 0$ for every H. By Proposition 1.48, f is n-compression rigid.

(2) If $\partial_n : C_n(L) \longrightarrow C_{n-1}(L)$ is injective, then $Z_n(L) = 0$ and $C_n(L) = Y_L$. Thus L/n has no cells z_β and

$$L/n = L^{n-1} \cup \bigcup_\alpha y_\alpha = L_{<n}.$$

We conclude that $\pi_{n+1}(L/n, L_{<n}) = 0$ and $\omega_n(H) = 0$ for every H also in this situation.

$\square$

The coefficient homotopy group $\pi_{n+1}(L/n, L_{<n})$ in the obstruction group can only be zero if $L_{<n} \hookrightarrow L/n$ is the identity. In fact:

Proposition 1.50. *Let $(L, L/n, h, L_{<n})$ be an n-truncation structure, $n \geq 3$, such that $H_n(L^n)$ has finite rank b. Then $\pi_{n+1}(L/n, L_{<n})$ maps onto $(\mathbb{Z}/2)^b$, and if $H_2(L) = 0$, then*

$$\pi_{n+1}(L/n, L_{<n}) \cong (\mathbb{Z}/2)^b.$$

Proof. The n-segmentation L/n has the form

$$L/n = L^{n-1} \cup \bigcup_\alpha y_\alpha \cup z_1 \cup \cdots \cup z_b,$$

where $\{z_1, \ldots, z_b\}$ is a basis of n-cells for $Z_n(L/n) = H_n(L/n) \cong H_n(L^n)$. The CW pair $(L/n, L_{<n})$ is $r = (n-1)$-connected, since all cells in $L/n - L_{<n}$ have dimension

$n > r$. The complex $L_{<n}$ is $s = 1$-connected as $n \geq 3$. Thus, as $n + 1 \leq r + s + 1 = (n - 1) + 2$, the quotient map $L/n \to (L/n)/L_{<n}$ induces a surjection $\pi_{n+1}(L/n, L_{<n}) \twoheadrightarrow \pi_{n+1}((L/n)/L_{<n})$. As $L_{<n} = L^{n-1} \cup \bigcup_\alpha y_\alpha$, we have $(L/n)/L_{<n} \cong S_1^n \vee \cdots \vee S_b^n$, where the sphere S_j^n corresponds to the cell z_j, $j = 1, \ldots, b$. Thus, from the proof of Proposition 1.18 (concerning virtual cell groups),

$$\pi_{n+1}((L/n)/L_{<n}) \cong \pi_{n+1}(S_1^n \vee \cdots \vee S_b^n) \cong (\mathbb{Z}/2)^b.$$

If $H_2(L) = 0$, then $H_2(L_{<n}) \cong H_2(L) = 0$ and since $L_{<n}$ is simply connected, it follows from the Hurewicz theorem that $L_{<n}$ is $s = 2$-connected. Therefore, as now $n + 1 \leq r + s = (n - 1) + 2$, the quotient map induces an isomorphism $\pi_{n+1}(L/n, L_{<n}) \cong \pi_{n+1}((L/n)/L_{<n})$. $\qquad\square$

1.3 Case Studies of Compression Rigid Categories

Proposition 1.51. *A morphism*

$$F = ([f], [f^n], [f/n], [f_{<n}]) : (K, K/n, h_K, K_{<n}) \to (L, L/n, h_L, L_{<n})$$

in **HoCW**$_{\supset <n}$ *is n-compression rigid if either $n = 3$ and $L^1 = \mathrm{pt}$, or $n \geq 4$ and*

$$\mathrm{im}(\pi_n(L^n, L^{n-1}) \to \pi_{n-1}(L^{n-1})) \cap \ker(\pi_{n-1}(L^{n-1}) \to \pi_{n-1}(L^{n-1}, L^{n-2})) = 0.$$

(The latter condition is in particular satisfied when $\pi_{n-1}(L^{n-2}) = 0$.)

Proof. Let $g_1, g_2 : K_{<n} \to L_{<n}$ be two cellular maps such that the square

commutes up to homotopy rel K^{n-1} for $i = 1, 2$. By Remark 1.7, the n-segmented space L/n can be written as a wedge sum

$$L/n = L_{<n} \vee \bigvee_\beta S_\beta^n.$$

The essential ingredient that facilitates the proof is the canonical retraction

$$r : L/n \longrightarrow L_{<n}, \quad r i_L = 1,$$

which maps the spheres S_β^n to a point. Then

$$i_L g_1 \simeq (f/n) i_K \simeq i_L g_2$$

rel K^{n-1} and thus

$$g_1 = r i_L g_1 \simeq r i_L g_2 = g_2$$

rel K^{n-1}. By Proposition 1.34, F is n-compression rigid.   □

Proposition 1.52. *Let K be a simply connected CW-complex having precisely one n-cell. Then any morphism $F : (K, K/n, h_K, K_{<n}) \to (K, K/n, h_K, K_{<n})$ in $\mathbf{HoCW}_{\supset <n}$ is n-compression rigid.*

Proof. Any homomorphism $\mathbb{Z} \to G$, where G is a torsion-free abelian group, is either zero or injective. Thus the boundary operator $\partial_n : C_n(K) = \mathbb{Z}e^n \to C_{n-1}(K)$ is either zero or injective. By Corollary 1.49, F is n-compression rigid.   □

Proposition 1.53. *If M is a closed, simply connected n-manifold with one n-cell, then any morphism $F : (M, M/n, h_M, M_{<n}) \to (L, L/n, h_L, L_{<n})$ in $\mathbf{HoCW}_{\supset <n}$ is n-compression rigid.*

Proof. Since M is simply connected, it is orientable and thus $H_n(M) \cong \mathbb{Z}$. On the other hand $H_n(M) = Z_n(M)$. The boundary operator $\partial_n : C_n(M) = \mathbb{Z}e^n \to C_{n-1}(M)$ is either zero or injective. If it were injective, we would reach the contradiction $0 = Z_n(M) = H_n(M) \cong \mathbb{Z}$. Thus $\partial_n = 0$ and F is n-compression rigid by Corollary 1.49.
□

Proposition 1.54. *If M and N are closed, simply connected 4-manifolds, each having one 4-cell, then for any $n \geq 3$, any morphism $F : (M, M/n, h_M, M_{<n}) \to (N, N/n, h_N, N_{<n})$ in $\mathbf{HoCW}_{\supset <n}$ is n-compression rigid.*

Proof. For $n \geq 5$, there is of course nothing to show since then $M = M_{<n}$, $N = N_{<n}$. For $n = 4$ the assertion follows from Proposition 1.53. Let $n = 3$. Since N is orientable, Poincaré duality implies $H_3(N) = 0$. Consequently, the sequence

$$C_4(N) \xrightarrow{\partial_4} C_3(N) \xrightarrow{\partial_3} C_2(N)$$

is exact. By the proof of Proposition 1.53, $\partial_4 = 0$. By exactness, ∂_3 is injective. By Corollary 1.49 (2), F is 3-compression rigid.   □

Let $\mathbf{CW}_{\partial=0}^n$ be the full subcategory of $\mathbf{CW}_{n \supset \partial}$ whose objects are those pairs (K, Y) for which the cellular boundary map $\partial_n : C_n(K) \to C_{n-1}(K)$ vanishes. By Corollary 1.49 (1), $\mathbf{CW}_{\partial=0}^n$ is an n-compression rigid category. For objects in this category, the cellular subgroup Y is uniquely determined, namely $Y = 0$. Many spaces that arise in the intended fields of application for the truncation machine are objects of $\mathbf{CW}_{\partial=0}^n$:

Proposition 1.55. *Let X be a complex algebraic threefold. Then the link of an isolated node in X is an object of $\mathbf{CW}_{\partial=0}^n$ for all n.*

Proof. Such a link is homeomorphic to $S^2 \times S^3$.   □

1.4 Truncation of Homotopy Equivalences

The following proposition asserts that the truncation of a homotopy equivalence is again a homotopy equivalence without requiring any compression rigidity assumptions.

Proposition 1.56. *Let* $f : (K, Y_K) \to (L, Y_L)$ *be a morphism in* $\mathbf{CW}_{n \supset \partial}$ *with* $f : K \to L$ *a homotopy equivalence. Then*

$$t_{<n}(f) : t_{<n}(K, Y_K) \longrightarrow t_{<n}(L, Y_L)$$

is an isomorphism in **HoCW**, *that is, represented by a homotopy equivalence.*

Proof. We will use the natural transformation $\mathrm{emb}_n : t_{<n} \to t_{<\infty}$ from Theorem 1.41. The induced maps $\mathrm{emb}_n(K, Y_K)_* : H_r(t_{<n}(K, Y_K)) \to H_r(K)$ and $\mathrm{emb}_n(L, Y_L)_* : H_r(t_{<n}(L, Y_L)) \to H_r(L)$ are isomorphisms for $r < n$. The commutative diagram

$$
\begin{array}{ccc}
t_{<n}(K, Y_K) & \xrightarrow{\ \mathrm{emb}_n(K,Y_K)\ } & K = t_{<\infty}(K, Y_K) \\[2pt]
\Big\downarrow{\scriptstyle t_{<n}(f)} & & \Big\downarrow{\scriptstyle t_{<\infty}(f)=[f]} \\[2pt]
t_{<n}(L, Y_L) & \xrightarrow{\ \mathrm{emb}_n(L,Y_L)\ } & L = t_{<\infty}(L, Y_L)
\end{array}
$$

induces a commutative diagram on homology:

$$
\begin{array}{ccc}
H_r(t_{<n}(K, Y_K)) & \xrightarrow{\ \mathrm{emb}_n(K,Y_K)_*\ } & H_r(K) \\[2pt]
\Big\downarrow{\scriptstyle t_{<n}(f)_*} & & \cong \Big\downarrow{\scriptstyle f_*} \\[2pt]
H_r(t_{<n}(L, Y_L)) & \xrightarrow{\ \mathrm{emb}_n(L,Y_L)_*\ } & H_r(L)
\end{array}
$$

If $r < n$, then $\mathrm{emb}_n(K, Y_K)_*$ and $\mathrm{emb}_n(L, Y_L)_*$ are isomorphisms, whence $t_{<n}(f)_*$ is an isomorphism. If $r \geq n$, then both $H_r(t_{<n}(K, Y_K))$ and $H_r(t_{<n}(L, Y_L))$ are zero so that $t_{<n}(f)_*$ is an isomorphism in this range as well. Thus $t_{<n}(f)$ is represented by a map between simply connected CW-complexes which is an H_*-isomorphism and hence a homotopy equivalence by Whitehead's theorem. $\square$

Caveat. In the situation of Proposition 1.56, one may not infer that $\tau_{<n}(f) = ([f], [f^n], [f/n], [f_{<n}])$ is an isomorphism in $\mathbf{HoCW}_{\supset <n}$. For one thing, f was only assumed to be a homotopy equivalence, not a homotopy equivalence *rel* K^{n-1}. Even if we made the assumption that f be a homotopy equivalence rel K^{n-1}, it does not in general follow that f^n is an equivalence. For example, let $f : D^{n+1} \to D^{n+1}$ be the map obtained by radially extending a map of degree 2 from ∂D^{n+1} to

∂D^{n+1}. (Here, D^{n+1} has the CW-structure $D^{n+1} = e^0 \cup e^n \cup_1 e^{n+1}$.) The $(n-1)$-skeleton of D^{n+1} is a point and f is a homotopy equivalence rel this point. However, $f^n = f| : \partial D^{n+1} \to \partial D^{n+1}$ is not an equivalence, since it has degree 2. Thus it is interesting to observe that, while the intermediary components f^n and f/n of a morphism do not preserve the property of being an equivalence, this property *is* preserved by the final component $f_{<n}$.

1.5 Truncation of Inclusions

In view of the fact that, up to homotopy equivalence, every map is an inclusion, it is worthwhile to investigate when an inclusion can be compressed into the spatial homology truncations of its domain and codomain. Here, we are starting with a "naked" inclusion map, not a morphism in $\mathbf{CW}_{n \supset \partial}$ whose underlying map is an inclusion. The goal is to state conditions under which an inclusion can be promoted to a morphism in $\mathbf{CW}_{n \supset \partial}$. The desired compression is then obtained by applying $t_{<n}$ to the morphism.

Proposition 1.57. *Let K be a simply connected CW-complex and $L \subset K$ a simply connected subcomplex. If $H_{n-1}(L)$ is free abelian, then the subcomplex-inclusion $f : L \hookrightarrow K$ is compressible into spatial homology n-truncations of L and K.*

Proof. Let $B_{r-1}(L) = \operatorname{im} \partial_r^L$ and $B_{r-1}(K) = \operatorname{im} \partial_r^K$ be the $(r-1)$-dimensional boundaries in L and K, respectively. Let $s : B_{n-1} \to C_n(L)$ be a splitting of $\partial_n^L| :$ $C_n(L) \twoheadrightarrow B_{n-1}(L)$, $\partial_n^L s = \operatorname{id}$. Let $u : B_{n-2} \to C_{n-1}(L)$ be a splitting of $\partial_{n-1}^L| :$ $C_{n-1}(L) \twoheadrightarrow B_{n-2}(L)$. The image of u determines a direct sum decomposition $C_{n-1}(L)$ $= Z_{n-1}(L) \oplus Y_{n-1}$ with $Y_{n-1} = \operatorname{im}(u)$. If $H_{n-1}(L)$ is free then the short exact sequence

$$0 \to B_{n-1}(L) \longrightarrow Z_{n-1}(L) \longrightarrow H_{n-1}(L) \to 0$$

splits and $Z_{n-1}(L) = B_{n-1}(L) \oplus H$, $H \cong H_{n-1}(L)$. Thus $C_{n-1}(L) = B_{n-1}(L) \oplus P$ with $P = H \oplus Y_{n-1}$. Let $R \subset C_{n-1}(K)$ be the subgroup generated by all $(n-1)$-cells of $K - L$. It follows that

$$C_{n-1}(K) = C_{n-1}(L) \oplus R = B_{n-1}(L) \oplus P \oplus R.$$

If $A \oplus B$ and A' are subgroups of some abelian group and $A \subset A'$, then the formula

$$(A \oplus B) \cap A' = A \oplus (B \cap A')$$

is available. It implies that

$$
\begin{aligned}
B_{n-1}(K) &= B_{n-1}(L) \oplus (P \oplus R) \cap B_{n-1}(K) \\
&= B_{n-1}(L) \oplus Q,
\end{aligned}
$$

since $B_{n-1}(L) \subset B_{n-1}(K)$, and where $Q = (P \oplus R) \cap B_{n-1}(K)$. Since Q is free abelian as a subgroup of the free abelian group $C_{n-1}(K)$, we can choose a basis $\{q_\alpha\}$ for Q. Since $Q \subset B_{n-1}(K)$, every q_α is a boundary, $q_\alpha = \partial_n^K(k_\alpha)$, $k_\alpha \in C_n(K)$. Define a map $t : Q \to C_n(K)$ by

$$t(\sum_i \lambda_{\alpha_i} q_{\alpha_i}) = \sum_i \lambda_{\alpha_i} k_{\alpha_i}.$$

Let $\sigma : B_{n-1}(K) \to C_n(K)$ be the map given by

$$\sigma(l + q) = s(l) + t(q), \ l \in B_{n-1}(L), \ q \in Q.$$

Then σ splits $\partial_n^K| : C_n(K) \twoheadrightarrow B_{n-1}(K)$ because

$$\begin{aligned}
\partial_n^K \sigma(l + \sum_i \lambda_{\alpha_i} q_{\alpha_i}) &= \partial_n^K s(l) + \partial_n^K t(\sum_i \lambda_{\alpha_i} q_{\alpha_i}) \\
&= \partial_n^L s(l) + \partial_n^K \sum_i \lambda_{\alpha_i} k_{\alpha_i} \\
&= l + \sum_i \lambda_{\alpha_i} \partial_n^K(k_{\alpha_i}) \\
&= l + \sum_i \lambda_{\alpha_i} q_{\alpha_i}.
\end{aligned}$$

Set $Y_L = \mathrm{im}(s)$ and $Y_K = \mathrm{im}(\sigma)$ so that $C_n(L) = Z_n(L) \oplus Y_L$, $C_n(K) = Z_n(K) \oplus Y_K$. If $y \in Y_L$, say $y = s(l)$, $l \in B_{n-1}(L)$, then $\sigma(l) = s(l) = y$ so that $y \in Y_K$. Hence, the chain map $f_* : C_n(L) \hookrightarrow C_n(K)$ induced by the inclusion $f : L \hookrightarrow K$ maps $f_*(Y_L) \subset Y_K$. This means that with these choices of Y_L and Y_K, f can be regarded as a morphism $f : (L, Y_L) \to (K, Y_K)$ in $\mathbf{CW}_{n \supset \partial}$. Thus $t_{<n}(f)$ is defined and yields the desired truncation $t_{<n}(f) : t_{<n}(L, Y_L) \to t_{<n}(K, Y_K)$. $\qquad \square$

1.6 Iterated Truncation

When you follow a truncation by a truncation in a lower degree, the resulting space is homotopy equivalent (rel relevant skeleton) to the result of truncating right away only in the lower degree.

Proposition 1.58. *Let $n > m \geq 3$ be integers, K a simply connected CW-complex and $(K, Y_n) \in Ob\,\mathbf{CW}_{n \supset \partial}$, $(K, Y_m) \in Ob\,\mathbf{CW}_{m \supset \partial}$. Then*

$$t_{<m}(t_{<n}(K, Y_n), Y_m) \cong t_{<m}(K, Y_m)$$

in $\mathbf{HoCW}_{m-1}$.

Proof. In the pair (K, Y_n), the second component Y_n is a subgroup $Y_n \subset C_n(K)$, and in the pair (K, Y_m), $Y_m \subset C_m(K)$. Carrying out the inner truncation, we obtain a space $t_{<n}(K, Y_n) = K_{<n}$, where $\tau_{<n}(K, Y_n) = (K, K/n, h_n, K_{<n})$, $h_{n*}i_{n*}C_n(K_{<n}) = Y_n$, $h_n : K/n \xrightarrow{\simeq} K^n$ rel K^{n-1}, $i_n : K_{<n} \hookrightarrow K/n$. Since

$$C_m(K_{<n}) = C_m((K_{<n})^{n-1}) = C_m(K^{n-1}) = C_m(K)$$

as $m < n$, the pair $(K_{<n}, Y_m)$ is indeed an object of $\mathbf{CW}_{m \supset \partial}$. Thus the outer truncation $t_{<m}(K_{<n}, Y_m)$ is defined and yields a space $t_{<m}(K_{<n}, Y_m) = (K_{<n})_{<m}$, where $\tau_{<m}(K_{<n}, Y_m) = (K_{<n}, K_{<n}/m, h_{nm}, (K_{<n})_{<m})$, $h_{nm*} i_{nm*} C_m((K_{<n})_{<m}) = Y_m$, $h_{nm} : K_{<n}/m \xrightarrow{\simeq} (K_{<n})^m = K^m$ rel K^{m-1}, $i_{nm} : (K_{<n})_{<m} \hookrightarrow K_{<n}/m$.

The right-hand side truncation yields $t_{<m}(K, Y_m) = K_{<m}$, where $\tau_{<m}(K, Y_m) = (K, K/m, h_m, K_{<m})$, $h_{m*} i_{m*} C_m(K_{<m}) = Y_m$, $h_m : K/m \xrightarrow{\simeq} K^m$ rel K^{m-1}, $i_m : K_{<m} \hookrightarrow K/m$. Since $(K_{<n})^m = K^m$, we may regard h_m as a homotopy equivalence $h_m : K/m \xrightarrow{\simeq} (K_{<n})^m$ rel K^{m-1}. Thus the quadruple $(K_{<n}, K/m, h_m, K_{<m})$ is an m-truncation structure completion of the pair $(K_{<n}, Y_m)$. So both

$$(K_{<n}, K_{<n}/m, h_{nm}, (K_{<n})_{<m}) \text{ and } (K_{<n}, K/m, h_m, K_{<m})$$

are m-truncation structure completions of $(K_{<n}, Y_m) \in Ob\mathbf{CW}_{m \supset \partial}$ satisfying

$$h_{nm*} i_{nm*} C_m((K_{<n})_{<m}) = Y_m = h_{m*} i_{m*} C_m(K_{<m}).$$

By Scholium 1.26, $(K_{<n})_{<m}$ and $K_{<m}$ are homotopy equivalent rel K^{m-1}. Consequently,

$$t_{<m}(t_{<n}(K, Y_n), Y_m) = t_{<m}(K_{<n}, Y_m) = (K_{<n})_{<m} \simeq K_{<m} = t_{<m}(K, Y_m)$$

rel K^{m-1}. $\square$

1.7 Localization at Odd Primes

Recall that $\mathbf{CW}^1$ denotes the category of simply connected CW-complexes and cellular maps. Let $G_{(odd)} = G \otimes \mathbb{Z}[\frac{1}{2}]$ denote the localization of an abelian group G at odd primes. Let $(-)_{(odd)} : \mathbf{CW}^1 \to \mathbf{CW}^1$ be the (Bousfield-Kan) localization functor at odd primes and let $loc : id \to (-)_{(odd)}$ be the localization natural transformation. The functor assigns to a simply connected CW-complex X a simply connected CW-complex $X_{(odd)}$ and to a map $f : X \to Y$ a map $f_{(odd)} : X_{(odd)} \to Y_{(odd)}$ such that

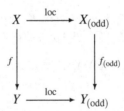

commutes. The localization map induces natural isomorphisms

$$\pi_*(X)_{(odd)} \cong \pi_*(X_{(odd)}), \ H_*(X; \mathbb{Z}[\tfrac{1}{2}]) \cong H_*(X)_{(odd)} \cong H_*(X_{(odd)}).$$

This localization preserves homotopy fibrations and cofibrations.

Lemma 1.59. *A homotopy between two maps $f, g : X \to Y$ induces a homotopy between the localized maps $f_{(\mathrm{odd})}, g_{(\mathrm{odd})} : X_{(\mathrm{odd})} \to Y_{(\mathrm{odd})}$.*

Proof. Let $H : X \times I \to Y$ be a homotopy between $f = H_0$ and $g = H_1$. The map $f_{(\mathrm{odd})}$ is an extension of $\mathrm{loc} \circ f : X \to Y_{(\mathrm{odd})}$ to $X_{(\mathrm{odd})}$ and $g_{(\mathrm{odd})}$ is an extension of $\mathrm{loc} \circ g : X \to Y_{(\mathrm{odd})}$ to $X_{(\mathrm{odd})}$:

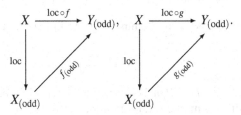

By [FHT01, Theorem 9.7(ii), p. 109], the homotopy $\mathrm{loc} \circ H : X \times I \to Y_{(\mathrm{odd})}$ extends to a homotopy $X_{(\mathrm{odd})} \times I \to Y_{(\mathrm{odd})}$ from $f_{(\mathrm{odd})}$ to $g_{(\mathrm{odd})}$.   □

Thus, $(-)_{(\mathrm{odd})} : \mathbf{CW}^1 \to \mathbf{CW}^1$ induces a functor on the corresponding homotopy categories, $(-)_{(\mathrm{odd})} : \mathbf{HoCW}^1 \to \mathbf{HoCW}^1$: If $[f] : X \to Y$ is a homotopy class represented by a cellular map $f : X \to Y$, then $[f]_{(\mathrm{odd})} := [f_{(\mathrm{odd})}]$ is well-defined. We define the odd-primary spatial homology truncation

$$t_{<n}^{(\mathrm{odd})} : \mathbf{CW}_{n \supset \partial} \longrightarrow \mathbf{HoCW}^1$$

to be the composition

$$t_{<n}^{(\mathrm{odd})} = (-)_{(\mathrm{odd})} \circ t_{<n}.$$

Explicitly, $t_{<n}^{(\mathrm{odd})}$ assigns to an object (K, Y_K) in $\mathbf{CW}_{n \supset \partial}$ the localization

$$t_{<n}^{(\mathrm{odd})}(K, Y_K) = (t_{<n}(K, Y_K))_{(\mathrm{odd})} = (K_{<n})_{(\mathrm{odd})},$$

where $\tau_{<n}(K, Y_K) = (K, K/n, h_K, K_{<n})$, and to a morphism $f : (K, Y_K) \to (L, Y_L)$ the homotopy class

$$t_{<n}^{(\mathrm{odd})}(f) = (t_{<n}f)_{(\mathrm{odd})} = [f_{<n}]_{(\mathrm{odd})} = [f_{<n(\mathrm{odd})}],$$

where $\tau_{<n}(f) = ([f], [f^n], [f/n], [f_{<n}])$. (Thus, this definition forgets that the original homotopy classes were rel $(n-1)$-skeleta.)

Proposition 1.60. *Let $f : (K, Y_K) \to (L, Y_L)$ and $g : (L, Y_L) \to (P, Y_P)$ be morphisms in $\mathbf{CW}_{n \supset \partial}$. If $H_2(P) = 0$ and $H_n(P^n)$ has finite rank, then*

$$t_{<n}^{(\mathrm{odd})}(g \circ f) = t_{<n}^{(\mathrm{odd})}(g) \circ t_{<n}^{(\mathrm{odd})}(f)$$

in **HoCW**.

Proof. Set $h = gf$. If

$$\tau_{<n}(f) = ([f], [f^n], [f/n], [f_{<n}]), \quad \tau_{<n}(g) = ([g], [g^n], [g/n], [g_{<n}])$$

and

$$\tau_{<n}(h) = ([h], [h^n], [h/n], [h_{<n}])$$

then $g/n \circ f/n \simeq h/n$ rel K^{n-1} by Lemma 1.38. As in the proof of Theorem 1.39, we obtain homotopies

$$i_P h_{<n} \simeq (h/n) i_K \simeq i_P g_{<n} f_{<n}$$

rel K^{n-1}. Let $H : K_{<n} \times I \to P/n$ be a homotopy rel K^{n-1} between $i_P h_{<n}$ and $i_P g_{<n} f_{<n}$. Composition with the localization $\mathrm{loc} : P/n \to P/n_{(\mathrm{odd})}$ yields a homotopy $\mathrm{loc} \circ H : K_{<n} \times I \to P/n_{(\mathrm{odd})}$ rel K^{n-1} between $\mathrm{loc} \circ i_P h_{<n}$ and $\mathrm{loc} \circ i_P g_{<n} f_{<n}$. Using the commutative diagram

$$
\begin{array}{ccc}
P_{<n} & \xrightarrow{\;\mathrm{loc}\;} & P_{<n(\mathrm{odd})} \\[2pt]
\Big\downarrow{\scriptstyle i_P} & & \Big\downarrow{\scriptstyle i_{P(\mathrm{odd})}} \\[2pt]
P/n & \xrightarrow{\;\mathrm{loc}\;} & P/n_{(\mathrm{odd})},
\end{array}
$$

$\mathrm{loc} \circ H$ is a homotopy rel K^{n-1} between $i_{P(\mathrm{odd})} \circ \mathrm{loc}\, h_{<n}$ and $i_{P(\mathrm{odd})} \circ \mathrm{loc}\, g_{<n} f_{<n}$. By the obstruction theory Lemma 1.43,

$$i_{P(\mathrm{odd})*}\omega(\mathrm{loc}\, h_{<n}, \mathrm{loc}\, g_{<n} f_{<n}) = \omega(i_{P(\mathrm{odd})} \circ \mathrm{loc}\, h_{<n}, i_{P(\mathrm{odd})} \circ \mathrm{loc}\, g_{<n} f_{<n})$$
$$= 0 \in C^{n+1}(K_{<n} \times I; \pi_n(P/n_{(\mathrm{odd})})).$$

By Proposition 1.50, $\pi_{n+1}(P/n, P_{<n})$ is all 2-torsion, whence its odd-primary localization vanishes. Since $(-)_{(\mathrm{odd})}$ preserves homotopy fibrations,

$$\pi_{n+1}(P/n_{(\mathrm{odd})}, P_{<n(\mathrm{odd})}) = \pi_{n+1}(P/n, P_{<n})_{(\mathrm{odd})} = 0.$$

Thus the exactness of

$$\pi_{n+1}(P/n_{(\mathrm{odd})}, P_{<n(\mathrm{odd})}) \longrightarrow \pi_n(P_{<n(\mathrm{odd})}) \xrightarrow{\;i_{P(\mathrm{odd})*}\;} \pi_n(P/n_{(\mathrm{odd})})$$

implies that $i_{P(\mathrm{odd})*}$ is injective and hence

$$\omega(\mathrm{loc}\, h_{<n}, \mathrm{loc}\, g_{<n} f_{<n}) = 0 \in C^{n+1}(K_{<n} \times I; \pi_n(P_{<n(\mathrm{odd})})).$$

By Lemma 1.43, there exists a homotopy $G : K_{<n} \times I \to P_{<n(\text{odd})}$ rel K^{n-1} between $\text{loc}\, h_{<n}$ and $\text{loc}\, g_{<n} f_{<n}$. Consider the commutative diagrams

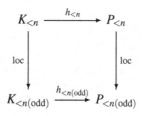

and

$$
\begin{array}{ccc}
K_{<n} & \xrightarrow{\; g_{<n} f_{<n}\;} & P_{<n} \\
\downarrow{\scriptstyle \text{loc}} & & \downarrow{\scriptstyle \text{loc}} \\
K_{<n(\text{odd})} & \xrightarrow{\; g_{<n(\text{odd})} f_{<n(\text{odd})}\;} & P_{<n(\text{odd})}.
\end{array}
$$

By [FHT01, Theorem 9.7(ii), p. 109], G extends to a homotopy between $h_{<n(\text{odd})}$ and $g_{<n(\text{odd})} f_{<n(\text{odd})}$. Thus

$$
t_{<n}^{(\text{odd})}(gf) = t_{<n}^{(\text{odd})}(h) = [h_{<n(\text{odd})}] = [g_{<n(\text{odd})}] \circ [f_{<n(\text{odd})}] = t_{<n}^{(\text{odd})}(g) \circ t_{<n}^{(\text{odd})}(f).
$$

$\square$

Let

$$
t_{<\infty}^{(\text{odd})} : \mathbf{CW}_{n \supset \partial} \longrightarrow \mathbf{HoCW}^1
$$

be the natural localization-followed-by-projection functor, that is, $t_{<\infty}^{(\text{odd})}(K, Y_K) = K_{(\text{odd})}$ for an object (K, Y_K) in $\mathbf{CW}_{n \supset \partial}$, and $t_{<\infty}^{(\text{odd})}(f) = [f_{(\text{odd})}]$ for a morphism $f : (K, Y_K) \to (L, Y_L)$ in $\mathbf{CW}_{n \supset \partial}$. (Here, $[f_{(\text{odd})}]$ denotes the absolute homotopy class of $f_{(\text{odd})}$, not the homotopy class rel some subspace.) Let $\mathbf{CW}_{n \supset \partial}^2$ be the full subcategory of $\mathbf{CW}_{n \supset \partial}$ having as objects all those pairs (K, Y) where K has vanishing second homology, i.e. is 2-connected, and $H_n(K^n)$ has finite rank.

Theorem 1.61. *Let $n \geq 3$ be an integer. There is an odd-primary spatial homology truncation functor* $t_{<n}^{(\text{odd})} : \mathbf{CW}_{n \supset \partial}^2 \longrightarrow \mathbf{HoCW}^1$ *together with a natural transformation* $\text{emb}_n^{(\text{odd})} : t_{<n}^{(\text{odd})} \to t_{<\infty}^{(\text{odd})}$ *such that for an object (K, Y) of $\mathbf{CW}_{n \supset \partial}^2$, one has $H_r(t_{<n}^{(\text{odd})}(K, Y)) = 0$ for $r \geq n$, and*

$$
\text{emb}_{n*}^{(\text{odd})} : H_r(t_{<n}^{(\text{odd})}(K, Y)) \xrightarrow{\;\cong\;} H_r(K; \mathbb{Z}[\tfrac{1}{2}])
$$

is an isomorphism for $r < n$.

Proof. The assignment $t_{<n}^{(\text{odd})}$ is a functor by Proposition 1.60. The natural transformation

$$\text{emb}_n^{(\text{odd})} : t_{<n}^{(\text{odd})} \longrightarrow t_{<\infty}^{(\text{odd})}$$

is defined by localizing emb_n:

$$\text{emb}_n^{(\text{odd})}(K, Y_K) = (\text{emb}_n(K, Y_K))_{(\text{odd})} : t_{<n}^{(\text{odd})}(K, Y_K) \longrightarrow t_{<\infty}^{(\text{odd})}(K, Y_K),$$

where $\text{emb}_n(K, Y_K) : t_{<n}(K, Y_K) \to K$. Given a morphism $f : (K, Y_K) \to (L, Y_L)$ in $\mathbf{CW}_{n \supset \partial}^2$, the square

$$
\begin{array}{ccc}
t_{<n}(K, Y_K) & \xrightarrow{\ \text{emb}_n(K, Y_K)\ } & K = t_{<\infty}(K, Y_K) \\
\downarrow{\scriptstyle t_{<n}(f)} & & \downarrow{\scriptstyle [f] = t_{<\infty}(f)} \\
t_{<n}(L, Y_L) & \xrightarrow{\ \text{emb}_n(L, Y_L)\ } & L = t_{<\infty}(L, Y_L)
\end{array}
$$

commutes in $\mathbf{HoCW}_{n-1}$, hence in $\mathbf{HoCW}^1$. So its localization

$$
\begin{array}{ccc}
(t_{<n}(K, Y_K))_{(\text{odd})} & \xrightarrow{\ (\text{emb}_n(K, Y_K))_{(\text{odd})}\ } & K_{(\text{odd})} \\
\downarrow{\scriptstyle (t_{<n}(f))_{(\text{odd})}} & & \downarrow{\scriptstyle [f_{(\text{odd})}]} \\
(t_{<n}(L, Y_L))_{(\text{odd})} & \xrightarrow{\ (\text{emb}_n(L, Y_L))_{(\text{odd})}\ } & L_{(\text{odd})}
\end{array}
$$

commutes in $\mathbf{HoCW}^1$. Consequently, $\text{emb}_n^{(\text{odd})}$ is a natural transformation. Given an object (K, Y) in $\mathbf{CW}_{n \supset \partial}^2$, let $(K, K/n, h, K_{<n}) = \tau_{<n}(K, Y)$. By definition of $\text{emb}_n(K, Y)$, the diagram

$$
\begin{array}{ccc}
t_{<n}(K, Y) = K_{<n} & \xrightarrow{\ \text{emb}_n(K, Y)\ } & K = t_{<\infty}(K, Y) \\
\downarrow{\scriptstyle [i_K]} & & \uparrow{\scriptstyle [j_K]} \\
K/n & \xrightarrow{\quad [h] \quad} & K^n
\end{array}
$$

commutes in $\mathbf{HoCW}_{n-1}$, hence in $\mathbf{HoCW}^1$. Thus its localization

$$
\begin{array}{ccc}
K_{<n(\text{odd})} & \xrightarrow{\ \text{emb}_n^{(\text{odd})}(K,Y)\ } & K_{(\text{odd})} \\[1mm]
{\scriptstyle [i_{K(\text{odd})}]}\Big\downarrow & & \Big\uparrow{\scriptstyle [j_{K(\text{odd})}]} \\[1mm]
K/n_{(\text{odd})} & \xrightarrow{\ [h_{(\text{odd})}]\ } & K^n_{(\text{odd})}
\end{array}
$$

commutes in $\mathbf{HoCW}^1$ and the induced map on homology,

$$
\text{emb}_{n*}^{(\text{odd})}(K,Y) : H_r(t_{<n}^{(\text{odd})}(K,Y)) \to H_r(K_{(\text{odd})}),
$$

factors as

$$
H_r(K_{<n(\text{odd})}) \xrightarrow{\ i_{K(\text{odd})*}\ } H_r(K/n_{(\text{odd})}) \xrightarrow{\ h_{(\text{odd})*}\ } H_r(K^n_{(\text{odd})}) \xrightarrow{\ j_{K(\text{odd})*}\ } H_r(K_{(\text{odd})}).
$$

For $r < n$, this is an isomorphism, since then each of the three maps

$$
H_r(K_{<n}) \xrightarrow{\ i_{K*}\ } H_r(K/n) \xrightarrow{\ h_*\ } H_r(K^n) \xrightarrow{\ j_{K*}\ } H_r(K)
$$

is an isomorphism, whence each of the three maps

$$
H_r(K_{<n})_{(\text{odd})} \xrightarrow{\ i_{K*}\otimes\text{id}_{(\text{odd})}\ } H_r(K/n)_{(\text{odd})} \xrightarrow{\ h_*\otimes\text{id}_{(\text{odd})}\ } H_r(K^n)_{(\text{odd})} \xrightarrow{\ j_{K*}\otimes\text{id}_{(\text{odd})}\ } H_r(K)_{(\text{odd})}
$$

is an isomorphism and the localization diagram

$$
\begin{array}{ccccccc}
H_r(K_{<n(\text{odd})}) & \xrightarrow{\ i_{K(\text{odd})*}\ } & H_r(K/n_{(\text{odd})}) & \xrightarrow{\ h_{(\text{odd})*}\ } & H_r(K^n_{(\text{odd})}) & \xrightarrow{\ j_{K(\text{odd})*}\ } & H_r(K_{(\text{odd})}) \\[1mm]
{\scriptstyle \cong}\Big\uparrow & & {\scriptstyle \cong}\Big\uparrow & & {\scriptstyle \cong}\Big\uparrow & & {\scriptstyle \cong}\Big\uparrow \\[1mm]
H_r(K_{<n})_{(\text{odd})} & \xrightarrow[\cong]{i_{K*}\otimes\text{id}_{(\text{odd})}} & H_r(K/n)_{(\text{odd})} & \xrightarrow[\cong]{h_*\otimes\text{id}_{(\text{odd})}} & H_r(K^n)_{(\text{odd})} & \xrightarrow[\cong]{j_{K*}\otimes\text{id}_{(\text{odd})}} & H_r(K)_{(\text{odd})}
\end{array}
$$

commutes. For $r \geq n$,

$$
H_r(t_{<n}^{(\text{odd})}(K,Y)) = H_r(K_{<n})_{(\text{odd})} = 0_{(\text{odd})} = 0.
$$

$\square$

1.8 Summary

Let us summarize spatial homology truncation as developed in the previous sections by displaying all assignments and functors constructed, together with all relevant categories, in one picture:

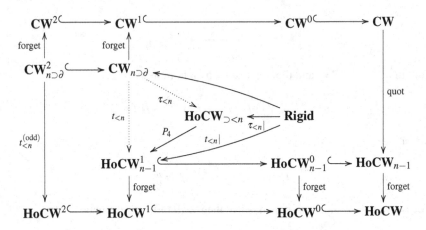

Arrows of the form $\hookrightarrow$ signify "full subcategory." The forgetful functor $\mathbf{CW}_{n\supset\partial} \to \mathbf{CW}^1$ sends an object (K,Y) to the simply connected space K and forgets the additional structure Y. This functor is surjective on objects and faithful, but not full. Dashed arrows mean assignments of objects and morphisms that need not be functors, whereas all fully drawn arrows are functors. The functor $\mathbf{CW} \to \mathbf{HoCW}_{n-1}$ is the natural quotient functor that is the identity on objects and sends a cellular map to its rel $(n-1)$-skeleton homotopy class. The category **Rigid** is any n-compression rigid subcategory of $\mathbf{CW}_{n\supset\partial}$, which need not be full. The arrow **Rigid** $\to \mathbf{CW}_{n\supset\partial}$ is the inclusion functor. The forgetful functor $\mathbf{HoCW}_{n-1} \to \mathbf{HoCW}$ is the identity on objects and sends a homotopy class rel $(n-1)$-skeleton to the absolute homotopy class of a representative map and thus forgets that the original class was rel $(n-1)$-skeleton. This functor is full but not faithful. The same holds for the functors $\mathbf{HoCW}^j_{n-1} \to \mathbf{HoCW}^j$, $j = 0,1,2,\ldots$.

1.9 The Interleaf Category

Many important spaces in topology and algebraic geometry have no odd-dimensional homology, see Examples 1.65 below. For such spaces, functorial spatial homology truncation simplifies considerably. On the theory side, the simplification arises as follows: To define general spatial homology truncation, we used intermediate auxiliary structures, the n-truncation structures. For spaces that lack odd-dimensional homology, these structures can be replaced by a much simpler structure (see Definition 1.67). Again every such space can be embedded in such a structure, see Proposition 1.68, which is the analogon of Proposition 1.6 for the general theory. On the application side, the crucial simplification is that the truncation functor $t_{<n}$ will not require that in truncating a given continuous map, the map preserve additional structure on the domain and codomain of the map. Recall that in general, $t_{<n}$ is defined on the category $\mathbf{CW}_{n \supset \partial}$, meaning that a map must preserve chosen subgroups "Y". We have seen that such a condition is generally necessary on maps, for otherwise no truncation exists. So what we will see in this section is that *arbitrary* continuous maps between spaces with trivial odd-dimensional homology can be functorially truncated. In particular the compression rigidity obstructions arising in the general theory will not arise for maps between such spaces.

Definition 1.62. Let $\mathbf{ICW}$ be the full subcategory of $\mathbf{CW}$ whose objects are simply connected CW-complexes K with finitely generated even-dimensional homology and vanishing odd-dimensional homology for any coefficient group. We call $\mathbf{ICW}$ the *interleaf category*.

Example 1.63. The space $K = S^2 \cup_2 e^3$ is simply connected and has vanishing integral homology in odd dimensions. However, $H_3(K; \mathbb{Z}/2) = \mathbb{Z}/2 \neq 0$.

Lemma 1.64. *Let X be a space whose odd-dimensional homology vanishes for any coefficient group. Then the even-dimensional integral homology of X is torsion-free.*

Proof. Taking the coefficient group $\mathbb{Q}/\mathbb{Z}$, we have

$$\mathrm{Tor}(H_{2k}(X), \mathbb{Q}/\mathbb{Z}) = H_{2k+1}(X) \otimes \mathbb{Q}/\mathbb{Z} \oplus \mathrm{Tor}(H_{2k}(X), \mathbb{Q}/\mathbb{Z}) = H_{2k+1}(X; \mathbb{Q}/\mathbb{Z}) = 0.$$

Thus $H_{2k}(X)$ is torsion-free, since the group $\mathrm{Tor}(H_{2k}(X), \mathbb{Q}/\mathbb{Z})$ is isomorphic to the torsion subgroup of $H_{2k}(X)$. □

Examples 1.65. (1) Any simply connected closed 4-manifold is in $\mathbf{ICW}$. Indeed, such a manifold is homotopy equivalent to a CW-complex of the form

$$\bigvee_{i=1}^{k} S_i^2 \cup_f e^4,$$

where the homotopy class of the attaching map $f : S^3 \to \bigvee_{i=1}^{k} S_i^2$ may be viewed as a symmetric $k \times k$ matrix with integer entries, as $\pi_3(\bigvee_{i=1}^{k} S_i^2) \cong M(k)$, with $M(k)$ the additive group of such matrices.

(2) Any simply connected closed 6-manifold with vanishing integral middle homology group is in $\mathbf{ICW}$. If G is any coefficient group, then $H_1(M; G) \cong H_1(M) \otimes G \oplus \mathrm{Tor}(H_0 M, G) = 0$, since $H_0(M) = \mathbb{Z}$. By Poincaré duality,

$$0 = H_3(M) \cong H^3(M) \cong \mathrm{Hom}(H_3 M, \mathbb{Z}) \oplus \mathrm{Ext}(H_2 M, \mathbb{Z}),$$

so that $H_2(M)$ is free. This implies that $\mathrm{Tor}(H_2M, G) = 0$ and hence $H_3(M;G) \cong H_3(M) \otimes G \oplus \mathrm{Tor}(H_2M, G) = 0$. Finally, by G-coefficient Poincaré duality,

$$H_5(M;G) \cong H^1(M;G) \cong \mathrm{Hom}(H_1M, G) \oplus \mathrm{Ext}(H_0M, G) = \mathrm{Ext}(\mathbb{Z}, G) = 0.$$

(3) Complex projective spaces are in **ICW**. This class will be vastly generalized in example (5).

(4) Any smooth, compact toric variety X is in **ICW**: Danilov's Theorem 10.8. in [Dan78] implies that $H^*(X;\mathbb{Z})$ is torsion-free and the map $A^*(X) \to H^*(X;\mathbb{Z})$ given by composing the canonical map from Chow groups to homology, $A^k(X) = A_{n-k}(X) \to H_{2n-2k}(X;\mathbb{Z})$, where n is the complex dimension of X, with Poincaré duality $H_{2n-2k}(X;\mathbb{Z}) \cong H^{2k}(X;\mathbb{Z})$, is an isomorphism. Since the odd-dimensional cohomology of X is not in the image of this map, this asserts in particular that $H^{\mathrm{odd}}(X;\mathbb{Z}) = 0$. By Poincaré duality, $H_{\mathrm{even}}(X;\mathbb{Z})$ is free and $H_{\mathrm{odd}}(X;\mathbb{Z}) = 0$. These two statements allow us to deduce from the universal coefficient theorem that $H_{\mathrm{odd}}(X;G) = 0$ for any coefficient group G. If we only wanted to establish $H_{\mathrm{odd}}(X;\mathbb{Z}) = 0$, then it would of course have been enough to know that the canonical, degree-doubling map $A_*(X) \to H_*(X;\mathbb{Z})$ is onto. One may then immediately reduce to the case of projective toric varieties because every complete fan Δ has a projective subdivision Δ', the corresponding proper birational morphism $X(\Delta') \to X(\Delta)$ induces a surjection $H_*(X(\Delta');\mathbb{Z}) \to H_*(X(\Delta);\mathbb{Z})$ (use the Umkehrmap) and the diagram

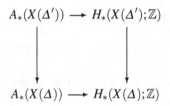

commutes, see [Dan78].

(5) Let G be a complex, simply connected, semisimple Lie group and $P \subset G$ a connected parabolic subgroup. Then the homogeneous space G/P is in **ICW**. It is simply connected, since the fibration $P \to G \to G/P$ induces an exact sequence

$$1 = \pi_1(G) \to \pi_1(G/P) \to \pi_0(P) \to \pi_0(G) = 0,$$

which shows that $\pi_1(G/P) \to \pi_0(P)$ is a bijection. According to [BGG73], there exist elements $s_w(P) \in H_{2l(w)}(G/P;\mathbb{Z})$ ("Schubert classes," given geometrically by Schubert cells), indexed by w ranging over a certain subset of the Weyl group of G, that form a basis for $H_*(G/P;\mathbb{Z})$. (For w in the Weyl group, $l(w)$ denotes the length of w when written as a reduced word in certain specified generators of the Weyl group.) In particular $H_{\mathrm{even}}(G/P;\mathbb{Z})$ is free and $H_{\mathrm{odd}}(G/P;\mathbb{Z}) = 0$. Thus $H_{\mathrm{odd}}(G/P;G) = 0$ for any coefficient group G.

The linear groups $SL(n,\mathbb{C})$, $n \geq 2$, and the subgroups $Sp(2n,\mathbb{C}) \subset SL(2n,\mathbb{C})$ of transformations preserving the alternating bilinear form

$$x_1 y_{n+1} + \cdots + x_n y_{2n} - x_{n+1} y_1 - \cdots - x_{2n} y_n$$

on $\mathbb{C}^{2n} \times \mathbb{C}^{2n}$ are examples of complex, simply connected, semisimple Lie groups. A parabolic subgroup is a closed subgroup that contains a Borel group B. For $G = SL(n,\mathbb{C})$, B is the group of all upper-triangular matrices in $SL(n,\mathbb{C})$. In this case, G/B is the complete flag manifold

$$G/B = \{0 \subset V_1 \subset \cdots \subset V_{n-1} \subset \mathbb{C}^n\}$$

of flags of subspaces V_i with $\dim V_i = i$. For $G = Sp(2n,\mathbb{C})$, the Borel subgroups B are the subgroups preserving a half-flag of isotropic subspaces and the quotient G/B is the variety of all such flags. Any parabolic subgroup P may be described as the subgroup that preserves some partial flag. Thus (partial) flag manifolds are in **ICW**. A special case is that of a maximal parabolic subgroup, preserving a single subspace V. The corresponding quotient $SL(n,\mathbb{C})/P$ is a Grassmannian $G(k,n)$ of k-dimensional subspaces of $\mathbb{C}^n$. For $G = Sp(2n,\mathbb{C})$, one obtains Lagrangian Grassmannians of isotropic k-dimensional subspaces, $1 \leq k \leq n$. So Grassmannians are objects in **ICW**.

The interleaf category is closed under forming fibrations.

Proposition 1.66. *Let F, E, B be CW-complexes that fit into a fibration $F \to E \to B$ with base B, total space E and fiber F. If B and F are objects in the interleaf category **ICW**, then so is E.*

Proof. Assume $B, F \in Ob\mathbf{ICW}$. Since B and F are in particular simply connected, the exactness of

$$\pi_1(F) \to \pi_1(E) \to \pi_1(B)$$

implies that E is simply connected as well. With G any coefficient group, we will first show that $H_{\mathrm{odd}}(E;G) = 0$. In degree 1, we have $H_1(E;G) = H_1(E) \otimes G \oplus \mathrm{Tor}(H_0(E),G) = 0$ since E is simply connected. In higher degrees, the claim follows from the spectral sequence of the fibration: Since the base is simply connected,

$$E^2_{p,q} \cong H_p(B; H_q(F;G))$$

and the latter term vanishes when p is odd (as B is in **ICW**) or q is odd (as F is in **ICW**). Since the differential d^2 has bidegree $(-2,1)$, either its domain $E^2_{p,q}$ is zero or else p and q are both even and its codomain $E^2_{p-2,q+1}$ is zero because $q+1$ is odd. Thus $d^2 = 0$ and $E^2 \cong E^3$. It follows by induction that all differentials d^r are zero, $r \geq 2$, using that d^r has bidegree $(-r, r-1)$ and one of these two numbers must be odd. Thus

$$E^2_{p,q} \cong E^3_{p,q} \cong \cdots \cong E^\infty_{p,q}.$$

On the other hand, E^∞ is isomorphic to the bigraded module $GH_*(E;G)$ associated to the filtration $F_pH_*(E;G) = \mathrm{im}(H_*(E_p;G) \to H_*(E;G))$, where $E_p \subset E$ is the preimage of the p-skeleton of B under the fibration. We conclude that

$$H_p(B;H_q(F;G)) \cong \frac{\mathrm{im}(H_{p+q}(E_p;G) \to H_{p+q}(E;G))}{\mathrm{im}(H_{p+q}(E_{p-1};G) \to H_{p+q}(E;G))}.$$

Let $n \geq 3$ be odd. The restricted fibration $F \to E_n \to B^n$ induces an exact sequence $\pi_1(F) \to \pi_1(E_n) \to \pi_1(B^n)$, which shows that E_n is simply connected since $n \geq 3$ implies $\pi_1(B^n) \cong \pi_1(B) = 1$. Using the homotopy lifting property, we can deduce that the pair (E,E_n) is n-connected from the fact that (B,B^n) is n-connected. Thus $H_i(E,E_n) = 0$ for $i \leq n$; in particular, $H_n(E_n;G) \to H_n(E;G)$ is surjective and

$$H_n(E;G) = \mathrm{im}(H_n(E_n;G) \to H_n(E;G)).$$

Then

$$0 = H_n(B;H_0(F;G)) \cong \frac{\mathrm{im}(H_n(E_n;G) \to H_n(E;G))}{\mathrm{im}(H_n(E_{n-1};G) \to H_n(E;G))},$$

whence

$$\mathrm{im}(H_n(E_n;G) \to H_n(E;G)) = \mathrm{im}(H_n(E_{n-1};G) \to H_n(E;G)).$$

From

$$0 = H_{n-1}(B;H_1(F;G)) \cong \frac{\mathrm{im}(H_n(E_{n-1};G) \to H_n(E;G))}{\mathrm{im}(H_n(E_{n-2};G) \to H_n(E;G))},$$

we find

$$\mathrm{im}(H_n(E_{n-1};G) \to H_n(E;G)) = \mathrm{im}(H_n(E_{n-2};G) \to H_n(E;G)).$$

Continuing in this manner, observing that for $p+q = n$, one of p or q must be odd and thus $H_p(B;H_q(F;G)) = 0$, we arrive at

$$H_n(E;G) = 0.$$

To see that the even homology of E is finitely generated, one may for instance argue as follows. By Lemma 1.64, the homology of B and F is torsion-free, hence free, since $H_*(B)$ and $H_*(F)$ are finitely generated. Thus all groups $E^2_{p,q} \cong E^\infty_{p,q}$ are free abelian and

$$H_n(E) \cong \bigoplus_{p+q=n} E^2_{p,q} \cong \bigoplus_{p+q=n} H_p(B) \otimes H_q(F).$$

This formula implies that $H_*(E)$ is finitely generated. $\square$

A multitude of spaces in algebraic geometry arise via fibrations that way. Let us give but one example. Let X be a smooth Schubert subvariety, "defined by inclusions" in the sense of [GR02], inside of

$$G/P = \{0 \subset V_{d_1} \subset V_{d_2} \subset \cdots \subset V_{d_r} \subset \mathbb{C}^n\},$$

where $G = GL(n, \mathbb{C})$ and P is the subgroup that stabilizes the standard partial flag with V_{d_i} spanned by the first d_i standard basis vectors in $\mathbb{C}^n$. Then, according to [GR02], X is fibered by Grassmannians. Since Grassmannians are in **ICW**, Proposition 1.66 shows that all such smooth Schubert varieties X are in **ICW**.

Definition 1.67. The *moduli category* **M(ICW)** of **ICW** consists of the following objects and morphisms: Objects are homotopy classes $[h_K]$ of cellular homotopy equivalences $h_K : K \to E(K)$, where K is an object of **ICW** and $E(K)$ is a CW-complex that has only even-dimensional cells. Morphisms are commutative diagrams

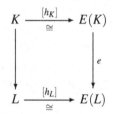

in **HoCW**. Composition is defined in the obvious way.

Proposition 1.68. *Any object of the interleaf category can be completed to an object of the moduli category M(ICW).*

Proof. Let K be an object of **ICW**. By Lemma 1.64, $H_{2k}(K)$ is torsion-free. Choose a decomposition of every homology group $H_{2k}(K)$ as a direct sum of infinite cyclic groups with specified generators g. Then, by minimal cell structure theory (which is applicable because K is simply connected; see e.g. [Hat02]), there is a CW-complex $E(K)$ and a cellular homotopy equivalence $h'_K : E(K) \to K$ such that each cell of $E(K)$ is a generator $2k$-cell e_g^{2k}, which is a cycle in cellular homology mapped by f to a cellular cycle representing the specified generator g of one of the cyclic summands of $H_{2k}(K)$. (There are no relator $(2k+1)$-cells since no g has finite order.) Thus $E(K)$ has only even-dimensional cells. Let $[h_K]$ be the inverse of $[h'_K]$ in **HoCW**. $\qquad\square$

Remark 1.69. Since objects K in **ICW** have finitely generated homology, the space $E(K)$ is a *finite* CW-complex.

With the help of this proposition, we construct a functor

$$M : \textbf{ICW} \longrightarrow \textbf{M(ICW)}.$$

Given an object K in **ICW**, use Proposition 1.68 to choose, once and for all, a cellular homotopy equivalence $h_K : K \to E(K)$ representing an object $[h_K]$ in **M(ICW)**. In addition, choose, once and for all, a cellular homotopy inverse $h'_K : E(K) \to K$ for h_K. (If K already has only cells of even dimension, then we take h_K and h'_K to be the identity maps.) Set

$$M(K) = [h_K].$$

Let $f : K \to L$ be a cellular map. If f is the identity map, set $E(f) = [\mathrm{id}_{E(K)}]$. Otherwise, set

$$E(f) = M(L) \circ [f] \circ M(K)^{-1} : E(K) \longrightarrow E(L).$$

Define

$$
M(f) = \qquad
\begin{array}{ccc}
K & \xrightarrow[\cong]{M(K)} & E(K) \\
\Big\downarrow{\scriptstyle [f]} & & \Big\downarrow{\scriptstyle E(f)} \\
L & \xrightarrow[\cong]{M(L)} & E(L)
\end{array}
$$

This is a morphism in **M(ICW)** as $E(f) \circ M(K) = M(L) \circ [f] \circ M(K)^{-1} \circ M(K) = M(L) \circ [f]$. We have $M(\mathrm{id}_K) = \mathrm{id}_{M(K)}$ and for a composition

$$K \xrightarrow{f} L \xrightarrow{g} P$$

we compute

$$
\begin{aligned}
E(g) \circ E(f) &= M(P)[g]M(L)^{-1} \circ M(L)[f]M(K)^{-1} \\
&= M(P)[g] \circ [f]M(K)^{-1} \\
&= M(P)[gf]M(K)^{-1} \\
&= E(gf).
\end{aligned}
$$

This shows that

$$M(gf) = M(g)M(f),$$

so that M is indeed a covariant functor. Next, we shall construct a preliminary truncation functor

$$T_{<n} : \mathbf{M(ICW)} \longrightarrow \mathbf{HoCW}$$

for any integer n. If $n \leq 0$, then we define $T_{<n}$ on objects to be the empty space, which is the initial object of **HoCW**. On morphisms, $T_{<n}$ is defined as the unique morphism from the initial object. We will henceforth assume that n is positive. On a class of a homotopy equivalence $h : K \to E(K)$, we set

$$T_{<n}[h] = E(K)^{n-1},$$

the $(n-1)$-skeleton of $E(K)$. If E is any space without odd-dimensional cells, then $H_r(E) = C_r(E)$, the cellular chain group in degree r, since all cellular boundary maps are zero. Thus for $r < n$,

$$H_r(T_{<n}[h]) = C_r(E(K)^{n-1}) = C_r(E(K)) = H_r(E(K)) \overset{h_*}{\cong} H_r(K),$$

while for $r \geq n$,

$$H_r(T_{<n}[h]) = C_r(E(K)^{n-1}) = 0.$$

This shows that $T_{<n}$ implements spatial homology truncation on K. Let $F : [h_1] \to [h_2]$ be a morphism in the moduli category, that is, F is a diagram

$$
\begin{array}{ccc}
K & \xrightarrow[\cong]{[h_1]} & E(K) \\
\downarrow & & \downarrow{\scriptstyle e_F} \\
L & \xrightarrow[\cong]{[h_2]} & E(L).
\end{array}
$$

Choose a cellular representative f_0 for the homotopy class e_F and put

$$T_{<n}(F) = [E(K)^{n-1} \xrightarrow{f_0^{n-1}} E(L)^{n-1}].$$

The following lemma shows that this is well-defined.

Lemma 1.70. *Let $f_0, f_1 : E(K) \to E(L)$ be two cellular maps. If $f_0 \simeq f_1$, then $f_0^{n-1} \simeq f_1^{n-1}$.*

Proof. Let $H : E(K) \times I \to E(L)$ be a cellular homotopy with $H(-,0) = f_0$ and $H(-,1) = f_1$. We will distinguish two cases according to whether n is even or odd. Suppose n is even. Since H is cellular, it restricts to a map $H| : (E(K) \times I)^{n-1} \to E(L)^{n-1}$ between $(n-1)$-skeleta. The $(n-1)$-skeleton of $E(K) \times I$ contains $E(K)^{n-2} \times I$, so that further restriction yields

$$H| : E(K)^{n-2} \times I \to E(L)^{n-1}.$$

Since $n-1$ is odd, we have $E(K)^{n-1} = E(K)^{n-2}$. Thus

$$H| : E(K)^{n-1} \times I \to E(L)^{n-1}$$

is a homotopy from f_0^{n-1} to f_1^{n-1}.

Now assume n is odd. In this case, we restrict H to the n-skeleton to get $H| : (E(K) \times I)^n \to E(L)^n$, and, by restricting further,

$$H| : E(K)^{n-1} \times I \to E(L)^n.$$

Since n is odd, we have $E(L)^n = E(L)^{n-1}$. Thus

$$H| : E(K)^{n-1} \times I \to E(L)^{n-1}$$

is a homotopy from f_0^{n-1} to f_1^{n-1}.   □

We have $T_{<n}(\mathrm{id}_{[h]}) = \mathrm{id}_{T_{<n}[h]}$. Furthermore, if $G : [h_2] \to [h_3]$ is another morphism

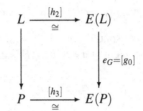

in **M(ICW)**, then

$$T_{<n}(G) \circ T_{<n}(F) = [g_0^{n-1}] \circ [f_0^{n-1}] = [(g_0 f_0)^{n-1}] = T_{<n}(G \circ F),$$

since $g_0 f_0$ is a representative of $e_G e_F$. Hence $T_{<n}$ is a functor.

Define the functor

$$t_{<n} : \mathbf{ICW} \longrightarrow \mathbf{HoCW}$$

to be the composition

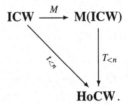

Let $t_{<\infty} : \mathbf{ICW} \longrightarrow \mathbf{HoCW}$ be the natural "inclusion-followed-by-quotient"-functor, that is, for objects K set $t_{<\infty}(K) = K$ and for morphisms f set $t_{<\infty}(f) = [f]$. There is an important natural transformation of functors

$$\mathrm{emb}_n : t_{<n} \longrightarrow t_{<\infty},$$

which we shall describe next. Given an object K of **ICW**, define $\mathrm{emb}_n(K)$ to be the composition

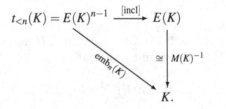

(Note that $\mathrm{emb}_n(K)$ has a canonical representative in **CW**, namely $E(K)^{n-1} \overset{\mathrm{incl}}{\hookrightarrow}$ $E(K) \overset{h'_K}{\longrightarrow} K$.) Given a morphism $f : K \to L$ in **ICW**, we have to show that the square

$$
\begin{array}{ccc}
t_{<n}(K) & \overset{\mathrm{emb}_n(K)}{\longrightarrow} & t_{<\infty}(K) \\
{\scriptstyle t_{<n}(f)} \big\downarrow & & \big\downarrow {\scriptstyle t_{<\infty}(f)} \\
t_{<n}(L) & \overset{\mathrm{emb}_n(L)}{\longrightarrow} & t_{<\infty}(L)
\end{array}
$$

commutes in **HoCW**. Using the morphism $M(f)$, given by the commutative diagram

$$
\begin{array}{ccc}
K & \overset{M(K)}{\underset{\cong}{\longrightarrow}} & E(K) \\
{\scriptstyle [f]} \big\downarrow & & \big\downarrow {\scriptstyle E(f)=M(L)\circ[f]\circ M(K)^{-1}} \\
L & \overset{M(L)}{\underset{\cong}{\longrightarrow}} & E(L),
\end{array}
$$

let f_0 be a cellular representative of $E(f)$, for example $f_0 = h_L \circ f \circ h'_K$, and consider the diagram

$$
\begin{array}{ccccc}
E(K)^{n-1} & \overset{[\mathrm{incl}]}{\longrightarrow} & E(K) & \overset{M(K)^{-1}}{\longrightarrow} & K \\
{\scriptstyle t_{<n}(f)=[f_0^{n-1}]} \big\downarrow & & {\scriptstyle [f_0]=E(f)} \big\downarrow & & \big\downarrow {\scriptstyle [f]=t_{<\infty}(f)} \\
E(L)^{n-1} & \overset{[\mathrm{incl}]}{\longrightarrow} & E(L) & \overset{M(L)^{-1}}{\longrightarrow} & L
\end{array}
$$

The left square commutes in **HoCW** by construction and the right square commutes in **HoCW** since

$$
M(L)^{-1} \circ E(f) = M(L)^{-1} \circ M(L) \circ [f] \circ M(K)^{-1} = [f] \circ M(K)^{-1} = t_{<\infty}(f) \circ M(K)^{-1}.
$$

Thus emb_n is a natural transformation.

Let us move on to implementing functorial spatial homology cotruncation on the interleaf category. Given an object K in **ICW**, we have the homotopy inverse cellular homotopy equivalences $h_K : K \rightleftarrows E(K) : h'_K$. If $n \leq 0$, then $t_{\geq n} : \mathbf{ICW} \to \mathbf{HoCW}$ will be the identity on objects and will be defined as $t_{\geq n}(f) = [f]$ for morphisms $f : K \to L$ in **ICW**. We will henceforth assume that n is positive. Define

$$
t_{\geq n}(K) = E(K)/E(K)^{n-1},
$$

that is, $t_{\geq n}(K)$ is the cofiber of the skeletal cofibration $E(K)^{n-1} \hookrightarrow E(K)$. Given a morphism $f : K \to L$ in **ICW**, the morphism $M(f)$ is represented by the homotopy commutative diagram

$$
\begin{array}{ccc}
K & \xrightarrow[\simeq]{h_K} & E(K) \\
\downarrow f & & \downarrow f_0 = h_L \circ f \circ h'_K \\
L & \xrightarrow[\simeq]{h_L} & E(L),
\end{array}
$$

$[f_0] = E(f)$. The square

$$
\begin{array}{ccc}
E(K)^{n-1} & \hookrightarrow & E(K) \\
\downarrow f_0^{n-1} & & \downarrow f_0 \\
E(L)^{n-1} & \hookrightarrow & E(L)
\end{array}
$$

commutes in **CW**. Thus f_0 induces a unique map

$$
\overline{f}_0 : t_{\geq n}(K) \longrightarrow t_{\geq n}(L)
$$

between the cofibers such that

$$
\begin{array}{ccccc}
E(K)^{n-1} & \hookrightarrow & E(K) & \twoheadrightarrow & t_{\geq n}(K) \\
\downarrow f_0^{n-1} & & \downarrow f_0 & & \downarrow \overline{f}_0 \\
E(L)^{n-1} & \hookrightarrow & E(L) & \twoheadrightarrow & t_{\geq n}(L)
\end{array}
$$

commutes in **CW**. We define

$$
t_{\geq n}(f) = [\overline{f}_0].
$$

(Note that we do not have to prove that this is well-defined, since no choices have been made: the map f_0 is at this point a canonical representative of the homotopy class $E(f)$.)

Lemma 1.71. *Let $h : X \to Y$ be a continuous map between topological spaces. Let $\sim$ be an equivalence relation on X. Then there exists a unique continuous map $\overline{h} : X/\sim \; \to Y$ such that*

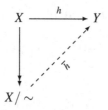

commutes iff $h(x) = h(x')$ *whenever* $x \sim x'$.

Lemma 1.72. *Let* E_1, E_2 *be two CW-complexes without odd-dimensional cells. If* $g, h : E_1 \to E_2$ *are two homotopic cellular maps, then* $\overline{g}$ *and* $\overline{h}$ *are homotopic, where* $\overline{g}, \overline{h} : E_1/E_1^k \to E_2/E_2^k$ *are induced by* g *and* h, *respectively.*

Proof. Let $H : E_1 \times I \to E_2$ be a cellular homotopy with $H(-, 0) = g$ and $H(-, 1) = h$. Since both E_1 and E_2 have only even-dimensional cells, we have

$$H(E_1^k \times I) \subset E_2^k.$$

(The details of that argument can be found in the proof of Lemma 1.70.) We shall apply Lemma 1.71 with $X = E_1 \times I$, $Y = E_2/E_2^k$, and h given by the composition

$$X \xrightarrow{H} E_2 \xrightarrow{\pi} E_2/E_2^k,$$

where π is the natural quotient projection. The equivalence relation $\sim$ on X is given as follows: $(e, t) \sim (e', t')$ iff $t = t'$ and either e, e' are both in E_1^k, or, if not, $e = e'$. It follows that $X/\sim = (E_1/E_1^k) \times I$. Suppose $(e, t) \sim (e', t')$. Let us check that then $h(e, t) = h(e', t')$. We have $t = t'$ and if one of e, e' does not lie in E_1^k, then $e = e'$ so that $(e', t') = (e, t)$ and therefore $h(e', t') = h(e, t)$. If e, e' both lie in E_1^k, then both $H(e', t)$ and $H(e, t)$ lie in E_2^k. Thus, in this case,

$$h(e', t') = \pi H(e', t) = [E_2^k] = \pi H(e, t) = h(e, t).$$

Hence, by Lemma 1.71, there exists a unique map

$$\overline{H} : (E_1/E_1^k) \times I = X/\sim \longrightarrow Y = E_2/E_2^k$$

such that

$$
\begin{array}{ccc}
E_1 \times I & \xrightarrow{\ H\ } & E_2 \\
\downarrow & & \downarrow{\scriptstyle \pi} \\
(E_1/E_1^k) \times I & \xrightarrow{\ \overline{H}\ } & E_2/E_2^k
\end{array}
$$

commutes and $\overline{H}(-, 0) = \overline{g}$, $\overline{H}(-, 1) = \overline{h}$. $\qquad\square$

Proposition 1.73. *For an object K in **ICW**, we have*

$$t_{\geq n}(\mathrm{id}_K) = \mathrm{id}_{t_{\geq n}(K)}$$

*in **HoCW**.*

Proof. The morphism $M(\mathrm{id}_K)$ is represented by the homotopy commutative square

$$
\begin{array}{ccc}
K & \xrightarrow{\;\;h_K\;\;} & E(K) \\
\simeq & & \\
{\scriptstyle \mathrm{id}_K}\downarrow & & \downarrow{\scriptstyle f_0 = h_K \circ h'_K} \\
& & \\
K & \xrightarrow[\simeq]{\;\;h_K\;\;} & E(K).
\end{array}
$$

Since h_K and h'_K are homotopy inverses, we have $f_0 \simeq \mathrm{id}_{E(K)}$. By Lemma 1.72, $\overline{f}_0 \simeq \overline{\mathrm{id}}_{E(K)} : E(K)/E(K)^{n-1} \to E(K)/E(K)^{n-1}$. As $E(K)/E(K)^{n-1} = t_{\geq n}(K)$ and $\overline{\mathrm{id}}_{E(K)} = \mathrm{id}_{t_{\geq n}(K)}$, we obtain

$$t_{\geq n}(\mathrm{id}_K) = [\overline{f}_0] = [\mathrm{id}_{t_{\geq n}(K)}].$$

$\square$

Proposition 1.74. *Given morphisms $f : K \to L$ and $g : L \to P$ in **ICW**, the functoriality relation*

$$t_{\geq n}(g \circ f) = t_{\geq n}(g) \circ t_{\geq n}(f)$$

*holds in **HoCW**.*

Proof. With

$$f_0 = h_L f h'_K, \quad g_0 = h_P g h'_L$$

and

$$(gf)_0 = h_P gf h'_K,$$

we must show

$$[\overline{(gf)_0}] = [\overline{g}_0] \circ [\overline{f}_0].$$

The maps $(gf)_0$ and $g_0 f_0$ are homotopic, as

$$[g_0 f_0] = [h_P g h'_L h_L f h'_K] = [h_P gf h'_K] = [(gf)_0].$$

By Lemma 1.72, $\overline{(gf)}_0 \simeq \overline{g_0 f_0}$. Furthermore, since the square

$$
\begin{array}{ccc}
E(K) & \longrightarrow & t_{\geq n}(K) \\
\downarrow{\scriptstyle g_0 f_0} & \quad \Big\downarrow{\scriptstyle \overline{g_0 f_0}} \;\; \Big\downarrow{\scriptstyle \overline{g_0} \circ \overline{f_0}} & \\
E(P) & \longrightarrow & t_{\geq n}(P)
\end{array}
$$

commutes if we use $\overline{g_0 f_0}$ and if we use $\overline{g_0} \circ \overline{f_0}$, uniqueness implies that

$$\overline{g_0 f_0} = \overline{g_0} \circ \overline{f_0}.$$

We conclude that $\overline{(gf)}_0$ is homotopic to $\overline{g_0} \circ \overline{f_0}$, as claimed. $\square$

Propositions 1.73 and 1.74 show that

$$t_{\geq n} : \mathbf{ICW} \longrightarrow \mathbf{HoCW}$$

is a covariant functor. Let us describe a natural transformation of functors

$$\mathrm{pro}_n : t_{<\infty} \longrightarrow t_{\geq n}.$$

Given an object K of $\mathbf{ICW}$, define $\mathrm{pro}_n(K)$ to be the composition

$$
\begin{array}{ccc}
t_{<\infty}(K) = K & \xrightarrow{\;\;M(K)\;\;} & E(K) \\
 & {\scriptstyle \mathrm{pro}_n(K)} \searrow & \downarrow{\scriptstyle [\mathrm{proj}]} \\
 & & E(K)/E(K)^{n-1} = t_{\geq n}(K).
\end{array}
$$

(Note that $\mathrm{pro}_n(K)$ has a canonical representative in $\mathbf{CW}$, namely

$$K \xrightarrow{\;h_K\;} E(K) \xrightarrow{\;\mathrm{proj}\;} E(K)/E(K)^{n-1}.)$$

Given a morphism $f : K \to L$ in $\mathbf{ICW}$, we have to show that the square

$$
\begin{array}{ccc}
t_{<\infty}(K) & \xrightarrow{\;\mathrm{pro}_n(K)\;} & t_{\geq n}(K) \\
\downarrow{\scriptstyle t_{<\infty}(f)} & & \downarrow{\scriptstyle t_{\geq n}(f)} \\
t_{<\infty}(L) & \xrightarrow{\;\mathrm{pro}_n(L)\;} & t_{\geq n}(L)
\end{array}
$$

commutes in **HoCW**. With $f_0 = h_L \circ f \circ h'_K$, we have $E(f) = [f_0]$ and

$$
\begin{array}{ccc}
E(K) & \longrightarrow & t_{\geq n}(K) \\
\downarrow f_0 & & \downarrow \bar{f}_0 \\
E(L) & \longrightarrow & t_{\geq n}(L)
\end{array}
$$

commutes in **CW**. Thus both squares of the diagram

$$
\begin{array}{ccccc}
t_{<\infty}(K) = K & \xrightarrow{M(K)=[h_K]} & E(K) & \longrightarrow & t_{\geq n}(K) \\
\downarrow {\scriptstyle t_{<\infty}(f)=[f]} & & \downarrow {\scriptstyle E(f)=[f_0]} & & \downarrow {\scriptstyle [\bar{f}_0]=t_{\geq n}(f)} \\
t_{<\infty}(L) = L & \xrightarrow{M(L)=[h_L]} & E(L) & \longrightarrow & t_{\geq n}(L)
\end{array}
$$

commute in **HoCW**. Thus pro_n is a natural transformation.

Proposition 1.75. *The functor $t_{\geq n}$ implements spatial homology cotruncation, that is, if K is an object of **ICW**, then*

$$
\mathrm{pro}_{n*} : H_r(K) \longrightarrow H_r(t_{\geq n}(K))
$$

is an isomorphism for $r \geq n$ and $\widetilde{H}_r(t_{\geq n}(K)) = 0$ for $r < n$.

Proof. Since $(E(K), E(K)^{n-1})$ is a CW pair, the inclusion $E(K)^{n-1} \hookrightarrow E(K)$ is a closed cofibration, whence

$$
\widetilde{H}_*(t_{\geq n}(K)) = \widetilde{H}_*(E(K)/E(K)^{n-1}) \cong H_*(E(K), E(K)^{n-1}).
$$

For $r < n$, the exact sequence

$$
H_r(E(K)^{n-1}) \xrightarrow{\cong} H_r(E(K)) \xrightarrow{0} \widetilde{H}_r(t_{\geq n}(K)) \xrightarrow{\partial_* = 0} H_{r-1}(E(K)^{n-1}) \xrightarrow{\cong} H_{r-1}(E(K))
$$

of the pair $(E(K), E(K)^{n-1})$ shows that $\widetilde{H}_r(t_{\geq n}(K)) = 0$. For $r = n$, the commutative diagram with exact top row

$$
0 = H_r(E(K)^{n-1}) \longrightarrow H_r(E(K)) \xrightarrow[\cong]{\mathrm{proj}_*} \widetilde{H}_r(t_{\geq n}(K)) \xrightarrow{\partial_*} H_{r-1}(E(K)^{n-1}) \xrightarrow{\cong} H_{r-1}(E(K))
$$

$$
M(K)_* {\Big\downarrow} \cong \qquad \nearrow {\scriptstyle \mathrm{pro}_n(K)_*}
$$

$$
H_r(K)
$$

shows that proj_*, and hence $\text{pro}_n(K)_*$, is an isomorphism. For $r > n$, the claim follows from the exactness of the top row and the commutativity in the diagram

$$0 = H_r(E(K)^{n-1}) \longrightarrow H_r(E(K)) \xrightarrow[\cong]{\text{proj}_*} \widetilde{H}_r(t_{\geq n}(K)) \xrightarrow{\partial_*} H_{r-1}(E(K)^{n-1}) = 0.$$

$$M(K)_* \Big\uparrow \cong \qquad \nearrow \text{pro}_n(K)^*$$

$$H_r(K).$$

$\square$

1.10 Continuity Properties of Homology Truncation

Continuity of homology truncation refers to the question whether $\tilde{t}_{<n}(f)$ is close to $\tilde{t}_{<n}(g)$ when f is close to g in the compact-open topology. Here, $\tilde{t}_{<n}(f)$ and $\tilde{t}_{<n}(g)$ denote particular representatives of the homotopy classes $t_{<n}(f)$ and $t_{<n}(g)$, respectively. Our motivation for studying this question is the intention to apply the answers obtained in setting up fiberwise homology truncation, see Section 1.11: Suppose $E \to B$ is a fiber bundle with fiber F, structure group $G(F)$, and continuous transition functions $g_{\alpha\beta} : U_\alpha \cap U_\beta \to G(F)$, where $\{U_\alpha\}$ is an open cover of B over which the bundle trivializes. Let $G(t_{<n}F)$ be a topological group acting on the truncation $t_{<n}F$ of the fiber. Continuity of $\tilde{t}_{<n}$ would ideally mean the existence of a continuous homomorphism $\tau_n : G(F) \to G(t_{<n}F)$ such that

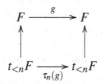

commutes for all $g \in G(F)$. Whenever such a τ_n exists, it can be used to form a fiber bundle $\text{ft}_{<n} E \to B$, the *fiberwise truncation* of E, with fiber $t_{<n}F$ and structure group $G(t_{<n}F)$ by gluing via the transition functions $\tau_n \circ g_{\alpha\beta} : U_\alpha \cap U_\beta \to G(t_{<n}F)$. The fact that τ_n is a group homomorphism ensures that the cocycle condition is again satisfied for the system $\{\tau_n \circ g_{\alpha\beta}\}$. Techniques in this direction will enable one to define intersection spaces for classes of pseudomanifolds that have nontrivial, twisted link bundles. On the other hand, it is to be noted that a fiberwise homology truncation

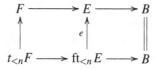

cannot generally be carried out for any fibration and any n because the morphism of the associated Serre spectral sequences induced by e, together with $H_q(t_{<n}F) \rightarrow H_q(F)$ being an isomorphism for $q < n$ and $H_q(t_{<n}F) = 0$ for $q \geq n$, places restrictions on the differentials in the spectral sequence of $E \rightarrow B$. Thus, suitable assumptions on the fibration need to be adopted.

For topological spaces X and Y, let $\text{Map}(X,Y)$ denote the set of all continuous maps $X \rightarrow Y$. We endow this set with the compact-open topology. If X is a locally compact, locally connected Hausdorff space, then the subspace $\text{Homeo}(X) \subset \text{Map}(X,X)$ consisting of all homeomorphisms $X \rightarrow X$ is a topological group, see [Are46]. If X and Y are CW-complexes, let $\text{Map}_{CW}(X,Y) \subset \text{Map}(X,Y)$ denote the subspace of all cellular maps and let $\text{Homeo}_{CW}(X) \subset \text{Homeo}(X) \cap \text{Map}_{CW}(X,X)$ denote the subspace of all homeomorphisms that are cellular with cellular inverse. The space $\text{Homeo}_{CW}(X)$ is a group under composition. Any CW-complex is Hausdorff and locally path connected, in particular locally connected. It is locally compact if, and only if, each point has a neighborhood that meets only finitely many cells. Thus for a finite CW-complex X, $\text{Homeo}(X)$ is a topological group. Every subgroup of a topological group is itself a topological group when given the subspace topology. Hence $\text{Homeo}_{CW}(X)$ is a topological group for a finite CW-complex X. A fiber bundle with fiber F a priori has structure group $\text{Homeo}(F)$. Let us mention but one example class that allows the structure group to take values in $\text{Homeo}_{CW}(F)$.

Proposition 1.76. *Suppose F is a smooth, compact manifold and ξ a smooth fiber bundle with fiber F and finite structure group G. Then the transition functions of ξ take values in $\text{Homeo}_{CW}(F)$ for a suitable CW-structure on F.*

Proof. The fiber F is a smooth G-space. By [Ill78], F is a G-CW-complex. For a finite group, a G-CW-complex is the same thing as an ordinary CW-complex with a cellular G-action. The latter means that G permutes the cells of F; in particular, G acts by cellular homeomorphisms that have cellular inverses and the map $G \rightarrow \text{Homeo}(F)$ factors through a map $G \rightarrow \text{Homeo}_{CW}(F)$. $\qquad\square$

Here is one way how bundles with finite structure group arise:

Proposition 1.77. *Let G be a Lie group and B a smooth path-connected manifold with finite fundamental group. Then any G-bundle over B having a connection with curvature zero ("flat" bundle) can be reduced to a finite structure group.*

Proof. By [Mil58, Lemma 1], the G-bundle ξ is induced from the universal covering bundle ξ' with projection $\widetilde{B} \rightarrow B$ (a $\pi = \pi_1(B)$-bundle) by a homomorphism $h : \pi \rightarrow G$. This means that the transition functions $g_{ij} : U_i \cap U_j \rightarrow G$ of ξ are $g_{ij} = hg'_{ij}$, where $g'_{ij} : U_i \cap U_j \rightarrow \pi$ are the transition functions of ξ'. Thus the g_{ij} take values in the holonomy group $\text{im}(h) \subset G$, which is finite, since π is finite. $\quad\square$

For a topological space X, let $G(X) \subset \text{Map}(X,X)$ be the subspace of all (unbased) self-homotopy equivalences of X. If X is compact and has the homotopy type of a finite CW-complex, then $G(X)$ is a grouplike topological monoid under composition of maps, see [Fuc71]. In other words, $G(X)$ is a strictly associative

H-space with strict unit and a global homotopy inverse, i.e. a map $v : G(X) \to G(X)$ such that the composition

$$G(X) \xrightarrow{\Delta} G(X) \times G(X) \xrightarrow{\mathrm{id} \times v} G(X) \times G(X) \xrightarrow{\mu} G(X)$$

is homotopic to the constant map at id_X, where μ is the composition of maps. Let $G[X] = \pi_0 G(X)$ denote the group of homotopy classes of self-homotopy equivalences of X.

Let K be an object of the interleaf category with finitely many cells and n an integer. To avoid a discussion of trivialities, we assume that n is positive. The functor

$$t_{<n} : \mathbf{ICW} \longrightarrow \mathbf{HoCW}$$

assigns to a homeomorphism $f \in \mathrm{Homeo}_{CW}(K)$ a morphism

$$t_{<n}(f) : t_{<n}K \longrightarrow t_{<n}K,$$

which is the homotopy class of some cellular map t. This t is a homotopy equivalence because the functoriality of $t_{<n}$ implies that any representative of $t_{<n}(f^{-1})$ is a homotopy inverse for t. Thus $t \in G(t_{<n}K)$ and $t_{<n}(f) = [t] \in G[t_{<n}K]$. The functor $t_{<n}$ thus defines a map

$$t_{<n} : \mathrm{Homeo}_{CW}(K) \longrightarrow G[t_{<n}K].$$

By the functoriality of $t_{<n}$, this map is a group homomorphism. We wish to construct a continuous lift

$$
\begin{array}{ccc}
 & & G(t_{<n}K) \\
 & \overset{\tilde{t}_{<n}}{\nearrow} & \downarrow \\
\mathrm{Homeo}_{CW}(K) & \xrightarrow{t_{<n}} & G[t_{<n}K]
\end{array}
$$

which will in fact be an H-map, but not in general a monoid homomorphism. (Note that $G(t_{<n}K)$ is indeed a grouplike topological monoid because $E(K)^{n-1} = t_{<n}(K)$ is a finite CW-complex by Remark 1.69.) Recall that we had associated homotopy inverse homotopy equivalences

$$h_K : K \rightleftarrows E(K) : h_K'$$

with K. The CW-complex $E(K)$ has only even-dimensional cells and we have

$$t_{<n}K = E(K)^{n-1}.$$

Set

$$\tilde{t}_{<n} : \mathrm{Homeo}_{CW}(K) \longrightarrow G(t_{<n}K)$$
$$f \mapsto (h_K \circ f \circ h'_K)^{n-1}.$$

Since

$$[(h_K \circ f \circ h'_K)^{n-1}] = t_{<n}(f),$$

the map $\tilde{t}_{<n}$ is indeed a lift of $t_{<n}$. It is not only continuous, but also respects, up to homotopy, the monoid multiplication:

Theorem 1.78. *The map $\tilde{t}_{<n}$ is an H-map.*

Proof. Let $Q : K \times I \to K$ be a cellular homotopy from $Q(-,0) = h'_K \circ h_K$ to $Q(-,1) = \mathrm{id}_K$. By cellularity, Q maps $K^{n-1} \times I \subset (K \times I)^n$ to K^n. Let us denote this restriction by $Q^n : K^{n-1} \times I \to K^n$. We will study the maps

$$H(f,g,t) = h_K^n g^n Q^n(-,t) f^{n-1} h_K'^{n-1} : E(K)^{n-1} \to E(K)^n$$

where $f,g \in \mathrm{Homeo}_{CW}(K)$ and $t \in I$. The following properties will be established for H:

(1) $H(f,g,0) = \tilde{t}_{<n}(g)\tilde{t}_{<n}(f)$,
(2) $H(f,g,1) = \tilde{t}_{<n}(gf)$,
(3) $H(f,g,t)(t_{<n}K) \subset t_{<n}K$,
(4) $H(f,g,t) : t_{<n}K \to t_{<n}K$ is a homotopy equivalence.

It follows from (3) and (4) that H is a map

$$H : \mathrm{Homeo}_{CW}(K) \times \mathrm{Homeo}_{CW}(K) \times I \longrightarrow G(t_{<n}K).$$

We will then show that

(5) H is continuous.

Thus H will be an explicit "sputnik homotopy" in the terminology of Stasheff.

(1) We have

$$Q^n(-,0) = Q|_{K^{n-1} \times \{0\}} = h'_K h_K|_{K^{n-1}} = h_K'^{n-1} h_K^{n-1}$$

and

$$h_K^n g^n h_K'^{n-1} = h_K^{n-1} g^{n-1} h_K'^{n-1}.$$

Thus

$$\begin{aligned}
H(f,g,0) &= h_K^n g^n Q^n(-,0) f^{n-1} h_K'^{n-1} \\
&= h_K^n g^n h_K'^{n-1} h_K^{n-1} f^{n-1} h_K'^{n-1} \\
&= h_K^{n-1} g^{n-1} h_K'^{n-1} h_K^{n-1} f^{n-1} h_K'^{n-1} \\
&= (h_K g h'_K)^{n-1} (h_K f h'_K)^{n-1} \\
&= \tilde{t}_{<n}(g)\tilde{t}_{<n}(f).
\end{aligned}$$

(2) Holds since

$$\begin{aligned}
H(f,g,1) &= h_K^n g^n Q^n(-,1) f^{n-1} h_K'^{n-1} \\
&= h_K^n g^n \operatorname{id}_K |_{K^{n-1}} f^{n-1} h_K'^{n-1} \\
&= h_K^{n-1} g^{n-1} f^{n-1} h_K'^{n-1} \\
&= (h_K g f h_K')^{n-1} \\
&= \tilde{t}_{<n}(gf).
\end{aligned}$$

(3) We distinguish two cases according to whether n is even or odd. For n even, $E(K)^{n-1} = E(K)^{n-2}$. Let $Q^{n-1} : K^{n-2} \times I \to K^{n-1}$ be the restriction of Q^n. The commutative diagram

$$\begin{array}{ccccccccccc}
E(F)^{n-1} & \xrightarrow{h_K'^{n-1}} & K^{n-1} & \xrightarrow{f^{n-1}} & K^{n-1} & \xrightarrow{Q^n(-,t)} & K^n & \xrightarrow{g^n} & K^n & \xrightarrow{h_K^n} & E(K)^n \\
{\scriptstyle =}\big\uparrow & & \big\uparrow & & \big\uparrow & & \big\uparrow & & \big\uparrow & & {\scriptstyle j}\big\uparrow \\
E(F)^{n-2} & \xrightarrow{h_K'^{n-2}} & K^{n-2} & \xrightarrow{f^{n-2}} & K^{n-2} & \xrightarrow{Q^{n-1}(-,t)} & K^{n-1} & \xrightarrow{g^{n-1}} & K^{n-1} & \xrightarrow{h_K^{n-1}} & E(K)^{n-1}
\end{array}$$

shows that for n even,

$$H(f,g,t) = j \circ h_K^{n-1} g^{n-1} Q^{n-1}(-,t) f^{n-2} h_K'^{n-2}$$

and so has an image that lies in $E(K)^{n-1} = t_{<n}K$.

For n odd, the statement follows from $H(f,g,t)(E(K)^{n-1}) \subset E(K)^n = E(K)^{n-1}$.

(4) Keeping f and g fixed, $H(f,g,-)$ defines a homotopy $H(f,g,-) : t_{<n}K \times I \to t_{<n}K$. Since $H(f,g,1) = \tilde{t}_{<n}(gf)$ is a homotopy equivalence, every $H(f,g,t)$ is homotopic to a homotopy equivalence, hence itself a homotopy equivalence.

(5) We will throughout avail ourselves of the following three basic properties of the compact-open topology:

(i) If $\phi : X' \to X$ and $\psi : Y \to Y'$ are continuous maps, then the map

$$\begin{aligned}
\operatorname{Map}(X,Y) &\longrightarrow \operatorname{Map}(X',Y') \\
f &\mapsto \psi \circ f \circ \phi
\end{aligned}$$

is continuous. (No point-set topological assumptions on the involved spaces.) In particular, if $A \subset X$ is any subspace of a topological space X then the restriction map

$$\begin{aligned}
\operatorname{Map}(X,Y) &\longrightarrow \operatorname{Map}(A,Y) \\
f &\mapsto f|_A
\end{aligned}$$

is continuous.

(ii) If X, Y, Z are topological spaces with Y locally compact Hausdorff, then composition of maps

$$\operatorname{Map}(X,Y) \times \operatorname{Map}(Y,Z) \xrightarrow{\circ} \operatorname{Map}(X,Z)$$

is continuous.

(iii) The exponential law (see e.g. [Bre93, Theorem VII.2.5]): If X, Y, Z are Hausdorff spaces with X, Z locally compact, then there is a homeomorphism

$$\text{Map}(Z \times X, Y) \cong \text{Map}(Z, \text{Map}(X, Y)).$$

The cartesian product of the continuous inclusion

$$\text{Homeo}_{CW}(K) \hookrightarrow \text{Map}_{CW}(K, K)$$

with itself defines a continuous map

$$c_1 : \text{Homeo}_{CW}(K) \times \text{Homeo}_{CW}(K) \longrightarrow \text{Map}_{CW}(K, K) \times \text{Map}_{CW}(K, K).$$

The restriction maps

$$\text{Map}_{CW}(K, K) \longrightarrow \text{Map}_{CW}(K^{n-1}, K^{n-1})$$

and

$$\text{Map}_{CW}(K, K) \longrightarrow \text{Map}_{CW}(K^n, K^n)$$

are continuous, since they are given by the composition

$$\text{Map}_{CW}(K, K) \hookrightarrow \text{Map}(K, K) \xrightarrow{\text{restr}} \text{Map}(K^{n-1} \text{ or } K^n, K).$$

Their product is a continuous map

$$c_2 : \text{Map}_{CW}(K, K) \times \text{Map}_{CW}(K, K) \longrightarrow \text{Map}_{CW}(K^{n-1}, K^{n-1}) \times \text{Map}_{CW}(K^n, K^n).$$

Composition with $h_K^{\prime n-1} : E(K)^{n-1} \to K^{n-1}$ yields a continuous map

$$\text{Map}_{CW}(K^{n-1}, K^{n-1}) \longrightarrow \text{Map}_{CW}(E(K)^{n-1}, K^{n-1}),$$

and composition with $h_K^n : K^n \to E(K)^n$ yields a continuous map

$$\text{Map}_{CW}(K^n, K^n) \longrightarrow \text{Map}_{CW}(K^n, E(K)^n).$$

Their product is a continuous map

$$c_3 : \text{Map}_{CW}(K^{n-1}, K^{n-1}) \times \text{Map}_{CW}(K^n, K^n)$$

$$\longrightarrow \text{Map}_{CW}(E(K)^{n-1}, K^{n-1}) \times \text{Map}_{CW}(K^n, E(K)^n).$$

By [Mun00, Theorem 46.11], the map $Q^n : K^{n-1} \times I \to K^n$ determines a continuous map $Q^n : K^{n-1} \to \mathrm{Map}(I, K^n)$. Composing with this map in the first factor and using the canonical inclusion on the second factor, we get a continuous map

$$\tilde{c}_4 : \mathrm{Map}_{CW}(E(K)^{n-1}, K^{n-1}) \times \mathrm{Map}_{CW}(K^n, E(K)^n)$$

$$\longrightarrow \mathrm{Map}(E(K)^{n-1}, \mathrm{Map}(I, K^n)) \times \mathrm{Map}(K^n, E(K)^n).$$

By the exponential law, we have a homeomorphism

$$\mathrm{Map}(E(K)^{n-1}, \mathrm{Map}(I, K^n)) \cong \mathrm{Map}(E(K)^{n-1} \times I, K^n),$$

since $E(K)^{n-1}$, K^n and I are all Hausdorff (being CW-complexes) and $E(K)^{n-1}$, I are locally compact because they have finitely many cells. Composing this homeomorphism (crossed with the identity) with $\tilde{c}_4$, we obtain a continuous map

$$c_4 : \mathrm{Map}_{CW}(E(K)^{n-1}, K^{n-1}) \times \mathrm{Map}_{CW}(K^n, E(K)^n)$$

$$\longrightarrow \mathrm{Map}(E(K)^{n-1} \times I, K^n) \times \mathrm{Map}(K^n, E(K)^n).$$

Composition is a continuous map

$$c_5 : \mathrm{Map}(E(K)^{n-1} \times I, K^n) \times \mathrm{Map}(K^n, E(K)^n)$$

$$\longrightarrow \mathrm{Map}(E(K)^{n-1} \times I, E(K)^n) \cong \mathrm{Map}(I, \mathrm{Map}(E(K)^{n-1}, E(K)^n)),$$

since K^n is locally compact Hausdorff. The composition $c_5 c_4 c_3 c_2 c_1$ is a continuous map

$$\mathrm{Homeo}_{CW}(K) \times \mathrm{Homeo}_{CW}(K) \longrightarrow \mathrm{Map}(I, \mathrm{Map}(E(K)^{n-1}, E(K)^n)).$$

By [Mun00, Theorem 46.11], this determines a continuous map

$$\mathrm{Homeo}_{CW}(K) \times \mathrm{Homeo}_{CW}(K) \times I \longrightarrow \mathrm{Map}(E(K)^{n-1}, E(K)^n),$$

since I is locally compact Hausdorff. The value of this map on a triple $(f, g, t) \in \mathrm{Homeo}_{CW}(K) \times \mathrm{Homeo}_{CW}(K) \times I$ equals $H(f, g, t)$ (and is in fact contained in $G(t_{<n}K)$). Thus H is continuous.

Restricting H to $g = \mathrm{id}_K$ and $t = 1$, we obtain the continuous map

$$H(-, \mathrm{id}_K, 1) : \mathrm{Homeo}_{CW}(K) \longrightarrow G(t_{<n}K).$$

Since $H(f, \mathrm{id}_K, 1) = \tilde{t}_{<n}(f)$, we conclude that $\tilde{t}_{<n}$ is continuous. The map H is a homotopy from $\tilde{t}_{<n}(-) \circ \tilde{t}_{<n}(-)$ to $\tilde{t}_{<n}(- \circ -)$. Therefore, the square

$$
\begin{array}{ccc}
\mathrm{Homeo}_{CW}(K) \times \mathrm{Homeo}_{CW}(K) & \xrightarrow{\ \circ\ } & \mathrm{Homeo}_{CW}(K) \\
\Big\downarrow{\scriptstyle \tilde{t}_{<n} \times \tilde{t}_{<n}} & & \Big\downarrow{\scriptstyle \tilde{t}_{<n}} \\
G(t_{<n}K) \times G(t_{<n}K) & \xrightarrow{\ \circ\ } & G(t_{<n}K)
\end{array}
$$

commutes up to homotopy and $\tilde{t}_{<n}$ is an H-map. $\square$

Let us discuss some observations concerning the problem of rectifying our truncation H-map into a strictly multiplicative map. An *H-equivalence* is a homotopy equivalence which is an H-map. By way of motivation, let us first mention the following simple fact.

Lemma 1.79. *Let X and Y be locally compact Hausdorff spaces. A homotopy equivalence*

$$ \phi : X \xrightleftharpoons{\qquad} Y : \psi $$

induces an H-equivalence

$$ \Phi : G(X) \xrightleftharpoons{\qquad} G(Y) : \Psi $$

by setting $\Phi(f) = \phi f \psi$, $\Psi(g) = \psi g \phi$.

Proof. The maps Φ, Ψ are continuous: The map

$$
\begin{array}{rcl}
\mathrm{Map}(X, X) & \longrightarrow & \mathrm{Map}(Y, Y) \\
f & \mapsto & \phi f \psi
\end{array}
$$

is continuous for the compact-open topology. Thus the composition

$$ G(X) \hookrightarrow \mathrm{Map}(X, X) \longrightarrow \mathrm{Map}(Y, Y) $$

is continuous. If $f : X \to X$ is a homotopy equivalence, then $\phi f \psi$ is a homotopy equivalence as well, whence the image of the composition lies in $G(Y)$. It follows that Φ is continuous. Similarly, or by symmetry, Ψ is continuous.

The maps Φ and Ψ are homotopy inverses of each other: We shall define a homotopy $\Psi\Phi \simeq \mathrm{id}_{G(X)}$. Let $P : X \times I \to X$ be a homotopy from $P(-, 0) = \psi\phi$ to $P(-, 1) = \mathrm{id}_X$. Define

$$ H : G(X) \times I \longrightarrow G(X) $$

by

$$ H(f, t)(x) = P(f(P(x, t)), t), \ f \in G(X), \ t \in I, \ x \in X. $$

Let us demonstrate that H is continuous. The map

$$P^* : I \longrightarrow \mathrm{Map}(X,X)$$
$$t \longmapsto P(-,t)$$

is continuous. Thus the product

$$c_1 = \mathrm{id}_{\mathrm{Map}(X,X)} \times (P^*,P^*) : \mathrm{Map}(X,X) \times I \longrightarrow \mathrm{Map}(X,X)^3$$

is continuous. Since X is locally compact Hausdorff, the composition map

$$c_2 = (\circ, \mathrm{id}) : \mathrm{Map}(X,X)^3 \longrightarrow \mathrm{Map}(X,X)^2,$$

sending (f,g,h) to $(f \circ g, h)$, as well as

$$c_3 = \circ : \mathrm{Map}(X,X)^2 \longrightarrow \mathrm{Map}(X,X),$$

sending (g,h) to $h \circ g$, is continuous. Thus the composition

$$G(X) \times I \hookrightarrow \mathrm{Map}(X,X) \times I \overset{c_3 c_2 c_1}{\longrightarrow} \mathrm{Map}(X,X)$$

is continuous. The value of this composition on a pair (f,t), with $f : X \to X$ a homotopy equivalence, is precisely $H(f,t)$. Thus H is continuous as a map $G(X) \times I \to \mathrm{Map}(X,X)$. The image $H(f,t)$ is again a homotopy equivalence, since it is homotopic, via H, to

$$H(f,1) = P(f(P(-,1)),1) = f.$$

Thus we get a continuous map $H : G(X) \times I \to G(X)$. Evaluating H at the other end of the cylinder, we obtain

$$H(f,0) = P(f(P(x,0)),0) = \psi \phi f \psi \phi = \Psi \Phi(f).$$

Consequently, H is a homotopy between $\Psi \Phi$ and $\mathrm{id}_{G(X)}$. Similarly, or by symmetry, one gets a homotopy $\Phi \Psi \simeq \mathrm{id}_{G(Y)}$.

It remains to be verified that Φ and Ψ are H-maps. We need to exhibit a sputnik homotopy

$$H : G(X) \times G(X) \times I \longrightarrow G(Y)$$

that establishes the homotopy commutativity of the diagram

$$
\begin{array}{ccc}
G(X) \times G(X) & \overset{\Phi \times \Phi}{\longrightarrow} & G(Y) \times G(Y) \\
\downarrow{\scriptstyle \circ} & & \downarrow{\scriptstyle \circ} \\
G(X) & \overset{\Phi}{\longrightarrow} & G(Y).
\end{array}
$$

Define such an H by

$$H(f,g,t)(y) = \phi f P(g\psi(y),t), \; f,g \in G(X), \; t \in I, \; y \in Y.$$

The continuity of H follows by the usual arguments already detailed a number of times in previous proofs, using that X and Y are locally compact Hausdorff. The fact that the image of H, a priori only known to lie in $\mathrm{Map}(Y,Y)$, really lies in $G(Y)$, follows from the fact that $H(f,g,t)$ is homotopic, via H, to $H(f,g,1) = \phi f P(g\psi(-),1) = \phi f g\psi$, which is a homotopy equivalence. We have

$$H(f,g,0)(y) = \phi f P(g\psi(y),0) = \phi f \psi \phi(g\psi(y)) = (\Phi(f)\Phi(g))(y)$$

and

$$H(f,g,1)(y) = \phi f g\psi(y) = \Phi(fg)(y).$$

Similarly, or by symmetry, Ψ is an H-map. $\qquad\qquad\qquad\qquad\qquad\qquad\square$

Let us contrast the above lemma with the analogous problem in the world of CW-complexes. For a CW-complex K, let $G_{CW}(K) = G(K) \cap \mathrm{Map}_{CW}(K,K)$ be the topological monoid of cellular self-homotopy equivalences of K. A map $r : X \to Y$ is called a *homotopy retraction* if there exists a map $s : Y \to X$ such that $rs \simeq \mathrm{id}_Y$. If such maps exist, one says that Y is a *homotopy retract of X*. (Sometimes the terminology "Y is dominated by X" is used.) If X,Y are H-spaces and s is an H-map, we say that Y is an *H-homotopy retract of X*.

Lemma 1.80. *Let K and E be locally compact CW-complexes (i.e. each point has a neighborhood that meets only finitely many cells). If E has no odd-dimensional cells and $K \simeq E$, then $G_{CW}(E)$ is an H-homotopy retract of $G_{CW}(K)$.*

Proof. Let $\phi : K \to E$ be a cellular homotopy equivalence with cellular homotopy inverse $\psi : E \to K$. Let $P : E \times I \to E$ be a cellular homotopy from $P(-,0) = \phi\psi$ to $P(-,1) = \mathrm{id}_E$. Let

$$R : G_{CW}(K) \longrightarrow G_{CW}(E)$$

be the map $R(f) = \phi f \psi$ and let

$$S : G_{CW}(E) \longrightarrow G_{CW}(K)$$

be the map $S(g) = \psi g \phi$.

The maps R,S are continuous: The map

$$\mathrm{Map}(K,K) \longrightarrow \mathrm{Map}(E,E)$$
$$f \mapsto \phi f \psi$$

is continuous, so the composition

$$G_{CW}(K) \hookrightarrow \mathrm{Map}(K,K) \longrightarrow \mathrm{Map}(E,E)$$

is continuous. Since its image lies in $G_{CW}(E)$, it follows that R is continuous. Similarly, S is continuous.

Let us define a homotopy $RS \simeq \mathrm{id}_{G_{CW}(E)}$. Define

$$H : G_{CW}(E) \times I \longrightarrow \mathrm{Map}(E,E)$$

to be

$$H(g,t)(x) = P(g(P(x,t)),t),\ g \in G_{CW}(E),\ t \in I,\ x \in E.$$

The continuity of H is demonstrated as in the proof of Lemma 1.79. The image $H(g,t)$ is again a homotopy equivalence, so we get a continuous map $H : G_{CW}(E) \times I \to G(E)$. We claim that $H(f,t) : E \to E$ is in fact a cellular map: This follows from the fact that P restricts to map

$$P| : E^k \times I \longrightarrow E^k,$$

a key observation that has already been used to prove Lemma 1.70: If k is even, then P restricts as

$$P| : E^k \times I \subset (E \times I)^{k+1} \to E^{k+1} = E^k,$$

while if k is odd, then P restricts as

$$P| : E^k \times I = E^{k-1} \times I \subset (E \times I)^k \to E^k.$$

Hence for a point $x \in E^k$, we have $P(x,t) \in E^k$, thus $g(P(x,t)) \in E^k$ and so $H(g,t)(x) = P(g(P(x,t)),t) \in E^k$. Therefore, H is a continuous map $H : G_{CW}(E) \times I \to G_{CW}(E)$. Evaluating H at time zero, we obtain

$$H(g,0)(x) = P(g(P(x,0)),0) = \phi\psi g\phi\psi(x) = RS(g)(x).$$

Evaluation at time one gives

$$H(g,1)(x) = P(g(P(x,1)),1) = g(x).$$

We conclude that H is a homotopy between RS and $\mathrm{id}_{G_{CW}(E)}$.

It remains to be verified that S is an H-map. We need to exhibit a sputnik homotopy

$$H : G_{CW}(E) \times G_{CW}(E) \times I \longrightarrow G_{CW}(K)$$

that establishes the homotopy commutativity of the diagram

$$
\begin{array}{ccc}
G_{CW}(E) \times G_{CW}(E) & \xrightarrow{S \times S} & G_{CW}(K) \times G_{CW}(K) \\
\downarrow{\scriptstyle\circ} & & \downarrow{\scriptstyle\circ} \\
G_{CW}(E) & \xrightarrow{\ \ S\ \ } & G_{CW}(K).
\end{array}
$$

Define such an H by

$$H(f,g,t)(y) = \psi f P(g\phi(y),t), \; f,g \in G_{CW}(E), \; t \in I, \; y \in K.$$

The continuity of H follows by the usual arguments already detailed a number of times in previous proofs, using also that K and E are locally compact and Hausdorff, being CW-complexes. The fact that the image of H, a priori only known to lie in $\mathrm{Map}(K,K)$, really lies in $G_{CW}(K)$, follows on the one hand from the fact that $H(f,g,t)$ is homotopic, via H, to $H(f,g,1) = \psi f P(g\phi(-),1) = \psi f g\phi$, which is a homotopy equivalence, and on the other hand from the fact that $H(f,g,t) : K \to K$ is cellular because $P(E^k \times I) \subset E^k$ as pointed out above. We have

$$H(f,g,0)(y) = \psi f P(g\phi(y),0) = \psi f \phi \psi g\phi(y) = (S(f)S(g))(y)$$

and

$$H(f,g,1)(y) = \psi f g\phi(y) = S(fg)(y).$$

$\square$

Remark 1.81. The key issue in the proof of the previous lemma is of course the construction of a homotopy *through cellular maps*. As we have seen, this works if the codomain has only even-dimensional cells. If there are cells of odd dimension as well, then the method of proof breaks down and does not yield an induced homotopy equivalence $G_{CW}(K) \simeq G_{CW}(E)$, unless one assumes for instance that the tracks of a homotopy $\psi\phi \simeq \mathrm{id}_K$ remain in the skeleton which they start out from.

Let us return to our finite CW-complex K, an object of the interleaf category. We can now improve the truncation H-map $\tilde{t}_{<n}$ to a strictly multiplicative map, in fact a monoid homomorphism, in the following manner.

Proposition 1.82. *There exists a topological monoid G, which is an H-homotopy retract of $G_{CW}(K)$, a homotopy retraction $R : G_{CW}(K) \to G$ and a monoid homomorphism $t : G \to G(t_{<n}K)$ such that the homology truncation H-map $\tilde{t}_{<n} :$ $\mathrm{Homeo}_{CW}(K) \to G(t_{<n}K)$ factors as*

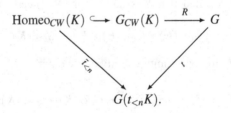

$$G(t_{<n}K).$$

Proof. Consider the homotopy equivalence

$$h_K : K \rightleftarrows E(K) : h'_K$$

The CW-complex $E(K)$ has only even-dimensional cells and is finite, so in particular locally compact. By Lemma 1.80, $G_{CW}(E(K))$ is an H-homotopy retract of $G_{CW}(K)$. In fact, a homotopy retraction

$$R : G_{CW}(K) \longrightarrow G_{CW}(E(K))$$

is given by $R(f) = h_K f h'_K$ and a homotopy section

$$S : G_{CW}(E(K)) \longrightarrow G_{CW}(K)$$

for R is given by the H-map $S(g) = h'_K g h_K$. Set $G = G_{CW}(E(K))$ and define

$$t : G \longrightarrow G(t_{<n}K)$$

by restricting a cellular homotopy equivalence to the $(n-1)$-skeleton, that is, $t(f) = f^{n-1}$. Observe that $E(K)^{n-1} = t_{<n}K$ and $t(f) : E(K)^{n-1} \to E(K)^{n-1}$ is indeed a homotopy equivalence by Lemma 1.70. The map t is continuous because the restriction map

$$\mathrm{Map}(E(K), E(K)) \longrightarrow \mathrm{Map}(E(K)^{n-1}, E(K))$$

is continuous, whence the composition

$$G_{CW}(E(K)) \hookrightarrow \mathrm{Map}(E(K), E(K)) \longrightarrow \mathrm{Map}(E(K)^{n-1}, E(K))$$

is continuous. The image of the composition, however, lies in $G(E(K)^{n-1})$ and its value on a map is the value of t. Furthermore, t is a monoid homomorphism, since $t(\mathrm{id}) = \mathrm{id}$ and $t(fg) = (fg)^{n-1} = f^{n-1}g^{n-1} = t(f)t(g)$. Lastly, we have indeed produced a factorization, as

$$tR(f) = t(h_K f h'_K) = (h_K f h'_K)^{n-1} = \tilde{t}_{<n}(f)$$

for $f \in \mathrm{Homeo}_{CW}(K)$. $\qquad\square$

We have so far discussed continuity properties of spatial homology truncation for spaces that have only even-dimensional cells. Let us now turn to the much harder problem of continuity for homology truncation of arbitrary (simply connected) complexes. We will not discuss low-dimensional truncation but immediately turn to degrees $n \geq 3$. (The category $\mathbf{CW}_{n \supset \partial}$ and the notion of n-compression rigidity have only been defined for $n \geq 3$ and are irrelevant for $n \leq 2$.) Let (K, Y) be an object of $\mathbf{CW}_{n \supset \partial}$ and let G be a discrete group.

Definition 1.83. A group homomorphism $\rho : G \to \mathrm{Homeo}(K)$ is an *n-compression rigid representation* (with respect to Y) if $\rho(G)$ consists of n-compression rigid morphisms $(K, Y) \to (K, Y)$ in $\mathbf{CW}_{n \supset \partial}$.

Example 1.84. Suppose B is the base space of a flat fiber bundle $\widetilde{B} \times_\rho F$ given by a holonomy representation $\rho : \pi_1(B) \to \mathrm{Homeo}_{CW}(F)$, where the fiber F is a simply connected CW-complex whose boundary operator ∂_n in its cellular chain complex is either zero or injective. Then by Corollary 1.49, any cellular map $F \to F$ is an n-compression rigid morphism $(F,Y) \to (F,Y)$ and ρ is an n-compression rigid representation. When $n = 3$ and the 1-skeleton of F is a point, then the condition on the boundary operator of the fiber is not even needed (by Proposition 1.51).

An n-compression rigid representation $\rho : G \to \mathrm{Homeo}(K)$ determines an n-compression rigid category $\mathbf{C}_\rho$ with one object (K,Y) and morphisms given by the image $\rho(G)$. By Corollary 1.40, one has a spatial homology truncation functor $t_{<n} : \mathbf{C}_\rho \to \mathbf{HoCW}_{n-1}$. Hence, for every $g \in G$, one gets a homotopy class $t_{<n}\rho(g) : t_{<n}(K,Y) \to t_{<n}(K,Y)$. Set $K_{<n} = t_{<n}(K,Y)$. If $g,h \in G$ are two group elements, then the functoriality of $t_{<n}$ on $\mathbf{C}_\rho$ implies

$$t_{<n}(\rho(gh)) = t_{<n}(\rho(g) \circ \rho(h)) = t_{<n}(\rho(g)) \circ t_{<n}(\rho(h)).$$

In particular, $t_{<n}\rho(g)$ is (the class of) a homotopy equivalence with homotopy inverse $t_{<n}\rho(g^{-1})$. The representation ρ determines thus a group homomorphism

$$\rho_{<n} = t_{<n}\rho : G \longrightarrow G[K_{<n}].$$

(A group homomorphism into a group of homotopy classes of self-homotopy equivalences of a space is called a *homotopy action*.) Using the result of [Coo78], where an obstruction theory for finding equivalent topological actions for given homotopy actions has been given, we derive:

Proposition 1.85. *Let $\rho : G \to \mathrm{Homeo}(K)$ be an n-compression rigid representation. If G has an Eilenberg–MacLane space $K(G,1)$ of dimension at most 2, for example if G is free, then there exists a homotopy equivalence $K_{<n} \simeq K'_{<n}$, inducing an isomorphism $G[K_{<n}] \cong G[K'_{<n}]$, and a lift $\tilde{\rho}_{<n} : G \to \mathrm{Homeo}(K'_{<n})$ such that*

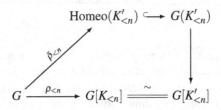

commutes.

Proof. The space $K_{<n}$ is a CW-complex. Thus the corollary to [Coo78, Theorem 1.1] applies and asserts that $\rho_{<n}$ is equivalent to a topological action. This means that there exists a homotopy equivalence $h : K_{<n} \to K'_{<n}$ with homotopy inverse $h' : K'_{<n} \to K_{<n}$ and a topological action $\tilde{\rho}_{<n} : G \to \mathrm{Homeo}(K'_{<n})$ such that

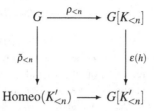

commutes, where $\varepsilon(h)[f] = [hfh']$ is conjugation by the homotopy equivalence. The map $\varepsilon(h)$ is a homomorphism as

$$\varepsilon(h)[fg] = [hfgh'] = [hfh'hgh'] = [hfh'] \circ [hgh'] = \varepsilon(h)[f] \circ \varepsilon(h)[g].$$

It is an isomorphism with inverse $\varepsilon(h') : G[K'_{<n}] \to G[K_{<n}]$, $\varepsilon(h')[f] = [h'fh]$. □

Examples 1.86. Here are some examples of groups G that have a $K(G,1)$ of dimension at most 2: Free groups were already mentioned. If G is the fundamental group of a connected closed surface Σ other than the sphere or the projective plane, then Σ itself is a two-dimensional $K(G,1)$. These surface groups are one-relator groups. More generally, a theorem of Lyndon asserts that any one-relator group G whose relator r is not a proper power $r = x^n, n \geq 2$, has a two-dimensional $K(G,1)$.

1.11 Fiberwise Homology Truncation

We will describe fiberwise homology truncation for the following three situations:

(1) Mapping tori, that is, fiber bundles over a circle,
(2) Flat bundles over spaces whose fundamental group G has a $K(G,1)$ of dimension at most 2 (for example flat bundles over closed surfaces other than $\mathbb{R}P^2$), and
(3) Fiber bundles over a sphere S^m, $m \geq 2$, where the fiber is a finite interleaf CW-complex.

1.11.1 Mapping Tori

Let F be a topological space and $f : F \to F$ a homeomorphism. The *mapping torus* E_f of f is the quotient space

$$E_f = (F \times I)/\sim,$$

where $(x,1) \sim (f(x),0)$ are identified. The factor projection $F \times I \to I$ induces a map $p : E_f \to S^1 = I/(0 \sim 1)$. Let us recall a well-known fact.

Lemma 1.87. *The map p is a locally trivial fiber bundle projection.*

Proof. Let $q : I \to S^1$ be the quotient map and $t_0 = q(0) = q(1) \in S^1$. It suffices to find a local chart near the point t_0. Let U be a small open neighborhood of t_0 in S^1 so that $q^{-1}(U)$ has two connected components V_0 and V_1 homeomorphic to half-open intervals, where V_0 is an open neighborhood of 0 in I and V_1 is an open neighborhood of 1 in I. Set $U_i = q(V_i)$, $i = 0, 1$, so that $U = U_0 \cup U_1$ and $U_0 \cap U_1 = \{t_0\}$. By definition of the mapping torus, the preimage space $p^{-1}(U)$ sits in a pushout square

$$
\begin{array}{ccc}
\{t_0\} \times F & \xrightarrow{\text{incl} \times f} & U_0 \times F \\
{\scriptstyle \text{incl} \times \text{id}_F} \downarrow & & \downarrow {\scriptstyle j_0} \\
U_1 \times F & \xrightarrow{\ \ j_1\ \ } & p^{-1}(U).
\end{array}
$$

The product $U \times F$ sits in the pushout square

$$
\begin{array}{ccc}
\{t_0\} \times F & \xrightarrow{\text{incl} \times \text{id}_F} & U_0 \times F \\
{\scriptstyle \text{incl} \times \text{id}_F} \downarrow & & \downarrow {\scriptstyle i_0} \\
U_1 \times F & \xrightarrow{\ \ i_1\ \ } & U \times F.
\end{array}
$$

By the universal property of the pushout, the commutative diagram

$$
\begin{array}{ccccc}
U_0 \times F & \xleftarrow{\text{incl} \times \text{id}_F} & \{t_0\} \times F & \xrightarrow{\text{incl} \times \text{id}_F} & U_1 \times F \\
{\scriptstyle \text{id} \times f} \downarrow & & {\scriptstyle \text{id}_F} \downarrow & & \downarrow {\scriptstyle \text{id} \times \text{id}_F} \\
U_0 \times F & \xleftarrow{\text{incl} \times f} & \{t_0\} \times F & \xrightarrow{\text{incl} \times \text{id}_F} & U_1 \times F
\end{array}
$$

induces a unique continuous map $\alpha : U \times F \to p^{-1}(U)$ which evidently lies over U such that

$$
\begin{array}{ccccc}
U_0 \times F & \xrightarrow{\ i_0\ } & U \times F & \xleftarrow{\ i_1\ } & U_1 \times F \\
{\scriptstyle \text{id} \times f} \downarrow & & {\scriptstyle \alpha} \downarrow & & \downarrow {\scriptstyle \text{id} \times \text{id}_F} \\
U_0 \times F & \xrightarrow{\ j_0\ } & p^{-1}(U) & \xleftarrow{\ j_1\ } & U_1 \times F
\end{array}
$$

commutes. The commutative diagram

$$
\begin{array}{ccccc}
U_0 \times F & \xleftarrow{\text{incl} \times f} & \{t_0\} \times F & \xrightarrow{\text{incl} \times \text{id}_F} & U_1 \times F \\
\downarrow{\scriptstyle \text{id} \times f^{-1}} & & \downarrow{\scriptstyle \text{id}_F} & & \downarrow{\scriptstyle \text{id} \times \text{id}_F} \\
U_0 \times F & \xleftarrow{\text{incl} \times \text{id}_F} & \{t_0\} \times F & \xrightarrow{\text{incl} \times \text{id}_F} & U_1 \times F
\end{array}
$$

induces a unique continuous map $\beta : p^{-1}(U) \to U \times F$ which lies over U such that

$$
\begin{array}{ccccc}
U_0 \times F & \xrightarrow{j_0} & p^{-1}(U) & \xleftarrow{j_1} & U_1 \times F \\
\downarrow{\scriptstyle \text{id} \times f^{-1}} & & \downarrow{\scriptstyle \beta} & & \downarrow{\scriptstyle \text{id} \times \text{id}_F} \\
U_0 \times F & \xrightarrow{i_0} & U \times F & \xleftarrow{i_1} & U_1 \times F
\end{array}
$$

commutes. Since α and β are inverse to each other, β is a homeomorphism and thus a local chart for p over U. $\qquad\square$

Let $f : (F,Y) \to (F,Y)$ be an isomorphism in $\mathbf{CW}_{n \supset \partial}$. We shall explain how one can perform fiberwise homological truncation on the fiber bundle $p : E = E_f \to S^1$. The result is a fiber bundle $\mathrm{ft}_{<n}(p) : \mathrm{ft}_{<n}(E) \to S^1$ whose fiber is homotopy equivalent to the truncation $F_{<n} = t_{<n}(F,Y)$. (Note that f is not required to be compression rigid here.)

Applying the covariant assignment $t_{<n} : \mathbf{CW}_{n \supset \partial} \to \mathbf{HoCW}_{n-1}$ to f, we obtain a homotopy class $t_{<n}(f)$. Choose a representative $f_{<n} : F_{<n} \to F_{<n}$ for $t_{<n}(f)$. Then $f_{<n}$ is a homotopy equivalence by Proposition 1.56. (We cannot deduce this from functoriality, since we did not require f to be n-compression rigid.) A construction due to Cooke [Coo78] will serve us at this point: Let $F'_{<n}$ be the infinite mapping telescope of $f_{<n}$,

$$
F'_{<n} = (\mathbb{Z} \times I \times F_{<n})/(n,1,x) \sim (n+1,0,f_{<n}(x)).
$$

A homotopy equivalence $h : F_{<n} \to F'_{<n}$ is given by $h(x) = (0,0,x)$. The shift

$$
f'_{<n} : F'_{<n} \longrightarrow F'_{<n}, \quad f'_{<n}(n,t,x) = (n-1,t,x),
$$

is a homeomorphism and the diagram

$$
\begin{array}{ccc}
F_{<n} & \xrightarrow[\simeq]{f_{<n}} & F_{<n} \\
\downarrow{\scriptstyle h}{\scriptstyle \simeq} & & {\scriptstyle \simeq}\downarrow{\scriptstyle h} \\
F'_{<n} & \xrightarrow[\simeq]{f'_{<n}} & F'_{<n}
\end{array}
$$

homotopy commutes. Set

$$\mathrm{ft}_{<n}(E) = E_{f'_{<n}},$$

the mapping torus of $f'_{<n}$, and let $\mathrm{ft}_{<n}(p) : \mathrm{ft}_{<n} E \to S^1$ be the mapping torus projection. By Lemma 1.87, $\mathrm{ft}_{<n}(p)$ is a locally trivial fiber bundle projection. The fiber is $F'_{<n}$, which is homotopy equivalent to $F_{<n}$ via h.

1.11.2 Flat Bundles

Let B be a connected space. Any flat fiber bundle $p : E \to B$ with fiber F over B has the form $E = \widetilde{B} \times_\rho F$, where $\rho : \pi_1(B) \to \mathrm{Homeo}(F)$ is the holonomy representation and $\widetilde{B}$ is the universal cover of B. The projection p is induced by projecting to the first component, followed by the covering projection $\widetilde{B} \to B$. Suppose that $\pi_1(B)$ has an Eilenberg–MacLane space $K(\pi_1 B, 1)$ of dimension at most 2. (For instance, B a closed surface other than $\mathbb{R}P^2$.) Let (F, Y) be an object of $\mathbf{CW}_{n \supset \partial}$ and $\rho : \pi_1(B) \to \mathrm{Homeo}(F)$ an n-compression rigid representation with respect to Y. We shall explain how to associate to the flat bundle $p : E = \widetilde{B} \times_\rho F \to B$ a fiberwise truncation $\mathrm{ft}_{<n}(p) : \mathrm{ft}_{<n}(E) \to B$, which is again a flat fiber bundle and has a fiber homotopy equivalent to the truncation $F_{<n} = t_{<n}(F, Y)$.

By Proposition 1.85, there exists a homotopy equivalence $F_{<n} \to F'_{<n}$ and a lift $\tilde{\rho}_{<n} : \pi_1(B) \to \mathrm{Homeo}(F'_{<n})$ such that

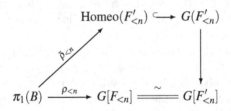

commutes. Set

$$\mathrm{ft}_{<n}(E) = \widetilde{B} \times_{\tilde{\rho}_{<n}} F'_{<n}$$

together with $\mathrm{ft}_{<n}(p) : \mathrm{ft}_{<n}(E) \to B$ induced as describe above, using the covering projection. Then $\mathrm{ft}_{<n}(p)$ is a flat fiber bundle with fiber $F'_{<n}$ homotopy equivalent to the truncation $F_{<n}$.

1.11.3 Remarks on Abstract Fiberwise Homology Truncation

As a thought experiment, an idealized, but motivational, abstract setup for fiberwise homology truncation might be formulated as follows. Let G be a topological group acting on a topological space F. Let $t_{<n}(F)$ be a spatial homological truncation of F.

Definition 1.88. An *abstract continuous homology truncation* for $(G, F, F_{<n})$ is a morphism $\tau_n : G \to G_n$ of topological groups together with an action of G_n on $F_{<n}$.

For example, if there existed a morphism of topological groups $\tau_n : \mathrm{Homeo}(F) \to \mathrm{Homeo}(F_{<n})$ truncating automorphisms of F in a continuous fashion, then one would obtain an abstract continuous homology truncation for $(\mathrm{Homeo}(F), F, F_{<n})$, taking the obvious action of $\mathrm{Homeo}(F_{<n})$ on $F_{<n}$.

Let $\xi_F = (E, p, B)$ be a numerable fiber bundle over B with fiber F and structure group G. Suppose an abstract continuous homology truncation for $(G, F, F_{<n})$ is given. The Milnor construction, among other such constructions, associates to G a numerable principal G-bundle $\omega_G = (EG, p_\omega, BG)$, with EG a free G-space weakly homotopy equivalent to a point. This bundle is universal in the sense that for each numerable principal G-bundle ξ, there exists a classifying map $f : B \to BG$ such that $\xi \cong f^*(\omega_G)$ as principal G-bundles. The morphism $\tau_n : G \to G_n$ induces a map $B\tau_n : BG \to BG_n$. Let ξ be the underlying numerable principal G-bundle of ξ_F. It is classified by a map $f : B \to BG$. Composition with $B\tau_n$ yields a map $f_n : B\tau_n \circ f : B \to BG_n$. Set $\xi_n = f_n^*(\omega_{G_n})$, a numerable principal G_n-bundle. Since G_n acts on $F_{<n}$, we obtain an associated fiber bundle $\mathrm{ft}_{<n}(\xi_F) = (\mathrm{ft}_{<n} E, \mathrm{ft}_{<n} p, B)$ with total space $\mathrm{ft}_{<n} E = E(\xi_n) \times_{G_n} F_{<n}$, fiber $F_{<n}$ and structure group G_n. We might call $\mathrm{ft}_{<n}(\xi_F)$ the *abstract fiberwise homology truncation* of ξ_F with respect to the given data.

As we have seen, however, it is in practice more realistic to take G_n to be a (grouplike) topological monoid. Moreover, the map $\tau_n : G \to G_n$ is usually not a monoid homomorphism, but only an H-map. For example, if a finite CW-complex F is an object of the interleaf category, then we have constructed an H-map $\tilde{t}_{<n} : \mathrm{Homeo}_{CW}(F) \longrightarrow G(t_{<n}F)$, see Theorem 1.78. In certain situations, the above general framework can be adapted to the monoid/H-map environment. We shall illustrate this in the case of a sphere as the base space.

1.11.4 Fiberwise Truncation Over Spheres

If G_n is the topological monoid of self-homotopy equivalences of a space, then the role of the Milnor construction will be played by Stasheff's classifying space BG_n. Given a space F, Stasheff [Sta63] associates to the monoid $H = G(F)$ a universal H-quasifibration

$$H \longrightarrow EH \xrightarrow{\ PH\ } BH.$$

The notion of a quasifibration was introduced by Dold and Thom in [DT58]. A continuous map $p : E \to B$ is a *quasifibration* if, for every point $b \in B$ and every $k \geq 0$, the induced map $p_* : \pi_k(E, p^{-1}(b)) \to \pi_k(B)$ is an isomorphism. The idea is that with respect to homotopy groups, quasifibrations should behave just like Hurewicz fibrations. In particular, the homotopy groups of each fiber $p^{-1}(b)$ fit into a long exact sequence

$$\cdots \to \pi_{k+1}(B) \longrightarrow \pi_k(p^{-1}(b)) \longrightarrow \pi_k(E) \xrightarrow{\ p_*\ } \pi_k(B) \longrightarrow \cdots.$$

The total space EH of the Stasheff quasifibration is aspherical, that is, $\pi_*(EH) = 0$. It follows from the long exact sequence that the homotopy boundary homomorphism induces an isomorphism

$$\pi_{k+1}(BH) \cong \pi_k(H). \tag{1.15}$$

Let $\xi_F = (E, p, S^m)$, $m \geq 2$, be a cellular topological fiber bundle over the m-sphere with fiber F. Assume that F is an object of the interleaf category and a finite CW-complex. Let n be a (positive) integer and

$$\phi : S^{m-1} \longrightarrow \text{Homeo}_{CW}(F) = G$$

be the clutching function for the bundle ξ_F. Set $G_n = H = G(t_{<n}F)$. In Section 1.10, we constructed an H-map

$$\tilde{i}_{<n} : G \longrightarrow G_n,$$

see Theorem 1.78. Composition yields a map

$$\psi = \tilde{i}_{<n} \circ \phi : S^{m-1} \longrightarrow G_n$$

and an element $[\psi] \in \pi_{m-1}(G_n)$. Under the above isomorphism (1.15),

$$\pi_m(BG_n) \cong \pi_{m-1}(G_n),$$

$[\psi]$ corresponds to a homotopy class $[\xi_n]$, where ξ_n is a map

$$\xi_n : S^m \longrightarrow BG_n.$$

Let $u : UE \to BG_n$ be Stasheff's universal fibration, a Hurewicz fibration that classifies Hurewicz fibrations with fibers of the homotopy type of $t_{<n}F$. Since $t_{<n}F$ is again a finite CW-complex by Remark 1.69, Stasheff's classification theorem applies and asserts that $[-, BG_n]$ and $L(t_{<n}F)(-)$ are naturally equivalent functors from the category of CW-complexes and homotopy classes of maps to the category of sets and functions, where $L(t_{<n}F)(X)$ is the set of fiber homotopy equivalence classes of Hurewicz fibrations with base space X and fibers of the homotopy type of $t_{<n}F$. The transformation $[-, BG_n] \to L(t_{<n}F)(-)$ is given by sending the homotopy class of a map $f : X \to BG_n$ to the pullback $f^*(u)$ of the universal fibration. Let $\text{ft}_{<n} \xi_F = (\text{ft}_{<n}(E), \text{ft}_{<n}(p), S^m)$ be the pullback Hurewicz fibration

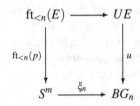

with fiber $F_{<n}$. This is the fiberwise truncation of ξ_F. Note that while we did start out with a bundle, we end up only with a fibration. This is to be expected, since G_n is not a group, only a monoid, and spatial homology truncation of a homeomorphism yields only a homotopy equivalence in general. However, whenever the base space of a Hurewicz fibration is a connected, locally finite polyhedron (such as in the present case), Fadell [Fad60] shows that the fibration can be replaced by a fiber homotopy equivalent fiber bundle. Thus, up to fiber homotopy equivalence, we end up with a bundle again.

1.12 Remarks on Perverse Links and Basic Sets

Let X^n be an even-dimensional PL stratified pseudomanifold that has no strata of odd dimension. In [MV86], the notion of a *perverse link* is introduced in order to obtain a more direct description of the category of (middle-)perverse sheaves on X. Let L be the link of a pure stratum S in X. A perverse link is a closed subspace $K \subset L$ such that for every perverse sheaf $\mathbf{P}^{\bullet}$ on $X - S$,

$$\mathcal{H}^k(K; \mathbf{P}^{\bullet}) = 0, \text{ for } k \geq -\tfrac{1}{2} \dim S, \text{ and}$$
$$\mathcal{H}^k(L, K; \mathbf{P}^{\bullet}) = 0, \text{ for } k < -\tfrac{1}{2} \dim S.$$

In a PL pseudomanifold such perverse links can always be constructed as certain simplicial subcomplexes. While perverse links thus provide some form of cohomological truncation, they cannot be used as a substitute for the spatial homology truncation machine built in Section 1.1, for the following reason: Let us consider the case of a space X having one isolated singular point c. Set $d = n/2$. On the complement $X - c$ of the singular point, the constant sheaf $\mathbf{P}^{\bullet} = \mathbb{R}_{X-c}[d]$ is a perverse sheaf in the indexing convention of [MV86]. Thus, the perverse link of the link of c satisfies

$$H^k(K) = 0, \text{ for } k \geq d, \text{ and } H^k(L, K) = 0 \text{ for } k < d.$$

The long exact sequence of the pair (L, K) shows that therefore

$$H^k(K) \cong H^k(L) \text{ for } k \leq d - 2.$$

For the missing degree $d - 1$, the sequence implies only an injection

$$H^{d-1}(L) \hookrightarrow H^{d-1}(K).$$

In the present case of an isolated singularity (or whenever the link happens to be a manifold), the perverse link is to be constructed as follows: Fix a triangulation T of L and let T' be the first barycentric subdivision of T. Let K be the union of all closed simplices in T', whose dimension is less than $(n-1)/2$. The following example shows that $H^{d-1}(K)$ can indeed be huge compared to $H^{d-1}(L)$, so that the above monomorphism is in general far from an isomorphism. Let $L = S^3$, the 3-sphere, so that $n = 4$, $d = 2$. Triangulate it for instance as the boundary of a standard 4-simplex. Then K is the union of all closed simplices of dimension at most 1 of the first barycentric subdivision. Thus K is a graph with a large number of cycles and the map $H^{d-1}(L) = H^1(S^3) = 0 \hookrightarrow H^1(K)$ is far away from being an isomorphism. Moreover, the example shows that the cohomology of the perverse link is not an invariant of the space L. Indeed, $H^{d-1}(K)$ depends on the triangulation of L: If we refine the triangulation of S^3 more and more, then the number of 1-cycles in K, and consequently the rank of $H^{d-1}(K)$, will increase beyond any bound and so the degree $(d-1)$-cohomology of K is in no way linked to the actual topology of S^3.

Chapter 2
Intersection Spaces

2.1 Reflective Algebra

For a given pseudomanifold, the homology of its intersection space is not isomorphic
to its intersection homology, but the two sets of groups are closely related. The reflective diagrams to be introduced in this section will be used to display the precise
relationship between the two theories in the isolated singularities case. This reflective nature of the relationship correlates with the fact that the two theories form a
mirror-pair for singular Calabi–Yau conifolds, see Section 3.8. Let R be a ring. If
M is an R-module, we will write M^* for the dual $\mathrm{Hom}(M, R)$. Let k be an integer.

Definition 2.1. Let H_*, H'_* and B_* be $\mathbb{Z}$-graded R-modules. Let A_- and A_+ be
R-modules. A *k-reflective diagram* is a commutative diagram of the form

$$(2.1)$$

M. Banagl, *Intersection Spaces, Spatial Homology Truncation, and String Theory*,
Lecture Notes in Mathematics 1997, DOI 10.1007/978-3-642-12589-8_2,
© Springer-Verlag Berlin Heidelberg 2010

containing the following exact sequences:

(1) $\cdots \to H'_{k+1} \to B_k \xrightarrow{\beta_-} A_- \xrightarrow{\alpha} A_+ \xrightarrow{\beta_+} B_{k-1} \to H'_{k-1} \to H_{k-1} \to \cdots$,

(2) $\cdots \to H'_{k+1} \to B_k \xrightarrow{\alpha_-\beta_-} H_k \xrightarrow{\alpha_+} A_+ \to 0$,

(3) $0 \to A_- \xrightarrow{\alpha_-} H_k \xrightarrow{\beta_+\alpha_+} B_{k-1} \to H'_{k-1} \to H_{k-1} \to \cdots$,

(4) $\cdots \to H'_{k+1} \to B_k \xrightarrow{\beta_-} A_- \xrightarrow{\alpha'_-} H'_k \to 0$,

(5) $0 \to H'_k \xrightarrow{\alpha'_+} A_+ \xrightarrow{\beta_+} B_{k-1} \to H'_{k-1} \to \cdots$.

The name derives from the obvious reflective symmetry of the diagram (2.1) across the vertical line through H_k and H'_k. The module H_* will eventually specialize to the reduced homology of the intersection space and H'_* will be intersection homology. The entire information of a reflective diagram may also be blown up into a braid diagram:

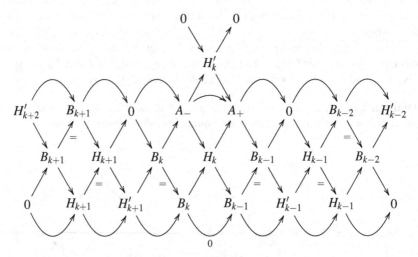

While a k-reflective diagram does not directly display the relation between H_k and H'_k, this relation can however be readily extracted from the diagram: Since

$$H_k/\operatorname{im}\alpha_- = H_k/\ker(\beta_+\alpha_+) \cong \operatorname{im}(\beta_+\alpha_+) = \operatorname{im}\beta_+ \quad \text{and} \quad \ker\alpha'_- = \operatorname{im}\beta_-,$$

we have the following T-diagram of two short exact sequences:

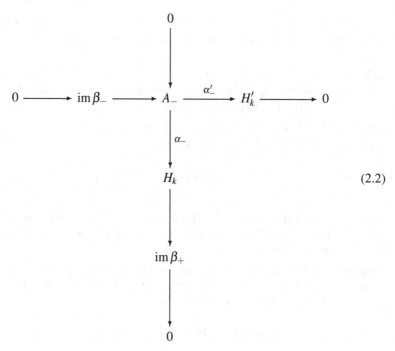

When R is a field, we can pick splittings and obtain a direct sum decomposition

$$H_k \cong \operatorname{im}\beta_- \oplus H'_k \oplus \operatorname{im}\beta_+.$$

Let l be an integer.

Definition 2.2. A *morphism* from a k-reflective to an l-reflective diagram is a commutative diagram of R-modules

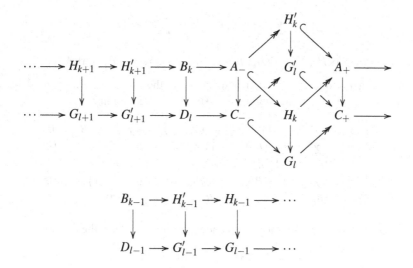

Reflective diagrams form a category, since the composition of two morphisms, defined by composing all the vertical arrows, is again a morphism of reflective diagrams.

Definition 2.3. A pair (H_*, H'_*) of $\mathbb{Z}$-graded modules is called *k-reflective across a $\mathbb{Z}$-graded module B_** if there exist modules A_- and A_+ such that the data $H_*, H'_*, B_*, A_\pm$ fits into a k-reflective diagram (2.1).

Definition 2.4. The *k-truncated Euler characteristic* $\chi_{<k}(B_*)$ of a finitely generated $\mathbb{Z}$-graded abelian group B_* is defined to be

$$\chi_{<k}(B_*) = \sum_{i<k} (-1)^i \operatorname{rk} B_i.$$

A reflective diagram for a pair (H_*, H'_*) implies in particular a relation between the Euler characteristics of H_* and H'_*, as well as a relation between the ranks of H_k and H'_k in the cut-off degree k.

Proposition 2.5. *The Euler characteristics of a k-reflective pair (H_*, H'_*) of finitely generated $\mathbb{Z}$-graded abelian groups fitting into a k-reflective diagram (2.1) with B_*, A_-, A_+ finitely generated obey the relation*

$$\chi(H_*) - \chi(H'_*) = \chi(B_*) - 2\chi_{<k}(B_*).$$

Furthermore, the identity

$$\operatorname{rk} H_k + \operatorname{rk} H'_k = \operatorname{rk} A_- + \operatorname{rk} A_+$$

holds in degree k.

Proof. Putting

$$\chi_{>k} = \sum_{i>k} (-1)^i \operatorname{rk} H_i, \ \chi_{<k} = \sum_{i<k} (-1)^i \operatorname{rk} H_i,$$

$$\chi'_{>k} = \sum_{i>k} (-1)^i \operatorname{rk} H'_i, \ \chi'_{<k} = \sum_{i<k} (-1)^i \operatorname{rk} H'_i,$$

$$h_k = \operatorname{rk} H_k, \ h'_k = \operatorname{rk} H'_k, \ b_k = \operatorname{rk} B_k, \ a_- = \operatorname{rk} A_-, \ a_+ = \operatorname{rk} A_+,$$

the five exact sequences (1)–(5) associated to the reflective diagram (2.1) in Definition 2.1 give the following linear system of five equations:

(1) $\chi_{>k} - \chi'_{>k} + (-1)^k a_- - (-1)^k a_+ + \chi'_{<k} - \chi_{<k} - \chi(B_*) = 0,$
(2) $\chi_{>k} - \chi'_{>k} - \chi_{>k}(B_*) - (-1)^k b_k + (-1)^k h_k - (-1)^k a_+ = 0,$
(3) $(-1)^k a_- - (-1)^k h_k - \chi_{<k}(B_*) + \chi'_{<k} - \chi_{<k} = 0,$
(4) $\chi_{>k} - \chi'_{>k} - \chi_{>k}(B_*) - (-1)^k b_k + (-1)^k a_- - (-1)^k h'_k = 0,$
(5) $(-1)^k h'_k - (-1)^k a_+ - \chi_{<k}(B_*) + \chi'_{<k} - \chi_{<k} = 0.$

These equations are not linearly independent because we have the relations

$$(2) + (3) = (1) = (4) + (5).$$

Thus equation (1) is redundant and one of the other four can be expressed in terms of the remaining three equations. The difference $(2) - (4)$ yields the equation

$$(6) \qquad\qquad h_k + h_k' - a_- - a_+ = 0.$$

The system $(1) - (5)$ is equivalent to the system (2), (3), (6). The latter three equations are linearly independent, since (3) and (6) are independent as (3) contains variables such as $\chi_{<k}$ that are absent from (6), and (2) is not in the span of $\{(3),(6)\}$, since (2) contains variables such as $\chi_{>k}$ that are absent from both (3) and (6). Using (2) and (5) (for example), we derive the formula for the difference of the Euler characteristics of H_* and H_*' as follows:

$$
\begin{aligned}
\chi(H_*) - \chi(H_*') &= (\chi_{>k} + (-1)^k h_k + \chi_{<k}) - (\chi_{>k}' + (-1)^k h_k' + \chi_{<k}') \\
&= (\chi_{>k} - \chi_{>k}') + (-1)^k h_k - (-1)^k h_k' + (\chi_{<k} - \chi_{<k}') \\
&= (\chi_{>k}(B_*) + (-1)^k b_k - (-1)^k h_k + (-1)^k a_+) + (-1)^k h_k \\
&\quad - (-1)^k h_k' + ((-1)^k h_k' - (-1)^k a_+ - \chi_{<k}(B_*)) \\
&= \chi_{>k}(B_*) + (-1)^k b_k - \chi_{<k}(B_*) \\
&= \chi(B_*) - 2\chi_{<k}(B_*).
\end{aligned}
$$

$\square$

We shall proceed to discuss duality for reflective diagrams over a field $\Bbbk = R$. Let Δ be the diagram (2.1).

Definition 2.6. The dual Δ^* of Δ is the k-reflective diagram

obtained by applying $\mathrm{Hom}(-, \Bbbk)$ to Δ.

Under this notion of duality, sequence (1) in Definition 2.1 is self-dual, sequences (2) and (3) are dual to each other, and sequences (4) and (5) are dual to each other.

Definition 2.7. Let (H_*, H_*') be k-reflective across B_* with reflective diagram Δ_H and let (G_*, G_*') be $(n-k)$-reflective across D_* with reflective diagram Δ_G. Then (H_*, H_*') and (G_*, G_*') are called *n-dual reflective pairs* if Δ_H and Δ_G are related by a duality isomorphism $\Delta_H^* \cong \Delta_G$.

2.2   The Intersection Space in the Isolated Singularities Case

Let $\bar{p}$ be a perversity. The intersection space of a stratified pseudomanifold M with one stratum is by definition $I^{\bar{p}}M = M$. (Such a space is a manifold, but a manifold is not necessarily a one-stratum space.) Let X be an n-dimensional compact oriented CAT pseudomanifold with isolated singularities $x_1, \ldots, x_w$, $w \geq 1$, and simply connected links $L_i = \mathrm{Link}(x_i)$, where CAT is PL or DIFF or TOP. (Pseudomanifolds whose links are all simply connected are sometimes called *supernormal* in the literature, see [CW91].) Thus X has two strata: the bottom pure stratum is $\{x_1, \ldots, x_w\}$ and the top stratum is the complement. By a DIFF pseudomanifold we mean a Whitney stratified pseudomanifold. By a TOP pseudomanifold we mean a topological stratified pseudomanifold as defined in [GM83]. In the present isolated singularities situation, this means that the L_i are closed topological manifolds and a small neighborhood of x_i is homeomorphic to the open cone on L_i. If CAT=TOP, assume for the moment $n \neq 5$. We shall define the *perversity $\bar{p}$ intersection space* $I^{\bar{p}}X$ for X.

Lemma 2.8. *Every link L_i, $i = 1, \ldots, w$, can be given the structure of a CW-complex.*

Proof. We begin with the case CAT=PL. Every link is then a closed PL manifold, which can be triangulated. The triangulation defines the CW-structure. For the case CAT=DIFF, i.e. the Whitney stratified case, we observe that links in Whitney stratified sets are again canonically Whitney stratified by intersecting with the strata of X. Since the links are contained in the top stratum, they are thus smooth manifolds. By the triangulation theorem of J. H. C. Whitehead, the link can then be smoothly triangulated. Again, the triangulation defines the desired CW-structure. Lastly, suppose CAT=TOP. If $n \leq 1$, then X has no singularities. If $n = 2$, the links are finite disjoint unions of circles. By the simple connectivity assumption, such unions must be empty. If $n = 3$, then by simple connectivity every link is a 2-sphere, so again X would be nonsingular. (Simple connectivity is of course not essential here, as circles and surfaces are certainly CW-complexes.) If $n = 4$, then the links are closed topological 3-manifolds. Since they are simply connected, the links must be 3-spheres according to the Poincaré conjecture, proved by Perelman. The space X would be nonsingular. (Simply connectivity is once more not essential for the existence of a CW-structure on the links because we could appeal to Moise's theorem [Moi52], asserting that every compact 3-manifold can be triangulated.) If $n \geq 6$, the links are closed topological manifolds of dimension at least 5. In this dimension range, topological manifolds have CW-structures by [KS77, FQ90]. $\square$

Remark 2.9. The preceding lemma makes a statement that is more refined than necessary for constructing the intersection space. CW-structures arising from triangulations for example, while having the virtue of being regular, typically are very large and have lots of cells that are not closely tied to the global topology of the space. To form the intersection space, it is enough to know that every link is homotopy equivalent to a CW-complex. Using such an equivalence, one is free to choose smaller CW-structures, indeed minimal cell structures consistent with the homology, or to obtain a CW-structure when it is not known to exist on the given link per se. This latter situation arises in the case TOP and $n = 5$, not covered by the lemma.

In this case, the links L_i are simply connected closed topological 4-manifolds. It is at present not known whether such a manifold possesses a CW-structure. It is not possible to obtain such a structure from a handlebody because a closed topological 4-manifold admits a topological handle decomposition if and only if it is smoothable, since the attaching maps can always the smoothed by an isotopy. For example, Freedman's closed simply connected 4-manifold with intersection form E_8 does not admit a handle decomposition. However, such links L_i are homotopy equivalent to a cell complex with one 0-cell, a finite number of 2-cells and one 4-cell. In the case TOP and $n = 5$, after having removed small open cone neighborhoods of the singularities, we glue in the mapping cylinders of these homotopy equivalences and now have CW-complexes sitting on the "boundary." The intersection space can then be defined, following the recipe below, in all dimensions, even when CAT=TOP.

We shall now invoke the spatial homology truncation machine of Section 1.1. If $k = n - 1 - \bar{p}(n) \geq 3$, we can and do fix completions (L_i, Y_i) of L_i so that every (L_i, Y_i) is an object in $\mathbf{CW}_{k \supset \partial}$. If $k \leq 2$, no groups Y_i have to be chosen and we simply apply the low-degree truncation of Section 1.1.5. Applying the truncation $t_{<k} : \mathbf{CW}_{k \supset \partial} \to \mathbf{HoCW}_{k-1}$ as defined on page 50, we obtain a CW-complex $t_{<k}(L_i, Y_i) \in Ob\,\mathbf{HoCW}_{k-1}$. The natural transformation $\mathrm{emb}_k : t_{<k} \to t_{<\infty}$ of Theorem 1.41 gives homotopy classes of maps

$$f_i = \mathrm{emb}_k(L_i, Y_i) : t_{<k}(L_i, Y_i) \longrightarrow L_i$$

such that for $r < k$,

$$f_{i*} : H_r(t_{<k}(L_i, Y_i)) \cong H_r(L_i),$$

while $H_r(t_{<k}(L_i, Y_i)) = 0$ for $r \geq k$. Let M be the compact manifold with boundary obtained by removing from X open cone neighborhoods of the singularities $x_1, \ldots, x_w$. The boundary is the disjoint union of the links,

$$\partial M = \bigsqcup_{i=1}^{w} L_i.$$

Let

$$L_{<k} = \bigsqcup_{i=1}^{w} t_{<k}(L_i, Y_i)$$

and define a homotopy class $g : L_{<k} \longrightarrow M$ by composing

$$L_{<k} \xrightarrow{f} \partial M \longrightarrow M,$$

where $f = \bigsqcup_i f_i$. The intersection space will be the homotopy cofiber of g:

Definition 2.10. The *perversity $\bar{p}$ intersection space* $I^{\bar{p}}X$ of X is defined to be

$$\boxed{I^{\bar{p}}X = \mathrm{cone}(g) = M \cup_g \mathrm{cone}(L_{<k}).}$$

More precisely, $I^{\bar{p}}X$ is a homotopy type of a space. If g_1 and g_2 are both representatives of the class g, then $\text{cone}(g_1) \simeq \text{cone}(g_2)$ by the following proposition.

Proposition 2.11. *If*

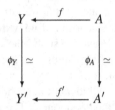

is a homotopy commutative diagram of continuous maps such that ϕ_Y and ϕ_A are homotopy equivalences, then there is a homotopy equivalence

$$Y \cup_f \text{cone} A \longrightarrow Y' \cup_{f'} \text{cone} A'$$

extending ϕ_Y.

This is Theorem 6.6 in [Hil65], where a proof can be found. The preceding construction of the intersection space $I^{\bar{p}}X$ depends on choices of cellular subgroups Y_i. If a link L_i is an object of the interleaf category **ICW**, then we may replace $t_{<k}(L_i, Y_i)$ in the construction by $t_{<k}L_i$, where $t_{<k} : \textbf{ICW} \rightarrow \textbf{HoCW}$ is the truncation functor of Section 1.9. The corresponding homotopy class f_i is to be replaced by the homotopy class $\text{emb}_k(L_i) : t_{<k}L_i \rightarrow L_i$ given by the natural transformation

$$\text{emb}_k : t_{<k} \longrightarrow t_{<\infty}$$

from Section 1.9. The construction of the intersection space thus becomes technically much simpler. The following theorem establishes generalized Poincaré duality for the rational reduced homology of intersection spaces and describes the relation to the intersection homology of Goresky and MacPherson.

Theorem 2.12. *Let X be an n-dimensional compact oriented supernormal singular CAT pseudomanifold with only isolated singularities. Let $\bar{p}$ and $\bar{q}$ be complementary perversities. Then:*

(1) The pair $(\widetilde{H}_*(I^{\bar{p}}X), IH_*^{\bar{p}}(X))$ is $(n-1-\bar{p}(n))$-reflective across the homology of the links, and

(2) $(\widetilde{H}_*(I^{\bar{p}}X;\mathbb{Q}), IH_*^{\bar{p}}(X;\mathbb{Q}))$ and $(\widetilde{H}_*(I^{\bar{q}}X;\mathbb{Q}), IH_*^{\bar{q}}(X;\mathbb{Q}))$ are n-dual reflective pairs.

Remark 2.13. Note that, as stated in the hypotheses, the theorem cannot formally be applied to a nonsingular X that is stratified with one stratum. The reason is simply that the reduced homology of a manifold $X = M$ does not possess Poincaré duality. If M is connected, then $\widetilde{H}_0(M) = 0$ but $\widetilde{H}_n(M) \cong \mathbb{Z}$ generated by the fundamental class.

We begin the proof of Theorem 2.12:

Proof. We prove statement (1) first. Put $L = \partial M$ and let $j : L \hookrightarrow M$ be the inclusion of the boundary. We will study the braid of the triple

Using the fact that f_* is an isomorphism in degrees less than k, as well as $H_r(L_{<k})=0$ for $r \geq k$, the braid becomes

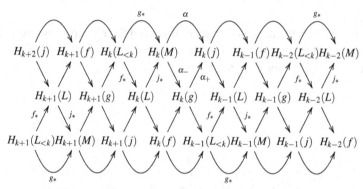

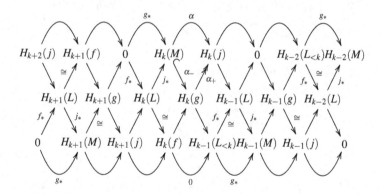

Since

$$H_*(g) = \widetilde{H}_*(\text{cone}(g)) = \widetilde{H}_*(I^{\bar{p}}X)$$

and

$$IH_r^{\bar{p}}(X) = \begin{cases} H_r(M,L) = H_r(j), & r > k \\ H_r(M), & r < k, \end{cases}$$

this can be rewritten as

$$IH^{\bar{p}}_{k+2}(X)\ H_{k+1}(f)\quad 0\qquad H_k(M)\quad H_k(j)\qquad 0\quad H_{k-2}(L_{<k})IH^{\bar{p}}_{k-2}(X)$$

$$H_{k+1}(L)\ \tilde{H}_{k+1}(I^{\bar{p}}X)\ H_k(L)\quad \tilde{H}_k(I^{\bar{p}}X)\quad H_{k-1}(L)\ \tilde{H}_{k-1}(I^{\bar{p}}X)\ H_{k-2}(L) \qquad (2.3)$$

$$0\qquad H_{k+1}(M)\ IH^{\bar{p}}_{k+1}(X)\quad H_k(f)\quad H_{k-1}(L_{<k})IH^{\bar{p}}_{k-1}(X)\ H_{k-1}(j)\qquad 0$$

By composing with the indicated isomorphisms and their inverses, we may replace $H_r(f)$ by $H_r(L)$ for $r \geq k$, $H_r(L_{<k})$ by $H_r(L)$ for $r < k$, $H_r(M)$ by $\tilde{H}_r(I^{\bar{p}}X)$ for $r > k$, and $H_r(j)$ by $\tilde{H}_r(I^{\bar{p}}X)$ for $r < k$ to obtain

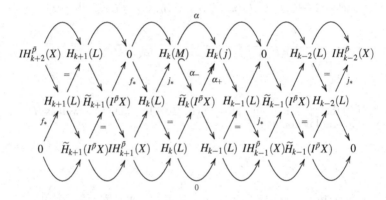

Finally, $IH^{\bar{p}}_k(X) = \operatorname{im} \alpha$, and we arrive at

where α'_- is given by regarding α as a map onto its image and α'_+ is the inclusion of $\operatorname{im}\alpha$ into $H_k(j)$. This braid contains the desired k-reflective diagram and all the required exact sequences.

For the remainder of the proof we will work with rational coefficients. To prove statement (2), we shall first construct duality isomorphisms

$$d : \widetilde{H}_r(I^{\bar{p}}X)^* \xrightarrow{\cong} \widetilde{H}_{n-r}(I^{\bar{q}}X).$$

There are three cases to consider: $r > k$, $r = k$, and $r < k$. For $r > k$, braid (2.3) contains the isomorphisms

$$H_r(M) \xrightarrow{\cong} \widetilde{H}_r(I^{\bar{p}}X).$$

For $I^{\bar{q}}X$, the cut-off degree k' is given by $k' = n - 1 - \bar{q}(n) = n - k$. Since $n - r < k'$, we have isomorphisms

$$\widetilde{H}_{n-r}(I^{\bar{q}}X) \xrightarrow{\cong} H_{n-r}(j)$$

by the braid of the $(n-k)$-reflective pair $(\widetilde{H}_*(I^{\bar{q}}X), IH_*^{\bar{q}}(X))$ analogous to braid (2.3). Using the Poincaré duality isomorphism $H_r(M)^* \cong H_{n-r}(j)$, we define d to be the unique isomorphism such that

$$
\begin{array}{ccc}
\widetilde{H}_r(I^{\bar{p}}X)^* & \xrightarrow{\cong} & H_r(M)^* \\
{\scriptstyle d}\downarrow{\scriptstyle \cong} & & {\scriptstyle PD}\downarrow{\scriptstyle \cong} \\
\widetilde{H}_{n-r}(I^{\bar{q}}X) & \xrightarrow{\cong} & H_{n-r}(j)
\end{array}
$$

commutes. Then

$$
\begin{array}{ccccc}
IH_r^{\bar{p}}(X)^* & \longrightarrow & \widetilde{H}_r(I^{\bar{p}}X)^* & \longrightarrow & H_r(L)^* \\
{\scriptstyle GMD}\downarrow{\scriptstyle \cong} & & {\scriptstyle d}\downarrow{\scriptstyle \cong} & & {\scriptstyle PD}\downarrow{\scriptstyle \cong} \\
IH_{n-r}^{\bar{q}}(X) & \longrightarrow & \widetilde{H}_{n-r}(I^{\bar{q}}X) & \twoheadrightarrow & H_{n-r-1}(L)
\end{array}
$$

commutes, where GMD denotes Goresky–MacPherson duality on intersection homology. Indeed, via the universal coefficient isomorphism (which is natural), this diagram is isomorphic to

$$
\begin{array}{ccccc}
H^r(M, \partial M) & \longrightarrow & H^r(M) & \xrightarrow{j^*} & H^r(\partial M) \\
{\scriptstyle -\cap[M,\partial M]}\downarrow{\scriptstyle \cong} & & {\scriptstyle -\cap[M,\partial M]}\downarrow{\scriptstyle \cong} & & {\scriptstyle -\cap[\partial M]}\downarrow{\scriptstyle \cong} \\
H_{n-r}(M) & \longrightarrow & H_{n-r}(M, \partial M) & \xrightarrow{\partial_*} & H_{n-r-1}(\partial M).
\end{array}
$$

It commutes on the nose, not only up to sign, because

$$\partial_*(\xi \cap [M,\partial M]) = j^*\xi \cap \partial_*[M,\partial M] = j^*\xi \cap [\partial M],$$

see [Spa66], Chapter 5, Section 6, 20, page 255. (Recall that we are using Spanier's sign conventions.) For $r < k$, we proceed by "reflecting the construction of the previous case." That is, using the isomorphisms

$$\widetilde{H}_r(I^{\bar{p}}X) \xrightarrow{\cong} H_r(j), \ H_{n-r}(M) \xrightarrow{\cong} \widetilde{H}_{n-r}(I^{\bar{q}}X), \ PD: H_r(j)^* \cong H_{n-r}(M),$$

we define d to be the unique isomorphism such that

$$
\begin{array}{ccc}
H_r(j)^* & \xrightarrow{\cong} & \widetilde{H}_r(I^{\bar{p}}X)^* \\
{\scriptstyle PD}\downarrow{\scriptstyle\cong} & & {\scriptstyle d}\downarrow{\scriptstyle\cong} \\
H_{n-r}(M) & \xrightarrow{\cong} & \widetilde{H}_{n-r}(I^{\bar{q}}X)
\end{array}
$$

commutes. It follows that

$$
\begin{array}{ccccc}
H_{r-1}(L)^* & \longrightarrow & \widetilde{H}_r(I^{\bar{p}}X)^* & \longrightarrow & IH_r^{\bar{p}}(X)^* \\
{\scriptstyle PD}\downarrow{\scriptstyle\cong} & & {\scriptstyle d}\downarrow{\scriptstyle\cong} & & {\scriptstyle GMD}\downarrow{\scriptstyle\cong} \\
H_{n-r}(L) & \longrightarrow & \widetilde{H}_{n-r}(I^{\bar{q}}X) & \longrightarrow & IH_{n-r}^{\bar{q}}(X)
\end{array}
$$

commutes as well. The remaining case $r = k$ is perhaps the most interesting one. Let

be the $(n - k)$-reflective diagram for the pair $(\widetilde{H}_*(I^{\bar{q}}X), IH_*^{\bar{q}}(X))$. The dual of the k-reflective diagram for $(\widetilde{H}_*(I^{\bar{p}}X), IH_*^{\bar{p}}(X))$ near k is

(2.4)

The following Poincaré duality isomorphisms will play a role in the construction of d:

$$d_M : H_k(M)^* \xrightarrow{\cong} H_{n-k}(j),$$
$$d'_M : H_k(j)^* \xrightarrow{\cong} H_{n-k}(M),$$
$$d_L : H_k(L)^* \xrightarrow{\cong} H_{n-k-1}(L).$$

Since the square

$$
\begin{array}{ccc}
H_k(M)^* & \xrightarrow{\beta^*_-} & H_k(L)^* \\
{\scriptstyle d_M} \downarrow {\scriptstyle \cong} & & {\scriptstyle d_L} \downarrow {\scriptstyle \cong} \\
H_{n-k}(j) & \xrightarrow{\delta_+} & H_{n-k-1}(L)
\end{array}
$$

commutes, d_L restricts to an isomorphism

$$d_L : \operatorname{im}\beta^*_- \xrightarrow{\cong} \operatorname{im}\delta_+.$$

Pick any splitting

$$s_{p\beta} : \operatorname{im}\beta^*_- \longrightarrow H_k(M)^*$$

for the surjection $\beta^*_- : H_k(M)^* \twoheadrightarrow \operatorname{im}\beta^*_-$. Set

$$s_{q\delta} = d_M s_{p\beta} d_L^{-1} : \operatorname{im}\delta_+ \longrightarrow H_{n-k}(j).$$

Then $s_{q\delta}$ splits $\delta_+ : H_{n-k}(j) \twoheadrightarrow \operatorname{im}\delta_+$ because

$$\delta_+ s_{q\delta} = \delta_+ d_M s_{p\beta} d_L^{-1} = d_L \beta^*_- s_{p\beta} d_L^{-1} = \operatorname{id}.$$

Pick any splitting

$$s_{p\alpha} : H_k(M)^* \longrightarrow \widetilde{H}_k(I^{\bar{p}}X)^*$$

for the surjection $\alpha^*_- : \widetilde{H}_k(I^{\bar{p}}X)^* \twoheadrightarrow H_k(M)^*$ and any splitting

$$s_{q\gamma} : H_{n-k}(j) \longrightarrow \widetilde{H}_{n-k}(I^{\bar{q}}X)$$

for the surjection $\gamma_+ : \widetilde{H}_{n-k}(I^{\bar{q}}X) \twoheadrightarrow H_{n-k}(j)$. The composition

$$s_p = s_{p\alpha} s_{p\beta} : \operatorname{im}\beta^*_- \longrightarrow \widetilde{H}_k(I^{\bar{p}}X)^*$$

is a splitting for $\beta^*_- \alpha^*_- : \widetilde{H}_k(I^{\bar{p}}X)^* \twoheadrightarrow \operatorname{im}\beta^*_-$. Similarly, the composition

$$s_q = s_{q\gamma} s_{q\delta} : \operatorname{im}\delta_+ \longrightarrow \widetilde{H}_{n-k}(I^{\bar{q}}X)$$

is a splitting for $\delta_+ \gamma_+ : \widetilde{H}_{n-k}(I^{\bar{q}}X) \twoheadrightarrow \operatorname{im}\delta_+$. Next, choose a splitting

$$t_p : IH_k^{\bar{p}}(X)^* \longrightarrow H_k(j)^*$$

for $\alpha_+^{\prime*} : H_k(j)^* \twoheadrightarrow IH_k^{\bar{p}}(X)^*$. Since duals of reflective diagrams are again reflective, diagram (2.4) has an associated T-diagram of type (2.2):

$$
\begin{array}{ccccccc}
 & & & & 0 & & \\
 & & & & \downarrow & & \\
0 & \longrightarrow & \mathrm{im}\,\beta_+^* & \longrightarrow & H_k(j)^* & \overset{\alpha_+^{\prime*}}{\longrightarrow} & IH_k^{\bar{p}}(X)^* \longrightarrow 0 \\
 & & & & \downarrow{\scriptstyle \alpha_+^*} & & \\
 & & & & \widetilde{H}_k(I^{\bar{p}}X)^* & & \\
 & & & & \downarrow{\scriptstyle \beta_-^* \alpha_-^*} & & \\
 & & & & \mathrm{im}\,\beta_-^* & & \\
 & & & & \downarrow & & \\
 & & & & 0 & &
\end{array}
$$

Thus we obtain a decomposition

$$\widetilde{H}_k(I^{\bar{p}}X)^* = \alpha_+^*(\mathrm{im}\,\beta_+^*) \oplus \alpha_+^* t_p IH_k^{\bar{p}}(X)^* \oplus s_p(\mathrm{im}\,\beta_-^*)$$

and every $v \in \widetilde{H}_k(I^{\bar{p}}X)^*$ can be written uniquely as

$$v = \alpha_+^*(b_+ + t_p(h)) + s_p(b_-)$$

with $b_+ \in \mathrm{im}\,\beta_+^*$, $h \in IH_k^{\bar{p}}(X)^*$ and $b_- \in \mathrm{im}\,\beta_-^*$. Write $x = b_+ + t_p(h)$. Setting

$$d(v) = \gamma_- d_M'(x) + s_q d_L(b_-)$$

defines a map

$$d : \widetilde{H}_k(I^{\bar{p}}X)^* \longrightarrow \widetilde{H}_{n-k}(I^{\bar{q}}X).$$

We claim that d is an isomorphism: By construction, the square

$$
\begin{array}{ccc}
H_k(j)^* & \overset{\alpha_+^*}{\longrightarrow} & \widetilde{H}_k(I^{\bar{p}}X)^* \\
{\scriptstyle d_M'}\downarrow{\scriptstyle \cong} & & \downarrow{\scriptstyle d} \\
H_{n-k}(M) & \overset{\gamma_-}{\longrightarrow} & \widetilde{H}_{n-k}(I^{\bar{q}}X)
\end{array}
$$

commutes. The square

$$
\begin{array}{ccc}
\widetilde{H}_k(I^{\bar{p}}X)^* & \overset{\beta_-^* \alpha_-^*}{\longrightarrow} & \mathrm{im}\,\beta_-^* \\
{\scriptstyle d}\downarrow & & {\scriptstyle \cong}\downarrow{\scriptstyle d_L} \\
\widetilde{H}_{n-k}(I^{\bar{q}}X) & \overset{\delta_+ \gamma_+}{\longrightarrow} & \mathrm{im}\,\delta_+
\end{array}
$$

commutes also, since

$$
\begin{aligned}
d_L\beta_-^*\alpha_-^*(v) &= d_L\beta_-^*\alpha^*(x) + d_L\beta_-^*\alpha_-^*s_p(b_-)\\
&= d_L(b_-)\\
&= \delta_+\gamma_+\gamma_-d_M'(x) + \delta_+\gamma_+s_qd_L(b_-)\\
&= \delta_+\gamma_+d(v).
\end{aligned}
$$

Hence we have a morphism of short exact sequences

$$
\begin{array}{ccccccccc}
0 & \longrightarrow & H_k(j)^* & \xrightarrow{\alpha_+^*} & \widetilde{H}_k(I^{\bar p}X)^* & \xrightarrow{\beta_-^*\alpha_-^*} & \mathrm{im}\,\beta_-^* & \longrightarrow & 0\\
& & \cong\big\downarrow{\scriptstyle d_M'} & & \big\downarrow{\scriptstyle d} & & \cong\big\downarrow{\scriptstyle d_L} & &\\
0 & \longrightarrow & H_{n-k}(M) & \xrightarrow{\gamma_-} & \widetilde{H}_{n-k}(I^{\bar q}X) & \xrightarrow{\delta_+\gamma_+} & \mathrm{im}\,\delta_+ & \longrightarrow & 0
\end{array}
$$

By the five-lemma, d is an isomorphism. It remains to be shown that the square

$$
\begin{array}{ccc}
\widetilde{H}_k(I^{\bar p}X)^* & \xrightarrow{\alpha_-^*} & H_k(M)^*\\
{\scriptstyle d}\big\downarrow{\scriptstyle \cong} & & {\scriptstyle \cong}\big\downarrow{\scriptstyle d_M}\\
\widetilde{H}_{n-k}(I^{\bar q}X) & \xrightarrow{\gamma_+} & H_{n-k}(j)
\end{array}
$$

commutes. This is established by the calculation

$$
\begin{aligned}
\gamma_+d(v) &= \gamma_+d_M'(x) + \gamma_+s_qd_L(b_-)\\
&= d_M\alpha^*(x) + \gamma_+s_{q\gamma}s_{q\delta}d_L(b_-)\\
&= d_M\alpha_-^*(\alpha_+^*(x)) + s_{q\delta}d_L(b_-)\\
&= d_M\alpha_-^*(\alpha_+^*(x)) + d_Ms_{p\beta}d_L^{-1}\circ d_L(b_-)\\
&= d_M\alpha_-^*(\alpha_+^*(x)) + d_Ms_{p\beta}(b_-)\\
&= d_M\alpha_-^*(\alpha_+^*(x)) + d_M(\alpha_-^*s_{p\alpha})s_{p\beta}(b_-)\\
&= d_M\alpha_-^*(\alpha_+^*(x)) + d_M\alpha_-^*s_p(b_-)\\
&= d_M\alpha_-^*(v).
\end{aligned}
$$

In summary, we have constructed the duality isomorphism

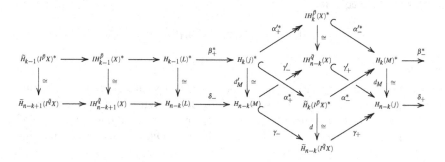

$$H_k(L)^* \longrightarrow IH^{\bar{p}}_{k+1}(X)^* \to \tilde{H}_{k+1}(I^{\bar{p}}X)^* \longrightarrow \cdots$$

$$d_L \Big\downarrow \cong \qquad \Big\downarrow \cong \qquad \Big\downarrow \cong$$

$$H_{n-k-1}(L) \to IH^{\bar{q}}_{n-k-1}(X) \twoheadrightarrow \tilde{H}_{n-k-1}(I^{\bar{q}}X) \longrightarrow \cdots$$

between the dual of the k-reflective diagram of the pair $(\tilde{H}_*(I^{\bar{p}}X), IH^{\bar{p}}_*(X))$ and the $(n-k)$-reflective diagram of the pair $(\tilde{H}_*(I^{\bar{q}}X), IH^{\bar{q}}_*(X))$. $\qquad \square$

Corollary 2.14. *If $n = \dim X$ is even, then the difference between the Euler characteristics of $\tilde{H}_*(I^{\bar{p}}X)$ and $IH^{\bar{p}}_*(X)$ is given by*

$$\chi(\tilde{H}_*(I^{\bar{p}}X)) - \chi(IH^{\bar{p}}_*(X)) = -2\chi_{<n-1-\bar{p}(n)}(L),$$

where L is the disjoint union of the links of all the isolated singularities of X. If $n = \dim X$ is odd, then

$$\chi(\tilde{H}_*(I^{\bar{n}}X)) - \chi(IH^{\bar{n}}_*(X)) = (-1)^{\frac{n-1}{2}} b_{(n-1)/2}(L),$$

where $b_{(n-1)/2}(L)$ is the middle dimensional Betti number of L and $\bar{n}$ is the upper middle perversity. Regardless of the parity of n, the identity

$$\mathrm{rk}\,\tilde{H}_k(I^{\bar{p}}X) + \mathrm{rk}\,IH^{\bar{p}}_k(X) = \mathrm{rk}\,H_k(M) + \mathrm{rk}\,H_k(M, L) \tag{2.5}$$

always holds in degree $k = n - 1 - \bar{p}(n)$, where M is the exterior of the singular set of X.

Proof. By Theorem 2.12, the pair $(H_*, H'_*) = (\tilde{H}_*(I^{\bar{p}}X), IH^{\bar{p}}_*(X))$ is $(n-1-\bar{p}(n))$-reflective across the homology of L. Therefore, Proposition 2.5 applies and we obtain

$$\chi(\tilde{H}_*(I^{\bar{p}}X)) - \chi(IH^{\bar{p}}_*(X)) = \chi(L) - 2\chi_{<n-1-\bar{p}(n)}(L).$$

If n is even, then L is an odd-dimensional closed oriented manifold and thus $\chi(L) = 0$ by Poincaré duality. If n is odd, then the cut-off value k for the upper middle perversity is $k = n - 1 - \bar{n}(n) = (n-1)/2$, the middle dimension of L. We have

$$\chi(L) = \chi_{<k}(L) + (-1)^k b_k(L) + \chi_{>k}(L) = 2\chi_{<k}(L) + (-1)^k b_k(L),$$

by Poincaré duality for L. Finally, as $A_- = H_k(M)$ and $A_+ = H_k(M, L)$, identity (2.5) follows from the equation

$$\mathrm{rk}\,H_k + \mathrm{rk}\,H'_k = \mathrm{rk}\,A_- + \mathrm{rk}\,A_+$$

of Proposition 2.5. $\qquad \square$

If a link of some singularity is not simply connected, so that the general construction of the intersection space as described above does not strictly apply, then one can in practice still often construct the intersection space provided one can find an ad hoc spatial homology truncation for this specific link. One then uses this truncation in place of the $t_{<k}L_i$ applied above; the rest of the construction remains the same. The simple connectivity assumption was adopted because our truncation machine required it (which in turn is due to the employment of the Hurewicz theorem). Inspection of the above proof on the other hand reveals that simple connectivity is nowhere necessary, only the existence of a spatial homology truncation of the link in the required dimension, dictated by the dimension of the pseudomanifold and the perversity. The following example illustrates this.

Example 2.15. Let us study Poincaré's own example of a three-dimensional space whose ordinary homology does not possess the duality that bears his name: $X^3 = \Sigma T^2$, the unreduced suspension of the 2-torus. This pseudomanifold has two singularities x_1, x_2, whose links are $L_1 = L_2 = T^2$, not simply connected. There are only two possible perversity functions to consider: $\bar{p}(3) = 0$ and $\bar{q}(3) = 1$. These two functions are complementary to each other.

Let us build the intersection space $I^{\bar{p}}X$ first. The cut-off value k is $k = n - 1 - \bar{p}(n) = 2$. We have spatial homology truncations

$$t_{<2}(L_1) = t_{<2}(L_2) = S^1 \vee S^1,$$

the 1-skeleton of T^2. The $\bar{p}$-intersection space is $I^{\bar{p}}X = \text{cone}(g)$, where g is the composition

$$L_{<2} = (S^1 \vee S^1) \times \{0,1\} \xrightarrow{\;\;f\;\;} L = T^2 \times \{0,1\}$$
$$g \searrow \qquad\qquad \downarrow j$$
$$M = T^2 \times I.$$

We shall proceed to work out its reduced homology. The braid utilized in the proof of Theorem 2.12 looks like this:

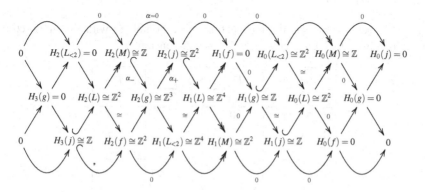

Therefore, the reduced homology of $I^{\bar{p}}X$,

$$\widetilde{H}_*(I^{\bar{p}}X) = H_*(g) = H_*(T^2 \times I, (S^1 \vee S^1) \times \{0,1\}),$$

is

$$\widetilde{H}_0(I^{\bar{p}}X) = 0,$$
$$\widetilde{H}_1(I^{\bar{p}}X) = \mathbb{Z}\langle \mathrm{pt} \times I \rangle,$$
$$\widetilde{H}_2(I^{\bar{p}}X) = \mathbb{Z}\langle T^2 \times \{\tfrac{1}{2}\}\rangle \oplus \mathbb{Z}\langle S^1 \times \mathrm{pt} \times I \rangle \oplus \mathbb{Z}\langle \mathrm{pt} \times S^1 \times I \rangle,$$
$$\widetilde{H}_3(I^{\bar{p}}X) = 0.$$

Let us now build the intersection space $I^{\bar{q}}X$. The cut-off value k is $k = n - 1 - \bar{q}(n) = 1$. The spatial homology truncations are

$$t_{<1}(L_1) = t_{<1}(L_2) = \mathrm{pt},$$

the 0-skeleton of T^2. The $\bar{q}$-intersection space is $I^{\bar{q}}X = \mathrm{cone}(g)$, where g is the composition

$$L_{<1} = \mathrm{pt} \times \{0,1\} \overset{f}{\hookrightarrow} L = T^2 \times \{0,1\}$$

with g to

$$M = T^2 \times I, \qquad j$$

Thus $I^{\bar{q}}X$ is obtained from a cylinder on the 2-torus by picking two points on it, one on each of the two boundary components, and then joining the two points by an arc outside of the cylinder. Its reduced homology

$$\widetilde{H}_*(I^{\bar{q}}X) = H_*(g) = H_*(T^2 \times I, \mathrm{pt} \times \{0,1\}),$$

can be determined from the long exact sequence of the pair and is given by

$$\widetilde{H}_0(I^{\bar{q}}X) = 0,$$
$$\widetilde{H}_1(I^{\bar{q}}X) = \mathbb{Z}\langle \mathrm{pt} \times I \rangle \oplus \mathbb{Z}\langle S^1 \times \mathrm{pt} \times \{\tfrac{1}{2}\}\rangle \oplus \mathbb{Z}\langle \mathrm{pt} \times S^1 \times \{\tfrac{1}{2}\}\rangle,$$
$$\widetilde{H}_2(I^{\bar{q}}X) = \mathbb{Z}\langle T^2 \times \{\tfrac{1}{2}\}\rangle,$$
$$\widetilde{H}_3(I^{\bar{q}}X) = 0.$$

The table below contrasts the intersection space homology with the intersection homology of X, listing the generators in each dimension.

r	$IH_r^{\bar{p}}(X)$	$IH_r^{\bar{q}}(X)$	$\widetilde{H}_r(I^{\bar{p}}X)$	$\widetilde{H}_r(I^{\bar{q}}X)$
0	pt	pt	0	0
1	$S^1 \times \mathrm{pt}$ $\mathrm{pt} \times S^1$	0	$\mathrm{pt} \times I$	$\mathrm{pt} \times I$ $S^1 \times \mathrm{pt}$ $\mathrm{pt} \times S^1$
2	0	$\Sigma(S^1 \times \mathrm{pt})$ $\Sigma(\mathrm{pt} \times S^1)$	$T^2 \times \{\tfrac{1}{2}\}$ $S^1 \times \mathrm{pt} \times I$ $\mathrm{pt} \times S^1 \times I$	$T^2 \times \{\tfrac{1}{2}\}$
3	$\Sigma(S^1 \times S^1)$	$\Sigma(S^1 \times S^1)$	0	0

The relative 2-cycle $S^1 \times \mathrm{pt} \times I$ in the $\bar{p}$-intersection space homology corresponds to the suspension $\Sigma(S^1 \times \mathrm{pt})$ in the $\bar{q}$-intersection homology, similarly $\mathrm{pt} \times S^1 \times I$ corresponds to $\Sigma(\mathrm{pt} \times S^1)$. In dimension 1, we have an analogous correspondence between the cycles $S^1 \times \mathrm{pt}, \mathrm{pt} \times S^1$. The fundamental class $\Sigma(S^1 \times S^1)$ is present in intersection homology but is not seen in the homology of the intersection spaces. This is a general phenomenon and explains why the duality holds for the reduced, not the absolute, homology. Except for this phenomenon, the homology of the intersection spaces sees more cycles than the intersection homology. The 2-cycle $T^2 \times \{\frac{1}{2}\}$, geometrically present in X, is recorded by both the homology of $I^{\bar{p}}X$ and $I^{\bar{q}}X$, but remains invisible to intersection homology, though an echo of it is the 3-cycle ΣT^2 in intersection homology. By the duality theorem, the 2-cycle $T^2 \times \{\frac{1}{2}\}$ must have a dual partner. Indeed, the intersection space homology automatically finds the geometrically dual partner as well: It is the suspension of a point, the relative cycle $\mathrm{pt} \times I$. The relative $\bar{p}$-cycle $S^1 \times \mathrm{pt} \times I$ is dual to the $\bar{q}$-cycle $\mathrm{pt} \times S^1$ and the relative $\bar{p}$-cycle $\mathrm{pt} \times S^1 \times I$ is dual to the $\bar{q}$-cycle $S^1 \times \mathrm{pt}$. In the table, one can also observe the reflective nature of the relationship between intersection homology and the homology of the intersection spaces. The example shows that in degrees other than $k = n - 1 - \bar{p}(n)$, the homology of $I^{\bar{p}}X$ need not contain a copy of intersection homology. (We shall return to this point in Section 3.7.) In degree k it always does, as the proof of the theorem shows.

Let us also illustrate Corollary 2.14, relating the Euler characteristics of $\tilde{H}_*(I^{\bar{p}}X)$ and $IH_*^{\bar{p}}(X)$, in the context of this example. In general, see also Proposition 2.5,

$$\chi(\tilde{H}_*(I^{\bar{p}}X)) - \chi(IH_*^{\bar{p}}(X)) = \chi(L) - 2\chi_{<n-1-\bar{p}(n)}(L).$$

We have $\chi(L) = \chi(T^2 \times \{0,1\}) = 0$ and, since $k = 2$ for perversity $\bar{p}$, $\chi_{<2}(L) = 2 - 4 = -2$, whence

$$\chi(\tilde{H}_*(I^{\bar{p}}X)) - \chi(IH_*^{\bar{p}}(X)) = 4.$$

Indeed, $\chi(\tilde{H}_*(I^{\bar{p}}X)) = 0 - 1 + 3 - 0 = 2$ and $\chi(IH_*^{\bar{p}}(X)) = 1 - 2 + 0 - 1 = -2$. Furthermore, since $a_- = \mathrm{rk}\, H_2(T^2 \times I) = 1$ and $a_+ = \mathrm{rk}\, H_2(T^2 \times I, \partial) = 2$, we have according to equation (2.5),

$$\mathrm{rk}\, \tilde{H}_2(I^{\bar{p}}X) + \mathrm{rk}\, IH_2^{\bar{p}}(X) = \mathrm{rk}\, H_2(T^2 \times I) + \mathrm{rk}\, H_2(T^2 \times I, \partial) = 1 + 2 = 3,$$

in concurrence with the ranks listed in the table. Since $\bar{q} = \bar{n}$ and the dimension $n = 3$ is odd, we have for $I^{\bar{q}}X$:

$$\chi(\tilde{H}_*(I^{\bar{q}}X)) - \chi(IH_*^{\bar{q}}(X)) = -\mathrm{rk}\, H_1(T^2 \times \{0,1\}) = -4,$$

consistent with $\chi(\tilde{H}_*(I^{\bar{q}}X)) = 0 - 3 + 1 - 0 = -2$ and $\chi(IH_*^{\bar{q}}(X)) = 1 - 0 + 2 - 1 = 2$. Formula (2.5) states that

$$\mathrm{rk}\, \tilde{H}_1(I^{\bar{q}}X) + \mathrm{rk}\, IH_1^{\bar{q}}(X) = \mathrm{rk}\, H_1(T^2 \times I) + \mathrm{rk}\, H_1(T^2 \times I, \partial) = 2 + 1 = 3,$$

again in agreement with the ranks listed in the table.

Example 2.16. (The intersection space construction applied to a manifold point.) The intersection space construction may in principle also be applied to a nonsingular, two-strata pseudomanifold. What happens when the construction is applied to a manifold point x? One must remove a small open neighborhood of x and gets a compact oriented manifold M with boundary $\partial M = S^{n-1}$. The open neighborhood of x is an open n-ball, that is, the open cone on the link S^{n-1}. For a perversity $\bar{p}$, the cut-off degree $k = n - 1 - \bar{p}(n)$ is at most equal to $n - 1$. Thus the spatial homology truncation is $t_{<k}S^{n-1} = $ pt. The fundamental class of the sphere is lost, no matter which $\bar{p}$ one takes. Thus $I^{\bar{p}}N$ is M together with a whisker attached to the 0-cell of the boundary sphere of M. This space is homotopy equivalent to M and to $N - \{x\}$. The reduced homology of M satisfies Poincaré duality since $\tilde{H}_n(M)$ is dual to $\tilde{H}_0(M)$ and $H_r(M) \to H_r(M, \partial M) = H_r(M, S^{n-1})$ is an isomorphism for $0 < r < n$.

Remark 2.17. There are two ways to truncate a chain complex C_* algebraically. The "good" truncation $\tau_{<k}C_*$ truncates the homology cleanly and corresponds to the spatial homology truncation as introduced in Chapter 1. The so-called "stupid" truncation $\sigma_{<k}C_*$, defined by $(\sigma_{<k}C_*)_i = C_i$ for $i < k$ and $(\sigma_{<k}C_*)_i = 0$ for $i \geq k$, does not truncate the homology cleanly. On spaces, the stupid truncation $\sigma_{<k}L$ of a CW-complex L would be $\sigma_{<k}L = L^{k-1}$, the $(k-1)$-skeleton of L, and is thus much easier to define and to handle than the good spatial truncation. In light of these advantages, one may wonder whether in the construction of the intersection space, one could replace the good spatial truncation $t_{<k}(L,Y)$ of the link L by the above stupid truncation $\sigma_{<k}L$ and still get a space that possesses generalized Poincaré duality. The following example will show that this is in fact not possible. Let X^n be the 4-sphere, thought of as a stratified space

$$X = S^4 = D^4 \cup_{S^3} D^4 = M^4 \cup_{L^3} \text{cone}(L^3),$$

where $M^4 = D^4$ and $L^3 = S^3$ is the link of the cone point, thought of as the bottom stratum. Suppose L is equipped with the CW-structure

$$L = e_1^0 \cup e_2^0 \cup e_1^1 \cup e_2^1 \cup e_1^2 \cup e_2^2 \cup e_1^3 \cup e_2^3,$$

so that the equatorial spheres $S^0 \subset S^1 \subset S^2 \subset L$ are all subcomplexes. Is cone(g), where g is the composition

$$\sigma_{<k}L = L^{k-1} \overset{f}{\hookrightarrow} L = \partial M$$
$$\underset{g}{\searrow} \qquad \downarrow j$$
$$M,$$

a viable candidate for an intersection space of X? Since $\tilde{H}_*(M) = \tilde{H}_*(D^4) = 0$, the exact sequence of the pair (M, L^{k-1}) shows

$$\tilde{H}_*(\text{cone}(g)) \cong \tilde{H}_{*-1}(L^{k-1}).$$

For the middle perversity, one would take $k = n/2 = 2$. Thus $\sigma_{<2}L = L^1 = S^1$ and the middle homology of $\mathrm{cone}(g)$,

$$\widetilde{H}_2(\mathrm{cone}(g)) \cong \widetilde{H}_1(S^1),$$

has rank one. If $\mathrm{cone}(g)$ had Poincaré duality, then the signature of the nondegenerate, symmetric intersection form on $\widetilde{H}_2(\mathrm{cone}(g))$ would have to be nonzero. (Zero signature would imply even rank.) But the signature of $X = S^4$ is zero. Thus $\widetilde{H}_*(\mathrm{cone}(g))$ is a meaningless theory, unrelated to the geometry of X. It is therefore necessary to choose a subgroup $Y \subset C_2(L) = \mathbb{Z}e_1^2 \oplus \mathbb{Z}e_2^2$ such that $(L, Y) \in Ob\mathbf{CW}_{2 \supset \partial}$ and apply the good spatial truncation $t_{<2}(L, Y)$, not the stupid truncation $\sigma_{<2}L$. (Using $\sigma_{<1}L$ or $\sigma_{<3}L$ does not yield self-dual homology groups either.) Any such Y arises as the image of a splitting $s : \mathrm{im}\,\partial_2 \to C_2(L)$ for $\partial_2 :$ $C_2(L) \twoheadrightarrow \mathrm{im}\,\partial_2 = \ker \partial_1 = \mathbb{Z}\langle e_1^1 - e_2^1 \rangle$. So we could for instance take $Y = \mathbb{Z}e_1^2$ or $Y = \mathbb{Z}e_2^2$ because $\partial_2(e_1^2) = e_1^1 - e_2^1 = \partial_2(e_2^2)$.

2.3 Independence of Choices of the Intersection Space Homology

The construction of the intersection spaces $I^{\bar{p}}X$ involves choices of subgroups $Y_i \subset C_k(L_i)$, where the L_i are the links of the singularities, such that (L_i, Y_i) is an object in $\mathbf{CW}_{k \supset \partial}$ with $k = n - 1 - \bar{p}(n)$, $n = \dim X$. Moreover, the chain complexes $C_*(L_i)$ depend on the CW-structures on the links and these structures are another element of choice. In this section we collect some results on the independence of these choices of the intersection space homology $\widetilde{H}_*(I^{\bar{p}}X)$.

Theorem 2.18. *Let X^n be a compact oriented pseudomanifold with isolated singularities and fixed, simply connected links L_i that can be equipped with CW-structures. Then*
(1) $\widetilde{H}_*(I^{\bar{p}}X; \mathbb{Q})$ is independent of the choices involved in the construction of the intersection space $I^{\bar{p}}X$,
(2) $\widetilde{H}_r(I^{\bar{p}}X; \mathbb{Z})$ is independent of choices for $r \neq n - 1 - \bar{p}(n)$, and
(3) $\widetilde{H}_k(I^{\bar{p}}X; \mathbb{Z})$, $k = n - 1 - \bar{p}(n)$, is independent of choices if either

$$\mathrm{Ext}\big(\mathrm{im}(H_k(M, L) \to H_{k-1}(L)), H_k(M)\big) = 0,$$

or

$$\mathrm{Ext}\big(H_k(M, L), \mathrm{im}(H_k(L) \to H_k(M))\big) = 0.$$

Proof. We shall first look at the integral homology groups. For $r > k$, the proof of Theorem 2.12 exhibits isomorphisms

$$H_r(M) \xrightarrow{\cong} \widetilde{H}_r(I^{\bar{p}}X).$$

Thus $\tilde{H}_r(I^{\bar{p}}X)$ is independent of the choices of Y_i for $r > k$. Similarly, the isomorphisms

$$\tilde{H}_r(I^{\bar{p}}X) \xrightarrow{\cong} H_r(j) = H_r(M,L)$$

for $r < k$ show that $\tilde{H}_r(I^{\bar{p}}X)$ does not depend on the choices of Y_i. This proves statement (2); it remains to investigate $r = k$. By Theorem 2.12, the pair $(\tilde{H}_*(I^{\bar{p}}X),$ $IH_*^{\bar{p}}(X))$ is k-reflective across the homology of the links. The associated reflective diagram near k is

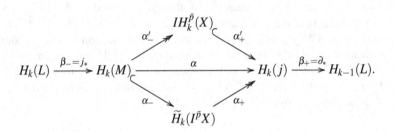

The sequence

$$0 \to H_k(M) \xrightarrow{\alpha_-} \tilde{H}_k(I^{\bar{p}}X) \xrightarrow{\beta_+\alpha_+} \operatorname{im}\beta_+ \to 0 \qquad (2.6)$$

is exact. If $\operatorname{Ext}(\operatorname{im}\beta_+, H_k(M)) = 0$, then the induced exact sequence

$$0 \to \operatorname{Hom}(\operatorname{im}\beta_+, H_k(M)) \longrightarrow \operatorname{Hom}(\tilde{H}_k(I^{\bar{p}}X), H_k(M)) \xrightarrow{-\circ\alpha_-} \operatorname{Hom}(H_k(M), H_k(M))$$
$$\longrightarrow \operatorname{Ext}(\operatorname{im}\beta_+, H_k(M)) = 0$$

shows that the sequence (2.6) splits. Thus

$$\tilde{H}_k(I^{\bar{p}}X) \cong H_k(M) \oplus \operatorname{im}\beta_+$$

is independent of the choice of Y_i. Similarly, if $\operatorname{Ext}(H_k(j), \operatorname{im}\beta_-) = 0$, then the exact sequence

$$0 \to \operatorname{im}\beta_- \longrightarrow \tilde{H}_k(I^{\bar{p}}X) \xrightarrow{\alpha_+} H_k(j) \to 0 \qquad (2.7)$$

splits and

$$\tilde{H}_k(I^{\bar{p}}X) \cong H_k(j) \oplus \operatorname{im}\beta_-$$

is again independent of the choice of Y_i. This establishes claim (3) of the theorem.

Finally, working with rational coefficients, the sequences (2.6) and (2.7) split without any assumption. This, together with

$$\tilde{H}_r(I^{\bar{p}}X;\mathbb{Q}) \cong \begin{cases} H_r(M;\mathbb{Q}), & r > k, \\ H_r(j;\mathbb{Q}), & r < k, \end{cases}$$

proves claim (1) of the theorem. $\qquad\qquad\qquad\qquad\qquad\qquad\qquad\qquad\qquad\square$

Remark 2.19. The assumption that the links be simply connected is adopted only to ensure the existence of $I^{\bar{p}}X$ and its omission does not invalidate the theorem, as the simple connectivity is not used during the proof. In practice, $I^{\bar{p}}X$ often exists even if the links are not simply connected, as illustrated by Example 2.15 above and Example 2.20 below.

Example 2.20. Let L_p be a three-dimensional lens space with fundamental group $\pi_1(L_p) \cong \mathbb{Z}/p$, $p \geq 2$. Let M^4 be the total space of a D^2-bundle over S^2 with $\partial M = L_p$. Let X^4 be the pseudomanifold obtained from M by coning off the boundary. Since this is a rational homology manifold, the ordinary rational homology of X enjoys Poincaré duality; nevertheless we shall investigate the intersection space $I^{\bar{m}}X$ of X. Here $k = n - 1 - \bar{m}(n) = 2$, so that we must determine a truncation $t_{<2}(L_p)$, as L_p is the link of the singularity. (Note that this is another example involving a non-simply connected link.) The standard cell structure for L_p is $L_p = e^0 \cup e^1 \cup_p e^2 \cup e^3$ with corresponding cellular chain complex

$$\mathbb{Z}e^3 \xrightarrow{0} \mathbb{Z}e^2 \xrightarrow{p} \mathbb{Z}e^1 \xrightarrow{0} \mathbb{Z}e^0.$$

Thus we may choose $t_{<2}(L_p) = e^0 \cup e^1 \cup_p e^2$, the 2-skeleton of L_p. Then $I^{\bar{m}}X = M/(S^1 \cup_p e^2)$, $S^1 \cup_p e^2 \hookrightarrow L_p = \partial M \hookrightarrow M$. The exact sequence of the pair (M, L_p),

$$H_2(L_p) \to H_2(M) \to H_2(M, L_p) \to H_1(L_p) \to H_1(M)$$

is

$$0 \to \mathbb{Z} \xrightarrow{p} \mathbb{Z} \to \mathbb{Z}/p \to 0,$$

whence

$$\mathrm{Ext}\big(\mathrm{im}(H_2(M, L_p) \to H_1(L_p)), H_2(M)\big) = \mathrm{Ext}(\mathbb{Z}/p, \mathbb{Z}) = \mathbb{Z}/p \neq 0,$$

but

$$\mathrm{Ext}\big(H_2(M, L_p), \mathrm{im}(H_2(L_p) \to H_2(M))\big) = \mathrm{Ext}(\mathbb{Z}, 0) = 0.$$

By Theorem 2.18, the integral homology $\tilde{H}_*(I^{\bar{m}}X; \mathbb{Z})$ is independent of choices.

Example 2.21. There are of course manifolds M with boundary $\partial M = L$ for which the hypothesis in (3) of Theorem 2.18 is not satisfied, that is, both Ext groups are nonzero. Consider for instance $M^9 = L_p \times S^4 \times D^2$, where L_p is a three-dimensional lens space with fundamental group $\pi_1(L_p) \cong \mathbb{Z}/p$, $p \geq 2$, and take $k = 3 = 8 - \bar{p}(9)$ for a perversity $\bar{p}$ with $\bar{p}(9) = 5$. The relevant homology groups are

$$H_2(\partial M) \cong \mathbb{Z}/p \langle \omega \times \mathrm{pt} \times [S^1] \rangle,$$
$$H_3(M) \cong \mathbb{Z} \langle [L_p] \times \mathrm{pt} \times \mathrm{pt} \rangle,$$
$$H_3(M, \partial M) \cong \mathbb{Z}/p \langle \omega \times \mathrm{pt} \times [D^2, \partial D^2] \rangle,$$

where $[-]$ denotes various fundamental classes and ω is the generating loop in L_p. The connecting homomorphism $\partial_* : H_3(M, \partial M) \to H_2(\partial M)$ is an isomorphism because it maps the generator $\omega \times \mathrm{pt} \times [D^2, \partial D^2]$ to

$$\partial_*(\omega \times \mathrm{pt} \times [D^2, \partial D^2]) = \omega \times \mathrm{pt} \times \partial_*[D^2, \partial D^2] = \omega \times \mathrm{pt} \times [\partial D^2],$$

which generates $H_2(\partial M)$. Thus the exact sequence of the pair $(M, \partial M)$ has the form

$$H_3(\partial M) \twoheadrightarrow H_3(M) \xrightarrow{0} H_3(M, \partial M) \xrightarrow{\cong} H_2(\partial M).$$

It follows that

$$\mathrm{Ext}(\mathrm{im}(H_3(M, \partial M) \to H_2(\partial M)), H_3(M)) = \mathrm{Ext}(\mathbb{Z}/p, \mathbb{Z}) = \mathbb{Z}/p \neq 0$$

and

$$\mathrm{Ext}(H_3(M, \partial M), \mathrm{im}(H_3(\partial M) \to H_3(M))) = \mathrm{Ext}(\mathbb{Z}/p, \mathbb{Z}) = \mathbb{Z}/p \neq 0.$$

As an application of Theorem 2.18, we shall see that even the integral homology of the (middle perversity) intersection space is well-defined independent of choices for large classes of isolated hypersurface singularities in complex algebraic varieties. Let $w_0 \leq w_1 \leq \cdots \leq w_n$ be a nondecreasing sequence of positive integers with $\gcd(w_0, \ldots, w_n) = 1$. We shall refer to the w_i as *weights*. Let $z_0, \ldots, z_n$ be complex variables. For each i, we assign the weight (or "degree") w_i to the variable z_i. This means that the *weighted degree* of a monomial $z_0^{u_0} z_1^{u_1} \cdots z_n^{u_n}$ is $w_0 u_0 + \cdots + w_n u_n$.

Definition 2.22. A polynomial $f \in \mathbb{C}[z_0, \ldots, z_n]$ is *weighted homogeneous* if

$$f(\lambda^{w_0} z_0, \ldots, \lambda^{w_n} z_n) = \lambda^d f(z_0, \ldots, z_n),$$

where d is the *weighted degree* of f.

For example, $f(z_0, z_1, z_2, z_3) = z_0^7 z_3 + z_1^4 + z_2^3 z_0 + z_3^2 z_1 + z_0^5 z_1 z_2$ is weighted homogeneous for weights $(w_0, w_1, w_2, w_3) = (5, 14, 17, 21)$ with weighted degree $d = 56$. If all weights are equal to one, then "weighted homogeneous" is synonymous with "homogeneous." We shall be specifically interested in threefolds, so we shall take $n = 3$.

Definition 2.23. A weight quadruple (w_0, w_1, w_2, w_3) is called *well-formed* if for any triple of distinct indices (i, j, k), $\gcd(w_i, w_j, w_k) = 1$ (see [BGN03]). We shall also refer to a polynomial $f \in \mathbb{C}[z_0, z_1, z_2, z_3]$ as well-formed if it is weighted homogeneous with respect to a well-formed weight quadruple.

The above example of a weighted homogeneous polynomial is well-formed in this sense.

Theorem 2.24. *Let X be a complex projective algebraic threefold with only isolated singularities. If all the singularities are hypersurface singularities that are weighted homogeneous and well-formed, then the integral homology $\widetilde{H}_*(I^{\bar{m}} X)$ is well-defined, independent of choices.*

Proof. Let x_i be one of the isolated hypersurface singularities of X. Since x_i is a hypersurface singularity, there exists a complex polynomial f_i in four variables z_0, z_1, z_2, z_3 such that an open neighborhood of x_i in X is homeomorphic to the intersection $V(f_i) \cap \text{int} D_\varepsilon^8$ of the hypersurface $V(f_i) = f_i^{-1}(0) \subset \mathbb{C}^4$ with an open ball $\text{int} D_\varepsilon^8 = \{z \in \mathbb{C}^4 \mid |z| < \varepsilon\}$ of suitably small radius $\varepsilon > 0$. Under the homeomorphism, x_i corresponds to the origin $0 \in V(f_i)$. The origin is the only singularity of $V(f_i)$. Set $S_\varepsilon^7 = \partial D_\varepsilon^8 = \{z \in \mathbb{C}^4 \mid |z| = \varepsilon\}$ and $L_i = V(f_i) \cap S_\varepsilon^7$. The space L_i has dimension

$$\dim L_i = \dim S_\varepsilon^7 + \dim_\mathbb{R} V(f_i) - \dim_\mathbb{R} \mathbb{C}^4 = 7 + 6 - 8 = 5.$$

For sufficiently small ε, L_i is a smooth manifold by [Mil68, Corollary 2.9]. Furthermore, [Mil68, Theorem 2.10] asserts that $V(f_i) \cap D_\varepsilon^8$ is homeomorphic to the (closed) cone on L_i. Thus an open neighborhood of x_i in X is homeomorphic to the (open) cone on L_i and thus L_i is the link of x_i in the sense of stratification theory. By [Mil68, Theorem 5.2], L_i is simply connected. By assumption, f_i is weighted homogeneous with well-formed weights. According to [BGN03, Proposition 7.1], see also [BG01, Lemma 5.8], this implies that $H_2(L_i; \mathbb{Z})$ is torsion-free. Since L_i is a compact manifold, $H_2(L_i; \mathbb{Z})$ is in addition finitely generated, hence free (abelian). Consequently,

$$H_2(L) = \bigoplus_{i=1}^{w} H_2(L_i)$$

is free, where w is the number of singularities of X. Thus the subgroup

$$\text{im}(H_3(M, L) \to H_2(L)) \subset H_2(L)$$

is also free and

$$\text{Ext}(\text{im}(H_3(M, L) \to H_2(L)), H_3(M)) = 0.$$

For the middle perversity $\bar{m}$, we have $\bar{m}(6) = 2$, so that $k = 3$. By the Independence-Theorem 2.18, $\widetilde{H}_*(I^{\bar{m}}X)$ is independent of choices. $\qquad\square$

If a link $L_i = V(f_i) \cap S_\varepsilon^7$ in the context of the above proof is in addition known to be spin, then Smale's classification [Sma62] of simply connected closed spin 5-manifolds implies that L_i is diffeomorphic to $S^5 \# m(S^2 \times S^3)$, since $H_2(L_i)$ is torsion-free. This geometric information allows us to work out the intersection space explicitly and to verify the independence of the intersection space homology rather directly. Indeed, if $m = 1$ so that the link is $S^2 \times S^3$, having the minimal CW-structure consistent with its homology, i.e. $S^2 \times S^3 = e^0 \cup e^2 \cup e^3 \cup e^5$, then the boundary operator $C_3(S^2 \times S^3) \to C_2(S^2 \times S^3)$ is zero. Therefore, $C_3 = Z_3$ (the cycle group) and $Y \subset C_3$ is forced to be zero. So there is in fact only one possible choice of Y.

Theorem 2.24 applies in particular to the case of nodal singularities: If a singular point x_i is a node, then the corresponding polynomial f_i is $z_0^2 + z_1^2 + z_2^2 + z_3^2$, which is homogeneous (and hence well-formed). In this case, the link is in fact $S^2 \times S^3$. The case of isolated nodal singularities is rather important in string theory. It arises there in the course of Calabi–Yau conifold transitions and will be discussed from this perspective in Chapter 3.

2.4 The Homotopy Type of Intersection Spaces for Interleaf Links

In the previous section we have seen that in general the rational homology of an intersection space of a given pseudomanifold is well-defined and that its integral homology is well-defined at least under certain homological assumptions on the exterior of the singular set. In the present section, we shall prove a stronger statement under stronger hypotheses on the links of the singular points: If the links lie in the interleaf category (Definition 1.62), then the homotopy type of the intersection space is well-defined.

Let A be a topological space and k a positive integer. Consider the following three properties for a map $f : K \to A$:

(T1) K is a simply connected CW-complex,
(T2) $f_* : H_r(K;\mathbb{Z}) \to H_r(A;\mathbb{Z})$ is an isomorphism for $r < k$ and
(T3) $H_r(K;\mathbb{Z}) = 0$ for $r \geq k$.

Lemma 2.25. *Let $f : K \to A$ be a map satisfying (T1)–(T3). If A is an object of the interleaf category, then so is K.*   •

Proof. By (T1), K is a simply connected CW-complex. By (T2) and (T3), the even-dimensional integral homology of K is finitely generated, since this is true for A. By Lemma 1.64, $H_{\text{even}}(A;\mathbb{Z})$ is torsion-free, hence free (abelian) because it is finitely generated. Thus, by (T2) and (T3), $H_{\text{even}}(K;\mathbb{Z})$ is finitely generated free (abelian). Since $H_{\text{odd}}(A;\mathbb{Z}) = 0$ implies, again by (T2) and (T3), that $H_{\text{odd}}(K;\mathbb{Z}) = 0$, we deduce by an application of the universal coefficient theorem that $H_{\text{odd}}(K;G) = 0$ for any coefficient group G.   □

Theorem 2.26. *Let X be an n-dimensional compact PL pseudomanifold with only isolated singularities $x_1,\ldots,x_w$ and links $L_i = \text{Link}(x_i)$, $i = 1,\ldots,w$. If all L_i, $i = 1,\ldots,w$, are objects of the interleaf category, then the homotopy type of the intersection space $I^{\bar{p}}X$ is well-defined independent of choices. More precisely: Let $k = n - 1 - \bar{p}(n)$. Given maps $f_i : (L_i)_{<k} \to L_i$, $i = 1,\ldots,w$, satisfying (T1)–(T3) and a second set of maps $\bar{f}_i : (L_i)_{<k} \to L_i$, $i = 1,\ldots,w$, satisfying (T1)–(T3) as well, there exists a homotopy equivalence*

$$I^{\bar{p}}X = \text{cone}(g) \simeq \text{cone}(\bar{g}) = \overline{I^{\bar{p}}X},$$

where g is the composition

$$\bigsqcup_i (L_i)_{<k} \xrightarrow{\sqcup f_i} \bigsqcup_i L_i = \partial M \xrightarrow{j} M$$

and $\bar{g}$ is the composition

$$\bigsqcup_i \overline{(L_i)}_{<k} \xrightarrow{\sqcup \bar{f}_i} \bigsqcup_i L_i = \partial M \xrightarrow{j} M.$$

Proof. Since L_i lies in **ICW**, there exists a homotopy equivalence $\varepsilon_i : L_i \to E(L_i)$, where $E(L_i)$ is a finite CW-complex that has only even-dimensional cells, see Proposition 1.68. Let $\varepsilon_i' : E(L_i) \to L_i$ be a homotopy inverse for ε_i. Set

$$L = \bigsqcup_i L_i, \ E(L) = \bigsqcup_i E(L_i),$$

$$\varepsilon = \bigsqcup_i \varepsilon_i : L \to E(L), \ \varepsilon' = \bigsqcup_i \varepsilon_i' : E(L) \to L.$$

Let $F : E(L)^{k-1} \to L$ be the restriction of ε' to the $(k-1)$-skeleton and let

$$G : E(L)^{k-1} \longrightarrow M$$

be the composition

$$E(L)^{k-1} \xrightarrow{F} L = \partial M \xhookrightarrow{j} M.$$

The space $\mathrm{cone}(G)$ will serve as a reference model for the perversity $\bar{p}$ intersection space of X. Indeed, we shall show that both $\mathrm{cone}(g)$ and $\mathrm{cone}(\bar{g})$ are homotopy equivalent to $\mathrm{cone}(G)$, hence they are in particular homotopy equivalent to each other.

Since $(L_i)_{<k}$ is a simply connected homology truncation of L_i (by (T1)–(T3) for f_i), it lies in **ICW** as well, by Lemma 2.25. Thus there exists a homotopy equivalence $(\varepsilon_i)_{<k} : (L_i)_{<k} \to E((L_i)_{<k})$, where $E((L_i)_{<k})$ is a finite CW-complex that has only even-dimensional cells. Let $(\varepsilon_i)_{<k}' : E((L_i)_{<k}) \to (L_i)_{<k}$ be a homotopy inverse for $(\varepsilon_i)_{<k}$.

We claim that $E((L_i)_{<k})$ has no cells of dimension k or higher. To verify this, let r_d denote the number of d-dimensional cells of $E((L_i)_{<k})$. If k is even, then the even cellular chain groups in degrees $\geq k$ are given by

$$C_{k+2m}(E((L_i)_{<k})) = \mathbb{Z}^{r_{k+2m}}, \ m \geq 0.$$

Since all boundary maps are trivial, we have

$$\mathbb{Z}^{r_{k+2m}} = C_{k+2m}(E((L_i)_{<k})) = H_{k+2m}(E((L_i)_{<k})) \cong H_{k+2m}((L_i)_{<k}) = 0$$

by (T3). Thus $r_{k+2m} = 0$ for all $m \geq 0$ and $E((L_i)_{<k})$ is a complex of dimension at most $k-1$. Similarly, if k is odd, then $C_k(E((L_i)_{<k})) = 0$ and $r_{k+2m+1} = 0$ for all $m \geq 0$ because the homology ranks equal the number of cells and the former vanish in dimensions $k+2m+1, m \geq 0$, as before – the claim is established.

Let $e_i : E((L_i)_{<k}) \to E(L_i)$ be a cellular approximation of (that is, cellular map homotopic to) the composition

$$E((L_i)_{<k}) \xrightarrow{(\varepsilon_i)_{<k}'} (L_i)_{<k} \xrightarrow{f_i} L_i \xrightarrow{\varepsilon_i} E(L_i).$$

As $E((L_i)_{<k})$ has no cells of dimension k or higher, the map e_i factors through the $(k-1)$-skeleton of $E(L_i)$,

$$E((L_i)_{<k}) \xrightarrow{\; e_i \;} E(L_i).$$

$$\downarrow{\hat{e}_i}$$

$$E(L_i)^{k-1}$$

Set

$$L_{<k} = \bigsqcup_i (L_i)_{<k}, \; E(L_{<k}) = \bigsqcup_i E((L_i)_{<k}),$$

$$\varepsilon_{<k} = \bigsqcup_i (\varepsilon_i)_{<k} : L_{<k} \to E(L_{<k}), \; \varepsilon'_{<k} = \bigsqcup_i (\varepsilon_i)'_{<k} : E(L_{<k}) \to L_{<k},$$

$$f = \bigsqcup_i f_i : L_{<k} \to L, \; e = \bigsqcup_i e_i : E(L_{<k}) \to E(L), \; \hat{e} = \bigsqcup_i \hat{e}_i : E(L_{<k}) \to E(L)^{k-1},$$

$$\tilde{e} = \hat{e}\varepsilon_{<k} : L_{<k} \to E(L)^{k-1}.$$

The diagram

$$
\begin{array}{ccc}
L_{<k} & \xrightarrow{\; g \;} & M \\
{\scriptstyle \tilde{e}}\big\downarrow & & \big\downarrow{\scriptstyle \mathrm{id}_M} \\
E(L)^{k-1} & \xrightarrow{\; G \;} & M
\end{array}
\tag{2.8}
$$

commutes up to homotopy because it factors as

$$
\begin{array}{ccccc}
L_{<k} & \xrightarrow{\; f \;} & L = \partial M & \xhookrightarrow{\; j \;} & M \\
{\scriptstyle \tilde{e}}\big\downarrow & & \big\downarrow{\scriptstyle \mathrm{id}_L} & & \big\downarrow{\scriptstyle \mathrm{id}_M} \\
E(L)^{k-1} & \xrightarrow{\; F \;} & L = \partial M & \xhookrightarrow{\; j \;} & M
\end{array}
$$

and the left-hand square homotopy commutes, since

$$
\begin{aligned}
F\tilde{e} &= (\varepsilon' \circ \mathrm{incl})(\hat{e}\varepsilon_{<k}) \\
&= \varepsilon' e \varepsilon_{<k} \\
&\simeq \varepsilon'(\varepsilon f \varepsilon'_{<k})\varepsilon_{<k} \\
&\simeq f.
\end{aligned}
$$

The map $\tilde{e}$ is a homotopy equivalence: If $r \geq k$, then $H_r(L_{<k}) = 0 = H_r(E(L)^{k-1})$, so that $\tilde{e}_* : H_r(L_{<k}) \to H_r(E(L)^{k-1})$ is an isomorphism in that range. Once we have shown that $\tilde{e}$ is a homology isomorphism in the complementary range $r < k$ as well, it will follow from Whitehead's theorem that $\tilde{e}$ is a homotopy equivalence, since

$L_{<k}$ and $E(L)^{k-1}$ are CW-complexes, $\tilde{e}$ induces a bijection between the connected components $(L_i)_{<k}$ of $L_{<k}$ and the connected components $E(L_i)^{k-1}$ of $E(L)^{k-1}$, and each of these components is simply connected. Suppose then that $r < k$. The skeletal inclusion incl : $E(L)^{k-1} \subset E(L)$ induces an isomorphism of cellular chain groups

$$\text{incl}_* : C_r(E(L)^{k-1}) \xrightarrow{\cong} C_r(E(L)).$$

As $H_r(E(L)^{k-1}) = C_r(E(L)^{k-1})$ and $H_r(E(L)) = C_r(E(L))$, we deduce that

$$\text{incl}_* : H_r(E(L)^{k-1}) \longrightarrow H_r(E(L))$$

is an isomorphism. By assumption, $f_* : H_r(L_{<k}) \to H_r(L)$ is an isomorphism (property (T2)). The commutativity of the pentagon

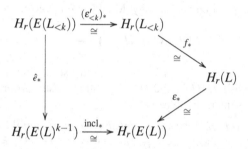

implies that $\hat{e}_*$ is an isomorphism. Hence, as $\tilde{e}$ is the composition of the homotopy equivalence $\varepsilon_{<k}$ and $\hat{e}$, $\tilde{e}_* : H_r(L_{<k}) \to H_r(E(L)^{k-1})$ is an isomorphism for $r < k$ (and thus for all r).

Proposition 2.11 applied to the diagram (2.8) yields a homotopy equivalence

$$\text{cone}(g) \simeq \text{cone}(G)$$

extending the identity map on M. Since the f_i and the $\overline{f}_i$ both satisfy properties (T1)–(T3), the same argument applied to the $\overline{f}_i$ instead of the f_i will produce a homotopy equivalence

$$\text{cone}(\overline{g}) \simeq \text{cone}(G).$$

$\square$

2.5 The Middle Dimension

Let X^n be a compact oriented pseudomanifold whose dimension n is divisible by 4 and which has only isolated singularities with simply connected links. We work exclusively with rational coefficients in this section. Since $\bar{n}(n) = \bar{m}(n)$, an upper middle perversity intersection space $I^{\bar{n}}X$ for X may be taken to be equal to a lower middle perversity intersection space $I^{\bar{m}}X$. Denote this space by $IX = I^{\bar{m}}X = I^{\bar{n}}X$. Let $m = n/2$ be the middle dimension. The compact manifold-with-boundary obtained by removing small open cone neighborhoods of the singularities is denoted by $(M, \partial M)$. Theorem 2.12 defines a nonsingular pairing

$$\widetilde{H}_r(IX) \otimes \widetilde{H}_{n-r}(IX) \longrightarrow \mathbb{Q}.$$

In the middle dimension, one obtains a nonsingular intersection form

$$\widetilde{H}_m(IX) \otimes \widetilde{H}_m(IX) \longrightarrow \mathbb{Q}.$$

We shall prove that this form is symmetric. In particular, it defines an element in the Witt group $W(\mathbb{Q})$ of the rationals. On middle perversity intersection homology, one has the symmetric, nonsingular Goresky–MacPherson intersection pairing

$$IH_m(X) \otimes IH_m(X) \longrightarrow \mathbb{Q},$$

which also defines an element in $W(\mathbb{Q})$. We shall show that these two elements coincide, so that while $\widetilde{H}_m(IX)$ and $IH_m(X)$ can be vastly different, they do yield essentially the same intersection theory.

Lemma 2.27. *Let $(M, \partial M)$ be an oriented compact manifold-with-boundary of dimension $2m$, with m even. Let $d_M : H_m(M) \to H^m(M, \partial M) = H_m(M, \partial M)^*$ be the Poincaré duality isomorphism inverse to capping with the fundamental class $[M, \partial M] \in H_{2m}(M, \partial M)$, $\alpha = j_* : H_m(M) \to H_m(M, \partial M)$ the canonical map, and let $w \in H_m(M)$. If $d_M(w)$ annihilates the image of α, then $\alpha(w) = 0$.*

Proof. Set $\omega = d_M(w)$, so that $\omega \cap [M, \partial M] = w$. If ω annihilates im α, then

$$\langle j^*(\omega), v \rangle = \langle \omega, j_*(v) \rangle = 0$$

for all $v \in H_m(M)$, $j^* : H^m(M, \partial M) \to H^m(M)$, which implies that $j^*(\omega) = 0$. The diagram

$$
\begin{array}{ccc}
H^m(M, \partial M) & \xrightarrow{-\cap[M,\partial M]} & H_m(M) \\
\downarrow{\scriptstyle j^*} & & \downarrow{\scriptstyle j_*} \\
H^m(M) & \xrightarrow{-\cap[M,\partial M]} & H_m(M, \partial M)
\end{array}
$$

commutes, whence

$$\alpha(w) = j_*(\omega \cap [M, \partial M]) = j^*(\omega) \cap [M, \partial M] = 0.$$

$\square$

Theorem 2.28. *The intersection form*

$$\Phi_{IX} : \tilde{H}_m(IX) \otimes \tilde{H}_m(IX) \longrightarrow \mathbb{Q}$$

is symmetric. Its Witt element $[\Phi_{IX}] \in W(\mathbb{Q})$ *is independent of choices. In fact, if*

$$\Phi_{IH} : IH_m(X) \otimes IH_m(X) \longrightarrow \mathbb{Q}$$

denotes the Goresky–MacPherson intersection form, then

$$[\Phi_{IX}] = [\Phi_{IH}] \in W(\mathbb{Q}).$$

Proof. We shall use the following description of the intersection form on $\tilde{H}_m(IX)$. Consider the commutative diagram (which is part of a self-duality isomorphism of an m-reflective diagram)

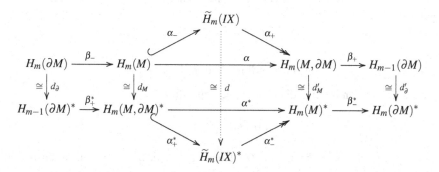

The dotted isomorphism d is to be described. It determines the intersection form Φ_{IX} by the formula

$$\Phi_{IX}(v \otimes w) = d(v)(w), \ v, w \in \tilde{H}_m(IX).$$

The Goresky–MacPherson intersection form $\Phi_{IH} : \operatorname{im}\alpha \otimes \operatorname{im}\alpha \to \mathbb{Q}$ is given by $\Phi_{IH}(v \otimes w) = d_M(v')(w)$ for any $v' \in H_m(M)$ with $\alpha(v') = v$. This is well-defined because if $\alpha(v'') = v$, then $v' - v''$ is in the image of β_-, $v' - v'' = \beta_-(u)$, $u \in H_m(\partial M)$, and

$$d_M(v' - v'')(w) = d_M\beta_-(u)(\alpha(w')) = \beta_+^* d_\partial(u)(\alpha(w')) = \alpha^*\beta_+^* d_\partial(u)(w') = 0,$$

where $\alpha(w') = w$.

Frequently, we shall make use of the symmetry identity $d_M(v)(w) = d'_M(w)(v)$, $v \in H_m(M)$, $w \in H_m(M, \partial M)$, which holds, since the cup product of m-dimensional cohomology classes commutes as m is even.

Choose a basis $\{e_1, \ldots, e_r\}$ for the subspace $\operatorname{im}\alpha \subset H_m(M, \partial M)$. Choose a subset $\{\bar{e}_1, \ldots, \bar{e}_r\} \subset H_m(M)$ with $\alpha(\bar{e}_i) = e_i$. In this basis, $\Phi_{IH}(e_i \otimes e_j) = d_M(\bar{e}_i)(e_j)$.

Define an annihilation subspace $Q \subset H_m(M, \partial M)$ by

$$Q = \{q \in H_m(M, \partial M) \mid d_M(\overline{e}_i)(q) = 0, \text{ for all } i\}.$$

We claim that $Q \cap \operatorname{im} \alpha = 0$: Let $v \in Q \cap \operatorname{im} \alpha$. Then $v = \alpha(w)$ for some $w \in H_m(M)$ and $d_M(\overline{e}_i)(\alpha(w)) = 0$ for all i. Consequently,

$$d_M(w)(e_i) = d_M'(e_i)(w) = d_M'(\alpha(\overline{e}_i))(w) = \alpha^* d_M(\overline{e}_i)(w) = d_M(\overline{e}_i)(\alpha(w)) = 0$$

for all i and we see that $d_M(w)$ annihilates $\operatorname{im} \alpha$. By Lemma 2.27, $v = \alpha(w) = 0$, which verifies the claim. Let us calculate the dimension of Q. The subspace $F \subset H_m(M, \partial M)^*$ spanned by $\{d_M(\overline{e}_1), \ldots, d_M(\overline{e}_r)\}$ has dimension $\dim F = r$, since $\{\overline{e}_1, \ldots, \overline{e}_r\}$ is a linearly independent set and d_M is an isomorphism. If V is any finite dimensional vector space and $F \subset V^*$ is a subspace, then the dimension of the corresponding annihilation space $W = \{v \in V \mid f(v) = 0 \text{ for all } f \in F\}$ is given by $\dim W = \dim V - \dim F$. Applying this dimension formula to $V = H_m(M, \partial M)$, we get

$$\dim Q = \dim H_m(M, \partial M) - r,$$

or

$$\dim \operatorname{im} \alpha + \dim Q = \dim H_m(M, \partial M).$$

Hence we have an internal direct sum decomposition

$$H_m(M, \partial M) = \operatorname{im} \alpha \oplus Q.$$

Let $\{q_1, \ldots, q_l\}$ be a basis for Q. Then $\{e_1, \ldots, e_r, q_1, \ldots, q_l\}$ is a basis for $H_m(M, \partial M)$. By construction, the formula

$$d_M(\overline{e}_i)(q_j) = 0 \tag{2.9}$$

holds for all i, j. For the dual $H_m(M, \partial M)^*$, we have the dual basis $\{e^1, \ldots, e^r, q^1, \ldots, q^l\}$. Since $d_M : H_m(M) \to H_m(M, \partial M)^*$ is an isomorphism, there are unique vectors $p_i \in H_m(M)$ such that $d_M(p_i) = q^i$. We claim that $\{p_1, \ldots, p_l, \overline{e}_1, \ldots, \overline{e}_r\}$ is a basis for $H_m(M)$. Since d_M is an isomorphism, the set $\{p_1, \ldots, p_l\}$ is linearly independent and spans an l-dimensional subspace $P \subset H_m(M)$. The linearly independent set $\{\overline{e}_1, \ldots, \overline{e}_r\}$ spans an r-dimensional subspace $\overline{E} \subset H_m(M)$. We will show that $P \cap \overline{E} = 0$. Let $v = \sum \pi_i p_i = \sum \varepsilon_j \overline{e}_j \in P \cap \overline{E}$, $\pi_i, \varepsilon_j \in \mathbb{Q}$. Then $d_M(v) = \sum \pi_i q^i$, $\alpha(v) = \sum \varepsilon_j e_j$, and

$$\sum \varepsilon_j d_M'(e_j) = d_M' \alpha(v) = \alpha^* d_M(v) = \alpha^* \sum \pi_i q^i.$$

Let $w \in H_m(M)$ be an arbitrary vector. Its image $\alpha(w)$ can be written as $\alpha(w) = \sum \omega_k e_k$, $\omega_k \in \mathbb{Q}$. Thus

$$(\sum \varepsilon_j d_M'(e_j))(w) = (\alpha^* \sum \pi_i q^i)(w) = (\sum \pi_i q^i)(\alpha(w))$$
$$= (\sum \pi_i q^i)(\sum \omega_k e_k) = \sum_{i,k} \pi_i \omega_k q^i(e_k) = 0.$$

It follows that $d'_M \sum \varepsilon_j e_j = 0$ and so $\sum \varepsilon_j e_j = 0$. Thus all coefficients ε_j vanish and $v = \sum \varepsilon_j \bar{e}_j = 0$. Consequently, $\{p_1, \ldots, p_l, \bar{e}_1, \ldots, \bar{e}_r\}$ is a linearly independent set in $H_m(M)$. It is a basis, as

$$\dim H_m(M) = \dim H_m(M, \partial M)^* = \dim H_m(M, \partial M) = r + l.$$

This finishes the verification of the claim. Since α_+ is surjective, we can choose $\bar{q}_i \in \tilde{H}_m(IX)$ with $\alpha_+(\bar{q}_i) = q_i$. We claim that

$$\mathcal{B} = \{\alpha_-(p_1), \ldots, \alpha_-(p_l), \alpha_-(\bar{e}_1), \ldots, \alpha_-(\bar{e}_r), \bar{q}_1, \ldots, \bar{q}_l\}$$

is a basis for $\tilde{H}_m(IX)$. Since α_- is injective, the set $\{\alpha_-(p_1), \ldots, \alpha_-(p_l), \alpha_-(\bar{e}_1), \ldots, \alpha_-(\bar{e}_r)\}$ is linearly independent and spans $\mathrm{im}\,\alpha_- \subset \tilde{H}_m(IX)$. The set $\{\bar{q}_1, \ldots, \bar{q}_l\}$ is linearly independent (since $\{q_1, \ldots, q_l\}$ is linearly independent) and spans an l-dimensional subspace $\bar{Q} \subset \tilde{H}_m(IX)$. We shall show that $\mathrm{im}\,\alpha_- \cap \bar{Q} = 0$. Suppose $v = \alpha_-(w) = \sum \lambda_i \bar{q}_i \in \mathrm{im}\,\alpha_- \cap \bar{Q}$. Since $\alpha_+(v) = \alpha(w) = \sum \lambda_i q_i$ and $\mathrm{im}\,\alpha \cap Q = 0$, we have $\sum \lambda_i q_i = 0$. Therefore, $\lambda_i = 0$ for all i and $v = 0$. It follows that $\mathcal{B}$ is a linearly independent set containing $r + 2l$ vectors. The exact sequence

$$H_m(\partial M) \xrightarrow{\alpha_- \beta_-} \tilde{H}_m(IX) \xrightarrow{\alpha_+} H_m(M, \partial M) \to 0$$

shows that

$$\dim \tilde{H}_m(IX) = \dim H_m(M, \partial M) + \mathrm{rk}(\alpha_- \beta_-).$$

Using $\mathrm{rk}(\alpha_- \beta_-) = \mathrm{rk}\,\beta_- = \dim \ker \alpha$ and

$$r = \mathrm{rk}\,\alpha = \dim H_m(M) - \dim \ker \alpha = r + l - \dim \ker \alpha,$$

we see that

$$\dim \tilde{H}_m(IX) = (r + l) + l = r + 2l.$$

Thus $\mathcal{B}$ is a basis for $\tilde{H}_m(IX)$. This basis yields a dual basis

$$\mathcal{B}^* = \{\alpha_-(p_1)^*, \ldots, \alpha_-(p_l)^*, \alpha_-(\bar{e}_1)^*, \ldots, \alpha_-(\bar{e}_r)^*, \bar{q}_1^*, \ldots, \bar{q}_l^*\}$$

for $\tilde{H}_m(IX)^*$. We observe that

$$\alpha^*(q^i) = 0 \tag{2.10}$$

and

$$\alpha(p_i) = 0 \tag{2.11}$$

for all i. Equality (2.10) holds, since on basis vectors,

$$\alpha^*(q^i)(\bar{e}_j) = q^i(\alpha(\bar{e}_j)) = q^i(e_j) = 0$$

and

$$\alpha^*(q^i)(p_j) = q^i(\alpha(p_j)) = q^i(\sum \varepsilon_k e_k) = \sum \varepsilon_k q^i(e_k) = 0.$$

Equality (2.11) follows from (2.10) by noting that, as d'_M is an isomorphism, $\alpha(p_i)$ vanishes if, and only if, $d'_M \alpha(p_i)$ vanishes, and

$$d'_M \alpha(p_i) = \alpha^* d_M(p_i) = \alpha^*(q^i) = 0.$$

Furthermore, the relation

$$\bar{q}_j^* = \alpha_+^*(q^j) \tag{2.12}$$

holds for all j. Its verification will be carried out by checking the identity on the three types of basis vectors of $\mathcal{B}$: On vectors of the type $\alpha_-(p_k)$, we have

$$\bar{q}_j^*(\alpha_-(p_k)) = 0 = q^j(\alpha(p_k)) = (\alpha_+^*(q^j))(\alpha_-(p_k)),$$

using equation (2.11). On vectors of the type $\alpha_-(\bar{e}_i)$, we have

$$\alpha_+^*(q^j)(\alpha_-(\bar{e}_i)) = q^j(\alpha(\bar{e}_i)) = q^j(e_i) = 0 = \bar{q}_j^*(\alpha_-(\bar{e}_i)).$$

Finally, on vectors of the type $\bar{q}_k$, we have

$$\bar{q}_j^*(\bar{q}_k) = \delta_{jk} = q^j(q_k) = q^j(\alpha_+(\bar{q}_k)) = \alpha_+^*(q^j)(\bar{q}_k),$$

which concludes the verification of (2.12).

Let us proceed to define the map $d : \tilde{H}_m(IX) \to \tilde{H}_m(IX)^*$. On the elements of the basis $\mathcal{B}$, we set

$$\begin{aligned} d(\alpha_-(\bar{e}_i)) &= \alpha_+^* d_M(\bar{e}_i), \\ d(\alpha_-(p_j)) &= \bar{q}_j^*, \\ d(\bar{q}_j) &= \alpha_-(p_j)^*. \end{aligned}$$

Set

$$\begin{aligned} IH &= \mathbb{Q}\langle \alpha_-(\bar{e}_1), \ldots, \alpha_-(\bar{e}_r) \rangle \subset \tilde{H}_m(IX), \\ IH^\dagger &= \mathbb{Q}\langle \alpha_-(\bar{e}_1)^*, \ldots, \alpha_-(\bar{e}_r)^* \rangle \subset \tilde{H}_m(IX)^*, \\ L_- &= \mathbb{Q}\langle \alpha_-(p_1), \ldots, \alpha_-(p_l) \rangle \subset \tilde{H}_m(IX), \\ L_-^\dagger &= \mathbb{Q}\langle \alpha_-(p_1)^*, \ldots, \alpha_-(p_l)^* \rangle \subset \tilde{H}_m(IX)^*, \\ L_+ &= \mathbb{Q}\langle \bar{q}_1, \ldots, \bar{q}_l \rangle \subset \tilde{H}_m(IX), \\ L_+^\dagger &= \mathbb{Q}\langle \bar{q}_1^*, \ldots, \bar{q}_l^* \rangle \subset \tilde{H}_m(IX)^*. \end{aligned}$$

We obtain thus corresponding internal direct sum decompositions

$$\tilde{H}_m(IX) = L_- \oplus IH \oplus L_+,$$

and

$$\tilde{H}_m(IX)^* = L_-^\dagger \oplus IH^\dagger \oplus L_+^\dagger.$$

Note that IH is isomorphic to the intersection homology of X. The isomorphism is given by

$$IH \xrightarrow{\cong} \mathrm{im}\,\alpha = IH_m(X),$$

$$\alpha_-(\bar{e}_i) \mapsto \alpha_+(\alpha_-(\bar{e}_i)) = \alpha(\bar{e}_i) = e_i.$$

We claim that $d(IH) \subset IH^\dagger$. To see this, we write $d(\alpha_-(\bar{e}_i))$ as a linear combination

$$d(\alpha_-(\bar{e}_i)) = \sum \pi_k \alpha_-(p_k)^* + \sum \varepsilon_j \alpha_-(\bar{e}_j)^* + \sum \lambda_s \bar{q}_s^*$$

with uniquely determined coefficients $\pi_k, \varepsilon_j, \lambda_s$. The coefficient π_k can be obtained by evaluating on the basis vector $\alpha_-(p_k)$:

$$\begin{aligned} \pi_k &= (\alpha_+^* d_M(\bar{e}_i))(\alpha_-(p_k)) \\ &= d_M(\bar{e}_i)(\alpha(p_k)) \\ &= 0, \end{aligned}$$

using (2.11). The coefficient λ_s can be obtained by evaluating on the basis vector $\bar{q}_s$:

$$\begin{aligned} \lambda_s &= (\alpha_+^* d_M(\bar{e}_i))(\bar{q}_s) \\ &= d_M(\bar{e}_i)(\alpha_+(\bar{q}_s)) \\ &= d_M(\bar{e}_i)(q_s) \\ &= 0 \end{aligned}$$

by (2.9). The claim is thus established. We claim next that the restriction $d| : IH \rightarrow IH^\dagger$ is injective: Let $v \in IH$ be a vector with $d(v) = 0$. Writing $v = \sum \varepsilon_i \alpha_-(\bar{e}_i)$, we have

$$\begin{aligned} \alpha_+^* d_M(\sum \varepsilon_i \bar{e}_i) &= \sum \varepsilon_i \alpha_+^* d_M(\bar{e}_i) \\ &= \sum \varepsilon_i d(\alpha_-(\bar{e}_i)) \\ &= d(v) \\ &= 0. \end{aligned}$$

Since α_+^* is injective and d_M is an isomorphism, it follows that $\sum \varepsilon_i \bar{e}_i = 0$. This can only happen when $\varepsilon_i = 0$ for all i, which implies that $v = 0$. This finishes the verification of the claim. From

$$\dim IH = \dim IH^\dagger = r$$

we conclude that

$$d| : IH \xrightarrow{\cong} IH^\dagger$$

is an isomorphism. Note that under the above isomorphism $IH \cong IH_m(X)$ to intersection homology, $d|_{IH}$ is just the Goresky–MacPherson duality isomorphism $IH_m(X) \xrightarrow{\cong} IH_m(X)^*$. By construction, the restrictions

$$d| : L_- \xrightarrow{\cong} L_+^\dagger, \ d| : L_+ \xrightarrow{\cong} L_-^\dagger$$

are isomorphisms as well. It follows from the above direct sum decompositions that d is an isomorphism.

Our next objective is to prove that the pairing Φ_{IX}, induced by d, is symmetric. A sequence of calculations will lead up to this. Pairing IH with itself is symmetric:

$$
\begin{aligned}
d(\alpha_-(\overline{e}_i))(\alpha_-(\overline{e}_k)) &= (\alpha_+^* d_M(\overline{e}_i))(\alpha_-(\overline{e}_k)) \\
&= d_M(\overline{e}_i)(\alpha(\overline{e}_k)) \\
&= d_M'(\alpha(\overline{e}_k))(\overline{e}_i) \\
&= (\alpha^* d_M(\overline{e}_k))(\overline{e}_i) \\
&= (\alpha_-^* \alpha_+^* d_M(\overline{e}_k))(\overline{e}_i) \\
&= (\alpha_+^* d_M(\overline{e}_k))(\alpha_-(\overline{e}_i)) \\
&= d(\alpha_-(\overline{e}_k))(\alpha_-(\overline{e}_i)).
\end{aligned}
$$

The pairing is zero between IH and L_-:

$$
\begin{aligned}
d(\alpha_-(\overline{e}_i))(\alpha_-(p_j)) &= (\alpha_+^* d_M(\overline{e}_i))(\alpha_-(p_j)) \\
&= d_M(\overline{e}_i)(\alpha(p_j)) \\
&= 0,
\end{aligned}
$$

by (2.11). The pairing is also zero between IH and L_+:

$$
\begin{aligned}
d(\alpha_-(\overline{e}_i))(\overline{q}_j) &= (\alpha_+^* d_M(\overline{e}_i))(\overline{q}_j) \\
&= d_M(\overline{e}_i)(\alpha_+(\overline{q}_j)) \\
&= d_M(\overline{e}_i)(q_j) \\
&= 0,
\end{aligned}
$$

by (2.9). The pairing vanishes between L_- and IH:

$$
d(\alpha_-(p_j))(\alpha_-(\overline{e}_k)) = \overline{q}_j^*(\alpha_-(\overline{e}_k)) = 0
$$

by definition of $\overline{q}_j^*$ as a dual basis element. Pairing L_- with itself yields another trivial block:

$$
d(\alpha_-(p_j))(\alpha_-(p_k)) = \overline{q}_j^*(\alpha_-(p_k)) = 0,
$$

again by definition of $\overline{q}_j^*$ as a dual basis element. Pairing L_- with L_+ and pairing L_+ with L_- both give the identity matrix in our chosen bases:

$$
\begin{aligned}
d(\alpha_-(p_j))(\overline{q}_k) &= \overline{q}_j^*(\overline{q}_k) \\
&= \delta_{jk} \\
&= \alpha_-(p_k)^*(\alpha_-(p_j)) \\
&= d(\overline{q}_k)(\alpha_-(p_j)).
\end{aligned}
$$

The pairing vanishes between L_+ and IH:

$$
d(\overline{q}_j)(\alpha_-(\overline{e}_k)) = \alpha_-(p_j)^*(\alpha_-(\overline{e}_k)) = 0
$$

by definition of $\alpha_-(p_j)^*$ as a dual basis element. Finally, pairing L_+ with itself yields a trivial block:

$$d(\bar{q}_j)(\bar{q}_k) = \alpha_-(p_j)^*(\bar{q}_k) = 0.$$

We have

$$\begin{aligned}
\Phi_{IX}(\alpha_-(\bar{e}_i) \otimes \alpha_-(\bar{e}_j)) &= (\alpha_+^* d_M(\bar{e}_i))(\alpha_-(\bar{e}_j)) \\
&= d_M(\bar{e}_i)(\alpha(\bar{e}_j)) \\
&= d_M(\bar{e}_i)(e_j) \\
&= \Phi_{IH}(e_i \otimes e_j).
\end{aligned}$$

In summary, we have shown that with respect to $\mathcal{B}$, Φ_{IX} has the matrix representation

$$(\Phi_{IX})_{\mathcal{B}} = \begin{array}{|c|c c|c|}
\hline
IH & L_- & L_+ & \\
\hline
(\Phi_{IH})_{\mathcal{B}} & 0 & 0 & IH \\
0 & 0 & \mathbf{1}_l & L_- \\
0 & \mathbf{1}_l & 0 & L_+ \\
\hline
\end{array}$$

with $(\Phi_{IH})_{\mathcal{B}}$ denoting the symmetric Goresky–MacPherson intersection matrix on $IH_m(X)$ with respect to the basis $\{e_1, \ldots, e_r\}$, and where $\mathbf{1}_l$ denotes the identity matrix of rank l. Thus Φ_{IX} defines an element $[\Phi_{IX}] \in W(\mathbb{Q})$ in the Witt group of the rationals. Set $S = L_- \oplus L_+ \subset \tilde{H}_m(IX)$. The subspace S is split ([MH73]) because it contains the Lagrangian subspace L_-, $\Phi_{IX}|_{L_-} = 0$, $\dim L_- = l = \frac{1}{2} \dim S$. Thus $[\Phi_{IX}|_S] = 0 \in W(\mathbb{Q})$ and we have

$$[\Phi_{IX}] = [\Phi_{IX}|_{IH}] + [\Phi_{IX}|_S] = [\Phi_{IH}] \in W(\mathbb{Q}).$$

It remains to be shown that the two squares

$$\begin{CD}
H_m(M) @>\alpha_->> \tilde{H}_m(IX) @>\alpha_+>> H_m(M, \partial M) \\
@V\cong V d_M V @V\cong V d V @V\cong V d_M' V \\
H_m(M, \partial M)^* @>\alpha_+^*>> \tilde{H}_m(IX)^* @>\alpha_-^*>> H_m(M)^*
\end{CD}$$

commute. The commutativity of the left-hand square can be checked on the basis $\{p_1, \ldots, p_l, \bar{e}_1, \ldots, \bar{e}_r\}$ of $H_m(M)$: For the vectors p_j, we find

$$d\alpha_-(p_j) = \bar{q}_j^* = \alpha_+^*(q^j) = \alpha_+^* d_M(p_j)$$

by (2.12), and for the vectors $\bar{e}_i$ we have

$$d\alpha_-(\bar{e}_i) = \alpha_+^* d_M(\bar{e}_i)$$

by definition. Thus the left-hand square commutes. The commutativity of the right-hand square will be verified on the elements of $\mathcal{B}$. For basis vectors $\alpha_-(\overline{e}_i)$,

$$
\begin{aligned}
\alpha_-^* d(\alpha_-(\overline{e}_i)) &= \alpha_-^* \alpha_+^* d_M(\overline{e}_i) \\
&= \alpha^* d_M(\overline{e}_i) \\
&= d_M' \alpha(\overline{e}_i) \\
&= d_M' \alpha_+(\alpha_-(\overline{e}_i)).
\end{aligned}
$$

For basis vectors $\alpha_-(p_j)$,

$$
\alpha_-^* d(\alpha_-(p_j)) = \alpha_-^*(\overline{q}_j^*) = \alpha_-^*(\alpha_+^*(q^j)) = \alpha^*(q^j) = 0 = d_M' \alpha(p_j) = d_M' \alpha_+(\alpha_-(p_j)),
$$

using (2.10), (2.11) and (2.12). For basis vectors $\overline{q}_j$, we need to verify the equality

$$
\alpha_-^*(\alpha_-(p_j)^*) = d_M'(q_j) \in H_m(M)^*.
$$

We will do this employing the basis $\{p_1, \ldots, p_l, \overline{e}_1, \ldots, \overline{e}_r\}$ of $H_m(M)$:

$$
\alpha_-^*(\alpha_-(p_j)^*)(p_k) = \alpha_-(p_j)^*(\alpha_-(p_k)) = \delta_{jk} = q^k(q_j) = d_M(p_k)(q_j) = d_M'(q_j)(p_k),
$$

$$
\alpha_-^*(\alpha_-(p_j)^*)(\overline{e}_k) = \alpha_-(p_j)^*(\alpha_-(\overline{e}_k)) = 0 = d_M(\overline{e}_k)(q_j) = d_M'(q_j)(\overline{e}_k)
$$

(using (2.9)). Hence the right-hand square commutes as well.   □

Example 2.29. Let $N^4 = S^2 \times T^2$. Drill out a small open 4-ball to obtain the compact 4-manifold $N_0 = N - \mathrm{int}\, D^4$ with boundary $\partial N_0 = S^3$. The manifold $M^8 = N_0 \times S^2 \times S^2$ is compact with simply connected boundary $L = \partial M = S^3 \times S^2 \times S^2$. The pseudomanifold

$$
X^8 = M \cup_L \mathrm{cone}\, L
$$

has one singular point of even codimension. Consequently, for classical intersection homology $IH_*^{\overline{m}}(X) = IH_*^{\overline{n}}(X)$ and for the intersection spaces $I^{\overline{m}} X = I^{\overline{n}} X$. We shall denote the former groups by $IH_*(X)$ and the middle perversity intersection space by IX. Our objective is to compute the intersection form on $\tilde{H}_4(IX)$ and compare it to the intersection form on $IH_4(X)$. We shall use the notation of the proof of the Duality Theorem 2.12.

Let

$$
a = [S^2 \times \mathrm{pt} \times \mathrm{pt}], \ b = [\mathrm{pt} \times S^1 \times \mathrm{pt}], \ c = [\mathrm{pt} \times \mathrm{pt} \times S^1]
$$

denote the three generating cycles of $H_*(N)$. Inspecting the long exact homology sequence of the pair $(N_0, \partial N_0)$, we see that $H_1(N_0) \cong H_1(N_0, \partial N_0) \cong H_1(N)$ is generated by the cycles b, c. We see furthermore that $H_2(N_0) \cong H_2(N)$ is generated by

$a, b \times c$ and $H_3(N_0) \cong H_3(N)$ is generated by $a \times b, a \times c$. The homology of N_0 is summarized in the following table:

$H_*(N_0)$	H_0	H_1	H_2	H_3	H_4
Generators	pt	b, c	$a, b \times c$	$a \times b, a \times c$	0

Let

$$x = [S^3 \times \text{pt} \times \text{pt}], \ y = [\text{pt} \times S^2 \times \text{pt}], \ z = [\text{pt} \times \text{pt} \times S^2]$$

be the indicated cycles in $H_*(L)$. By the Künneth theorem, the homology of L is given by:

$H_*(L)$	H_0	H_1	H_2	H_3	H_4	H_5	H_6	H_7
Generators	pt	0	y, z	x	$y \times z$	$x \times y, x \times z$	0	$x \times y \times z$

If V is a vector space with basis $e_1, \ldots, e_l$, then $e_1^*, \ldots, e_l^*$ will denote the dual basis for the linear dual V^*. The Poincaré duality isomorphism

$$d_L : H_4(L)^* \xrightarrow{\cong} H_3(L)$$

is given by

$$d_L(y \times z)^* = x.$$

We have

$$H_3(M) = H_3(N_0) \times H_0(S^2 \times S^2) \oplus H_1(N_0) \times H_2(S^2 \times S^2),$$

so that

$$H_3(M) = \mathbb{Q}\langle a \times b, a \times c, b \times y, c \times y, b \times z, c \times z \rangle.$$

The middle homology of M is given by

$$H_4(M) = H_2(N_0) \times H_2(S^2 \times S^2) \oplus H_0(N_0) \times H_4(S^2 \times S^2),$$

so that

$$H_4(M) = \mathbb{Q}\langle a \times y, b \times c \times y, a \times z, b \times c \times z, y \times z \rangle.$$

The map

$$\beta_- : H_4(L) \longrightarrow H_4(M)$$

maps the generator $y \times z$ to $y \times z \in H_4(M)$, in particular, β_- is injective. If v, w are two homology classes, we shall from now on briefly write vw for their cross product $v \times w$. The surjective dual map

$$\beta_-^* : H_4(M)^* \longrightarrow H_4(L)^*$$

maps $(yz)^*$ to $(yz)^*$ and all other basis elements to zero. Next, let us discuss the exact sequence

$$H_4(L) \xrightarrow{\beta_-} H_4(M) \xrightarrow{\alpha} H_4(j) \xrightarrow{\delta_+} H_3(L) \xrightarrow{\varepsilon} H_3(M).$$

(Note that $\beta_+ = \delta_+$ in the present context.) Let us first calculate the middle intersection homology group from it:

$$IH_4(X) = \operatorname{im}\alpha \cong \frac{H_4(M)}{\ker\alpha} = \frac{H_4(M)}{\operatorname{im}\beta_-} = \frac{H_4(M)}{\mathbb{Q}\langle yz \rangle}.$$

Hence,

$$IH_4(X) = \mathbb{Q}\langle ay, bcy, az, bcz \rangle.$$

We claim that ε is the zero map. Since the boundary homomorphism $H_4(N_0, \partial N_0) \to H_3(S^3)$ maps the fundamental class $[N] = a \times b \times c$, which we may identify with the relative fundamental class $[N_0, \partial N_0]$, to the fundamental class $[\partial N_0]$, the latter is mapped to 0 under $H_3(S^3) \to H_3(N_0)$. Pick some point in $S^2 \times S^2$ to get a commutative diagram of inclusions

$$
\begin{array}{ccc}
S^3 = \partial N_0 & \longrightarrow & N_0 \\
\downarrow & & \downarrow \\
S^3 \times S^2 \times S^2 & \longrightarrow & N_0 \times S^2 \times S^2
\end{array}
$$

which induces a commutative square

$$
\begin{array}{ccc}
H_3(S^3) & \longrightarrow & H_3(N_0) \\
\downarrow & & \downarrow \\
H_3(L) & \xrightarrow{\varepsilon} & H_3(M)
\end{array}
$$

Since the left vertical arrow maps $[\partial N_0]$ to x and the upper horizontal arrow maps $[\partial N_0]$ to 0, it follows that $\varepsilon(x) = 0$. Since x generates $H_3(L)$, ε is indeed the zero map. Consequently, δ_+ is surjective,

$$\operatorname{im}\delta_+ = \mathbb{Q}\langle x \rangle,$$

and

$$\frac{H_4(j)}{IH_4(X)} = \frac{H_4(j)}{\operatorname{im}\alpha} = \frac{H_4(j)}{\ker\delta_+} \cong \operatorname{im}\delta_+ = \mathbb{Q}\langle x \rangle.$$

We have

$$H_4(j) = H_4(X) = \mathbb{Q}\langle ay, bcy, az, bcz, abc \rangle,$$

with

$$\delta_+(abc) = x.$$

The Poincaré duality isomorphism

$$d_M : H_4(M)^* \xrightarrow{\cong} H_4(j)$$

is given by

$$(ay)^* \mapsto bcz$$
$$(bcy)^* \mapsto az$$
$$(az)^* \mapsto bcy$$
$$(bcz)^* \mapsto ay$$
$$(yz)^* \mapsto abc.$$

Take $s_{p\beta} : \operatorname{im}\beta_-^* \to H_4(M)^*$ to be

$$s_{p\beta}(yz)^* = (yz)^*.$$

This determines $s_{q\delta} : \operatorname{im}\delta_+ \to H_4(j)$:

$$s_{q\delta}(x) = d_M s_{p\beta} d_L^{-1}(x) = d_M s_{p\beta}(yz)^* = d_M(yz)^* = abc.$$

The middle intersection space homology group is given by

$$\widetilde{H}_4(IX) = \mathbb{Q}\langle ay, bcy, az, bcz, yz, abc\rangle.$$

Note that $\alpha(yz) = 0$, since yz is in the image of β_-. The factorization

$$H_4(M) \xrightarrow{\alpha_-} \widetilde{H}_4(IX) \xrightarrow{\alpha_+} H_4(j)$$

of α is given by

$$\begin{aligned}
\alpha_-(ay) &= ay, & \alpha_+(ay) &= ay \\
\alpha_-(bcy) &= bcy, & \alpha_+(bcy) &= bcy \\
\alpha_-(az) &= az, & \alpha_+(az) &= az \\
\alpha_-(bcz) &= bcz, & \alpha_+(bcz) &= bcz \\
\alpha_-(yz) &= yz, & \alpha_+(yz) &= 0 \\
& & \alpha_+(abc) &= abc.
\end{aligned}$$

The map $\beta_+ : H_4(j) \to H_3(L)$ agrees with δ_+, i.e. maps abc to x and all other basis elements to zero. Take the splitting $s_{p\alpha} : H_4(M)^* \to \widetilde{H}_4(IX)^*$ for α_-^* to be

$$s_{p\alpha}(ay)^* = (ay)^*, \; s_{p\alpha}(bcy)^* = (bcy)^*, \; s_{p\alpha}(az)^* = (az)^*,$$

$$s_{p\alpha}(bcz)^* = (bcz)^*, \; s_{p\alpha}(yz)^* = (yz)^*.$$

Take the splitting $s_{q\gamma} : H_4(j) \to \widetilde{H}_4(IX)$ for $\gamma_+ = \alpha_+$ to be

$$s_{q\gamma}(ay) = ay, \; s_{q\gamma}(bcy) = bcy, \; s_{q\gamma}(az) = az,$$

$$s_{q\gamma}(bcz) = bcz, \; s_{q\gamma}(abc) = abc.$$

Thus s_p is

$$\operatorname{im}\beta_-^* \xrightarrow{s_p} \widetilde{H}_4(IX)^*$$
$$(yz)^* \mapsto (yz)^*$$

and s_q is

$$\operatorname{im}\delta_+ \xrightarrow{s_q} \widetilde{H}_4(IX)$$
$$x \mapsto abc.$$

The Poincaré duality isomorphism

$$d_M' : H_4(j)^* \xrightarrow{\cong} H_4(M)$$

is given by

$$(ay)^* \mapsto bcz$$
$$(bcy)^* \mapsto az$$
$$(az)^* \mapsto bcy$$
$$(bcz)^* \mapsto ay$$
$$(abc)^* \mapsto yz.$$

The following table calculates the duality isomorphism

$$d : \widetilde{H}_4(IX)^* \xrightarrow{\cong} \widetilde{H}_4(IX)$$

on the middle intersection space homology group:

v	$d(v)$
$(ay)^* = \alpha_+^*(ay)^*$	$\gamma_- d_M'(ay)^* = \gamma_-(bcz) = bcz$
$(bcy)^* = \alpha_+^*(bcy)^*$	$\gamma_- d_M'(bcy)^* = \gamma_-(az) = az$
$(az)^* = \alpha_+^*(az)^*$	$\gamma_- d_M'(az)^* = \gamma_-(bcy) = bcy$
$(bcz)^* = \alpha_+^*(bcz)^*$	$\gamma_- d_M'(bcz)^* = \gamma_-(ay) = ay$
$(abc)^* = \alpha_+^*(abc)^*$	$\gamma_- d_M'(abc)^* = \gamma_-(yz) = yz$
$(yz)^* = s_p(yz)^*$	$s_q d_L(yz)^* = s_q(x) = abc$

(Note $\gamma_- = \alpha_-$ here.) The intersection form on $\widetilde{H}_4(IX)$ with respect to the basis $\{ay, bcy, az, bcz, abc, yz\}$ is thus given by the matrix

$$\begin{pmatrix} 0\,0\,0\,1\,0\,0 \\ 0\,0\,1\,0\,0\,0 \\ 0\,1\,0\,0\,0\,0 \\ 1\,0\,0\,0\,0\,0 \\ 0\,0\,0\,0\,0\,1 \\ 0\,0\,0\,0\,1\,0 \end{pmatrix}$$

On the basis elements $\{ay, bcy, az, bcz\}$, this matrix contains the block

$$\begin{pmatrix} 0\,0\,0\,1 \\ 0\,0\,1\,0 \\ 0\,1\,0\,0 \\ 1\,0\,0\,0 \end{pmatrix}$$

which is the intersection form on $IH_4(X)$.

2.6 Cap Products for Middle Perversities

The intersection space cohomology trivially has internal (with respect to the space *and* with respect to the perversity) cup products

$$H^r(I^{\bar{p}}X) \otimes H^s(I^{\bar{p}}X) \xrightarrow{\,\cup\,} H^{r+s}(I^{\bar{p}}X),$$

given by the ordinary cup product. The ordinary cap product

$$\widetilde{H}^r(I^{\bar{m}}X) \otimes \widetilde{H}_i(I^{\bar{m}}X) \xrightarrow{\,\cap\,} \widetilde{H}_{i-r}(I^{\bar{m}}X)$$

is of little use in establishing duality isomorphisms, since $\widetilde{H}_*(I^{\bar{m}}X)$ never contains an orientation class, the reason being that $\widetilde{H}_n(I^{\bar{m}}X) \cong \widetilde{H}_0(I^{\bar{n}}X)^* = 0$ ($n = \dim X$, X connected). Orientation classes for singular spaces X are usually contained in $H_*(X)$, so what would be desirable would be cap products of the type

$$\widetilde{H}^r(I^{\bar{m}}X) \otimes \widetilde{H}_i(X) \xrightarrow{\,\cap\,} \widetilde{H}_{i-r}(I^{\bar{n}}X)$$

and

$$\widetilde{H}^r(I^{\bar{n}}X) \otimes \widetilde{H}_i(X) \xrightarrow{\,\cap\,} \widetilde{H}_{i-r}(I^{\bar{m}}X).$$

We shall construct such products, at least on the even cohomology H^{2*} of the middle perversity intersection spaces. Chern classes of a complex vector bundle, for instance, lie in the even cohomology of the underlying base space. The L-class of a pseudomanifold, when defined, generally lies in the ordinary homology of X. Thus the new product allows one to multiply such classes and get a result that is again a middle perversity intersection space homology class. In constructing the products, we shall concentrate on the important two middle perversities and leave the obvious modifications necessary to deal with other perversities to the reader. Similarly, it is possible to go beyond the even cohomology-degree assumption, but we do not work this out here.

2.6.1 Motivational Considerations

The existence of a cap product of the type

$$\widetilde{H}^r(I^{\bar{m}}X) \otimes \widetilde{H}_i(X) \xrightarrow{\cap} \widetilde{H}_{i-r}(I^{\bar{n}}X)$$

seems counterintuitive from the point of view of classical intersection homology. The product asserts that one may take a class in the cohomology of the middle perversity intersection space, pair it with an *arbitrary* homology class and one will end up with a class that lifts back to a class in the homology of the middle perversity intersection space again. An analogous statement for intersection homology is certainly false, as the following example shows. Suppose X is the pseudomanifold with one singularity obtained by coning off the boundary of a compact manifold M of dimension, say, 10. The codimension of the singularity is even, so $I^{\bar{m}}X = I^{\bar{n}}X = IX$ and $IH_*^{\bar{m}}(X) = IH_*^{\bar{n}}(X) = IH_*(X)$. There cannot generally be a cap product

$$\cap : IH^2(X) \otimes H_4(X) \longrightarrow IH_2(X),$$

for example. The reason is that since 2 is below the middle dimension 5, we have $IH^2(X) = H^2(M)$ and $IH_2(X) = H_2(M)$. Furthermore, $H_4(X) = H_4(M, \partial M)$ so that the existence of the above product would amount to a cap product

$$\cap : H^2(M) \otimes H_4(M, \partial M) \longrightarrow H_2(M).$$

Such a product cannot generally be defined. The evaluation of the absolute chain-level product

$$\cap : C^j(M) \otimes C_i(M) \longrightarrow C_{i-j}(M), \tag{2.13}$$

on the submodule $C^j(M) \otimes C_i(\partial M)$ will lead to chains in $C_{i-j}(\partial M)$, but these chains can be nontrivial, even homologically. Thus the product (2.13) induces only a product

$$\cap : H^j(M) \otimes H_i(M, \partial M) \longrightarrow H_{i-j}(M, \partial M)$$

and not a product

$$\cap : H^j(M) \otimes H_i(M, \partial M) \longrightarrow H_{i-j}(M).$$

Why, then, does the pairing of an intersection space homology class with an arbitrary homology class again yield an intersection space homology class? Let us give a systematic analysis of the behavior of intersection homology versus the homology of intersection spaces in this regard. The analysis continues to be framed in the context of the above ten-dimensional X. Let the symbol "a" denote the absolute (co)homology of M and let the symbol "r" denote the relative (co)homology of the pair $(M, \partial M)$. Since we always wish to cap with arbitrary homology classes, we only deal with cap products of the type $- \cap r \to -$. As we have seen, capping an absolute cohomology class with a relative homology class gives a relative homology class. Capping a relative cohomology class with a relative homology class gives an absolute homology class, since (2.13) restricts to zero on the submodule $C^j(M, \partial M) \otimes C_i(\partial M)$. Thus, cap type behaves like a logical negation operator

$$r \otimes r \to a,$$
$$a \otimes r \to r.$$

We shall first focus on intersection homology. To simplify our analysis, we shall leave aside the middle dimension. In the tables below, a field will be crossed out (receive an entry "$\times$") if either middle dimensional elements would be required to fill it or a cap product for this field would land in a negative dimension. We investigate in detail for which i and j one can define a pairing

$$\cap : IH^j(X) \otimes H_i(X) \longrightarrow IH_{i-j}(X).$$

In terms of the pair $(M, \partial M)$, the groups $IH^j(X)$ have the following types:

j	0	1	2	3	4	5	6	7	8	9	10
$IH_{\bar{m}}^j(X)$	a	a	a	a	a	$\times$	r	r	r	r	r

(2.14)

The entries of the following table show what the actual cap type of the result of $\cap$ on $IH^j(X) \otimes H_i(X)$ is. Since cap type is negation, the rows of this table are obtained by negating the row (2.14).

$i \diagdown^j$	0	1	2	3	4	5	6	7	8	9	10
0	r	$\times$	$\times$	$\times$	$\times$	$\times$	$\times$	$\times$	$\times$	$\times$	$\times$
1	r	r	$\times$	$\times$	$\times$	$\times$	$\times$	$\times$	$\times$	$\times$	$\times$
2	r	r	r	$\times$	$\times$	$\times$	$\times$	$\times$	$\times$	$\times$	$\times$
3	r	r	r	r	$\times$	$\times$	$\times$	$\times$	$\times$	$\times$	$\times$
4	r	r	r	r	r	$\times$	$\times$	$\times$	$\times$	$\times$	$\times$
5	r	r	r	r	r	$\times$	$\times$	$\times$	$\times$	$\times$	$\times$
6	r	r	r	r	r	$\times$	a	$\times$	$\times$	$\times$	$\times$
7	r	r	r	r	r	$\times$	a	a	$\times$	$\times$	$\times$
8	r	r	r	r	r	$\times$	a	a	a	$\times$	$\times$
9	r	r	r	r	r	$\times$	a	a	a	a	$\times$
10	r	r	r	r	r	$\times$	a	a	a	a	a

(2.15)

The next table contains the dimensions $i - j$ of the results of $\cap$ on $IH^j(X) \otimes H_i(X)$ or on $\widetilde{H}^j(IX) \otimes H_i(X)$.

$_i\diagdown{}^j$	0	1	2	3	4	5	6	7	8	9	10
0	0	×	×	×	×	×	×	×	×	×	×
1	1	0	×	×	×	×	×	×	×	×	×
2	2	1	0	×	×	×	×	×	×	×	×
3	3	2	1	0	×	×	×	×	×	×	×
4	4	3	2	1	0	×	×	×	×	×	×
5	5	4	3	2	1	0	×	×	×	×	×
6	6	5	4	3	2	1	0	×	×	×	×
7	7	6	5	4	3	2	1	0	×	×	×
8	8	7	6	5	4	3	2	1	0	×	×
9	9	8	7	6	5	4	3	2	1	0	×
10	10	9	8	7	6	5	4	3	2	1	0

(2.16)

The table below decodes the a/r-type of $IH_{i-j}(X)$.

$i - j$	0	1	2	3	4	5	6	7	8	9	10
$IH^{\bar{m}}_{i-j}(X)$	a	a	a	a	a	×	r	r	r	r	r

(2.17)

Putting the result of table (2.17) into table (2.16) we obtain:

$_i\diagdown{}^j$	0	1	2	3	4	5	6	7	8	9	10
0	a	×	×	×	×	×	×	×	×	×	×
1	a	a	×	×	×	×	×	×	×	×	×
2	a	a	a	×	×	×	×	×	×	×	×
3	a	a	a	a	×	×	×	×	×	×	×
4	a	a	a	a	a	×	×	×	×	×	×
5	×	a	a	a	a	×	×	×	×	×	×
6	r	×	a	a	a	×	a	×	×	×	×
7	r	r	×	a	a	×	a	a	×	×	×
8	r	r	r	×	a	×	a	a	a	×	×
9	r	r	r	r	×	×	a	a	a	a	×
10	r	r	r	r	r	×	a	a	a	a	a

(2.18)

Take table (2.15) and table (2.18) and perform the transformation

(2.15)	(2.18)	(2.19)
r	r	$\to$ white
r	a	$\to$ black ■
a	r	$\to$ white
a	a	$\to$ white

on it. (White fields mean that there is no inconsistency between the actual result (2.15) of the cap product and the putative target (2.18). For instance, a, r receives white because there is a canonical map from absolute to relative homology. The pair r, a receives black, since you cannot always lift a relative class to an absolute one.) The result is:

$i \diagdown{}^j$	0	1	2	3	4	5	6	7	8	9	10
0	■	×	×	×	×	×	×	×	×	×	×
1	■	■	×	×	×	×	×	×	×	×	×
2	■	■	■	×	×	×	×	×	×	×	×
3	■	■	■	■	×	×	×	×	×	×	×
4	■	■	■	■	■	×	×	×	×	×	×
5	×	■	■	■	■	■	×	×	×	×	×
6		×	■	■	■	×		×	×	×	×
7			×	■	■	×			×	×	×
8				×	■	×			×	×	
9					×	×					×
10						×					

(2.19)

The presence of the black fields is the reason that no general cap product $\cap : IH^j(X) \otimes H_i(X) \longrightarrow IH_{i-j}(X)$ can be defined.

Let us carry out the very same kind of analysis for the homology of the intersection space, asking for a cap product $\cap : \widetilde{H}^j(IX) \otimes \widetilde{H}_i(X) \to \widetilde{H}_{i-j}(IX)$. The groups $\widetilde{H}^j(IX)$ have the following a/r-types:

j	0	1	2	3	4	5	6	7	8	9	10
$\widetilde{H}^j(IX)$	r	r	r	r	r	×	a	a	a	a	a

(2.20)

The entries of the following table show what the actual cap type of the result of $\cap$ on $\widetilde{H}^j(IX) \otimes H_i(X)$ is. Since cap type is negation, the rows are obtained by negating the row (2.20).

$i \diagdown{}^j$	0	1	2	3	4	5	6	7	8	9	10
0	a	×	×	×	×	×	×	×	×	×	×
1	a	a	×	×	×	×	×	×	×	×	×
2	a	a	a	×	×	×	×	×	×	×	×
3	a	a	a	a	×	×	×	×	×	×	×
4	a	a	a	a	a	×	×	×	×	×	×
5	a	a	a	a	a	×	×	×	×	×	×
6	a	a	a	a	a	×	r	×	×	×	×
7	a	a	a	a	a	×	r	r	×	×	×
8	a	a	a	a	a	×	r	r	r	×	×
9	a	a	a	a	a	×	r	r	r	r	×
10	a	a	a	a	a	×	r	r	r	r	r

(2.21)

The table below decodes the a/r-type of $\widetilde{H}_{i-j}(IX)$.

$$
\begin{array}{c|ccccccccccc}
i-j & 0 & 1 & 2 & 3 & 4 & 5 & 6 & 7 & 8 & 9 & 10 \\
\hline
\widetilde{H}_{i-j}(IX) & r & r & r & r & r & \times & a & a & a & a & a
\end{array}
\tag{2.22}
$$

Putting the result of table (2.22) into table (2.16) we obtain:

$i\backslash{}^j$	0	1	2	3	4	5	6	7	8	9	10
0	r	$\times$	$\times$	$\times$	$\times$	$\times$	$\times$	$\times$	$\times$	$\times$	$\times$
1	r	r	$\times$	$\times$	$\times$	$\times$	$\times$	$\times$	$\times$	$\times$	$\times$
2	r	r	r	$\times$	$\times$	$\times$	$\times$	$\times$	$\times$	$\times$	$\times$
3	r	r	r	r	$\times$	$\times$	$\times$	$\times$	$\times$	$\times$	$\times$
4	r	r	r	r	r	$\times$	$\times$	$\times$	$\times$	$\times$	$\times$
5	$\times$	r	r	r	r	$\times$	$\times$	$\times$	$\times$	$\times$	$\times$
6	a	$\times$	r	r	r	$\times$	r	$\times$	$\times$	$\times$	$\times$
7	a	a	$\times$	r	r	$\times$	r	r	$\times$	$\times$	$\times$
8	a	a	a	$\times$	r	$\times$	r	r	r	$\times$	$\times$
9	a	a	a	a	$\times$	$\times$	r	r	r	r	$\times$
10	a	a	a	a	a	$\times$	r	r	r	r	r

$$\tag{2.23}$$

Take table (2.21) and table (2.23) and perform the above transformation

(2.21)	(2.23)	(2.24)
r	r	$\rightarrow$ white
r	a	$\rightarrow$ black ■
a	r	$\rightarrow$ white
a	a	$\rightarrow$ white

on it to get

$i\backslash{}^j$	0	1	2	3	4	5	6	7	8	9	10
0		$\times$	$\times$	$\times$	$\times$	$\times$	$\times$	$\times$	$\times$	$\times$	$\times$
1			$\times$	$\times$	$\times$	$\times$	$\times$	$\times$	$\times$	$\times$	$\times$
2				$\times$	$\times$	$\times$	$\times$	$\times$	$\times$	$\times$	$\times$
3					$\times$	$\times$	$\times$	$\times$	$\times$	$\times$	$\times$
4						$\times$	$\times$	$\times$	$\times$	$\times$	$\times$
5	$\times$					$\times$	$\times$	$\times$	$\times$	$\times$	$\times$
6		$\times$	r			$\times$		$\times$	$\times$	$\times$	$\times$
7			$\times$			$\times$			$\times$	$\times$	$\times$
8				$\times$		$\times$				$\times$	$\times$
9					$\times$	$\times$					$\times$
10						$\times$					

$$\tag{2.24}$$

There are no blackouts for $\tilde{H}_*(IX)$. This explains why a cap product $\cap : \tilde{H}^j(IX) \otimes \tilde{H}_i(X) \to \tilde{H}_{i-j}(IX)$ can be defined.

2.6.2 Canonical Maps

Let X^n be a pseudomanifold with isolated singularities $x_1, \ldots, x_w$. Let $\hat{X}$ be the quotient of X obtained by identifying the points $x_1, \ldots, x_w$. Then $\hat{X}$ is again a pseudomanifold. It has one singular point whose link is the disjoint union of the links L_i of the points x_i. The quotient map $X \to \hat{X}$ is a normalization of $\hat{X}$ if all L_i are connected. If $w = 1$, then $\hat{X} = X$. For the homology we have the formula

$$\tilde{H}_r(\hat{X}) = H_r(M, \partial M).$$

If $j : \partial M \hookrightarrow M$ denotes the inclusion of the boundary, then $\hat{X}$ may also be described as $\hat{X} = \text{cone}(j)$. The reason why we introduce $\hat{X}$ here is that there will be canonical maps $c : I^{\bar{p}}X \to \hat{X}$, but if there is more than one singularity, i.e. $w \geq 2$, then there is no map from $I^{\bar{p}}X$ to X. However, as far as (co)homology is concerned, switching back and forth between X and $\hat{X}$ is no big deal, since the map $X \to \hat{X}$, for X connected, induces isomorphisms $H_r(X) \cong H_r(\hat{X})$ for $r \neq 1$ and $H_1(\hat{X}) \cong H_1(X) \oplus \mathbb{Z}^{w-1}$. The intersection homology does not change at all under normalization. Another interpretation of $\hat{X}$ is this: If a negative perversity value $\bar{p}(n) = -1$ were allowed (this would be one step below the zero perversity $\bar{0}$), then $k = n - 1 - \bar{p}(n) = n$, $L_{<k} = L_{<n} = L = \partial M$ (since L has dimension $n - 1$), $f = \text{id} : L_{<k} \to \partial M$ and $I^{\bar{p}}X = \text{cone}(g) = \text{cone}(jf) = \text{cone}(j) = \hat{X}$. So one may view $\hat{X}$, but not X, as an extreme case "$I^{-1}X$" of an intersection space of X and thus a canonical map c should have target $\hat{X}$, not X.

To a diagram of continuous maps

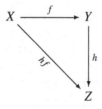

we can associate a commutative diagram

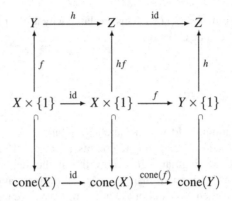

The pushout of the left column is $\mathrm{cone}(f)$, the pushout of the middle column is $\mathrm{cone}(hf)$ and the pushout of the right column is $\mathrm{cone}(h)$. Thus the horizontal maps of the diagram induce maps

$$\mathrm{cone}(f) \longrightarrow \mathrm{cone}(hf) \xrightarrow{\ c\ } \mathrm{cone}(h).$$

The braid of the triple (f, hf, h) contains the exact sequences

$$H_r(X) \xrightarrow{(hf)_*} H_r(Z) \xrightarrow{\ b_*\ } H_r(hf) \xrightarrow{\ \partial_*\ } H_{r-1}(X)$$

as well as

$$H_r(Y) \xrightarrow{\ h_*\ } H_r(Z) \xrightarrow{\ a_*\ } H_r(h) \xrightarrow{\ \partial_*\ } H_{r-1}(Y).$$

The diagram

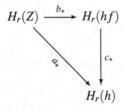

is contained in the braid and commutes. The corresponding diagram on cohomology

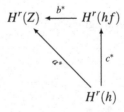

commutes also.

Applying this to the diagram

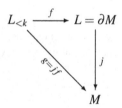

we obtain canonical maps

$$I^{\bar{p}}X = \mathrm{cone}(g) \xrightarrow{\ c\ } \mathrm{cone}(j) = \hat{X}$$

and

$$M \xrightarrow{\ b\ } I^{\bar{p}}X$$

(the latter is the canonical inclusion map from the target of a map to its mapping cone) such that

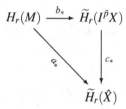

and

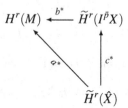

commute. The manifold M has the following interpretation as an intersection space: If $\bar{p}(n) = n - 1$ were allowed (it is actually one step above the top perversity $\bar{t}$), then $k = n - 1 - \bar{p}(n) = 0$, $L_{<k} = L_{<0} = \varnothing$ (the empty space) and $I^{\bar{p}}X = \mathrm{cone}(\varnothing \to M) = M^+$, the union of M with a disjoint point.

Remark 2.30. Due to the fact that the construction of intersection spaces requires (in general) certain choices in the kth cellular chain groups of the links, the existence of maps between $I^{\bar{p}}X$ and $I^{\bar{q}}X$ for different perversities $\bar{p}, \bar{q}$ is a somewhat delicate matter and will not be pursued in the present book. When such maps $I^{\bar{p}}X \to I^{\bar{q}}X$ exist, then certainly only for $\bar{p} \geq \bar{q}$. For such $\bar{p}, \bar{q}$ one has canonical maps $IH_*^{\bar{q}}(X) \to IH_*^{\bar{p}}(X)$ on intersection homology, once again documenting the reflective nature of the relationship between intersection space homology and intersection homology.

2.6.3 Construction of the Cap Products

We take rational coefficients for this section. With more care, integral products can also be defined, where possibly exceptional degrees are those close to k. The difficulty stems from the fact that for homology, $H_r(I^{\bar{m}}X;\mathbb{Z}) \cong H_r(M;\mathbb{Z})$ when $r > k$, while one need not have $H^r(I^{\bar{m}}X;\mathbb{Z}) \cong H^r(M;\mathbb{Z})$ for cohomology when $r > k$. Since $H^k(L_{<k}) \cong \text{Ext}(H_{k-1}(L),\mathbb{Z})$ (see Remark 1.42), one has in the borderline case $r = k+1$ the exact sequence

$$\text{Ext}(H_{k-1}(L),\mathbb{Z}) \longrightarrow H^{k+1}(I^{\bar{m}}X;\mathbb{Z}) \longrightarrow H^{k+1}(M;\mathbb{Z}) \xrightarrow{g^*=0} H^{k+1}(L_{<k};\mathbb{Z}) = 0,$$

which shows that for $r = k+1$, $H^r(I^{\bar{m}}X;\mathbb{Z}) \to H^r(M;\mathbb{Z})$ is only onto with kernel given by the image of the torsion subgroup of $H_{k-1}(L;\mathbb{Z})$. Over $\mathbb{Q}$, this group is zero, so we get an isomorphism. In order not to clutter up our statements with torsion-freeness assumptions in relevant degrees r close to k, we prefer to phrase them in this book for rational coefficients only.

Proposition 2.31. *Suppose* $n = \dim X \equiv 2 \mod 4$. *Then there exists a cap product*

$$\widetilde{H}^{2l}(I^{\bar{m}}X) \otimes \widetilde{H}_i(\hat{X}) \xrightarrow{\cap} \widetilde{H}_{i-2l}(I^{\bar{m}}X)$$

such that

$$
\begin{array}{ccc}
\widetilde{H}^{2l}(I^{\bar{m}}X) \otimes \widetilde{H}_i(\hat{X}) & \xrightarrow{\cap} & \widetilde{H}_{i-2l}(I^{\bar{m}}X) \\
\Big\uparrow{\scriptstyle c^*\otimes\text{id}} & & \Big\downarrow{\scriptstyle c_*} \\
\widetilde{H}^{2l}(\hat{X}) \otimes \widetilde{H}_i(\hat{X}) & \xrightarrow{\cap} & \widetilde{H}_{i-2l}(\hat{X})
\end{array}
$$

commutes, where the bottom arrow is the ordinary cap product.

Proof. Write $n = 4m+2$. If $i > n$, then $\widetilde{H}_i(\hat{X}) = 0$, so we may assume $i \leq n$. From $\bar{m}(2p) = p-1$ it follows that

$$k = n - 1 - \bar{m}(n) = 4m + 1 - 2m = 2m + 1 = n/2$$

is odd. Thus for $r = 2l$, we have either $r > k$ or $r < k$. Suppose $r > k$. In this case, the map $b^* : \widetilde{H}^r(I^{\bar{m}}X) \to \widetilde{H}^r(M)$ is an isomorphism. Since

$$i - r \leq n - r < n - k = n/2 = k,$$

the map

$$c_* : \widetilde{H}_{i-r}(I^{\bar{m}}X) = H_{i-r}(g) \to H_{i-r}(j) = H_{i-r}(M,\partial M) = \widetilde{H}_{i-r}(\hat{X})$$

is an isomorphism. Define

$$\widetilde{H}^r(I^{\bar{m}}X) \otimes \widetilde{H}_i(\hat{X}) \xrightarrow{\cap} \widetilde{H}_{i-r}(I^{\bar{m}}X)$$

through the diagram

$$
\begin{array}{ccc}
\widetilde{H}^r(I^{\bar{m}}X) \otimes \widetilde{H}_i(\hat{X}) & \dashrightarrow{\cap} & \widetilde{H}_{i-r}(I^{\bar{m}}X) \\
b^* \otimes \mathrm{id} \Big\downarrow \cong & & \cong \Big\downarrow c_* \\
\widetilde{H}^r(M) \otimes H_i(M, \partial M) & \xrightarrow{\cap} & H_{i-r}(M, \partial M).
\end{array}
$$

Since the diagrams

$$
\begin{array}{ccc}
\widetilde{H}^r(\hat{X}) \otimes \widetilde{H}_i(\hat{X}) & \xrightarrow{\cap} & \widetilde{H}_{i-r}(\hat{X}) \\
a^* \otimes \mathrm{id} \Big\downarrow & & \Big\| \\
\widetilde{H}^r(M) \otimes H_i(M, \partial M) & \xrightarrow{\cap} & H_{i-r}(M, \partial M)
\end{array}
\tag{2.25}
$$

and

$$
\begin{array}{ccc}
\widetilde{H}^r(\hat{X}) & \xrightarrow{c^*} & \widetilde{H}^r(I^{\bar{m}}X) \\
& {\scriptstyle a_*} \searrow & \cong \Big\downarrow b^* \\
& & \widetilde{H}^r(M)
\end{array}
$$

commute, we have for $\xi \in \widetilde{H}^r(\hat{X})$ and $x \in \widetilde{H}_i(\hat{X})$:

$$
\begin{aligned}
c^*(\xi) \cap x &= (b^*)^{-1} a^*(\xi) \cap x \\
&= c_*^{-1}(a^*(\xi) \cap x) \quad \text{(by definition)} \\
&= c_*^{-1}(\xi \cap x) \qquad \text{(by the commutativity of (2.25))}
\end{aligned}
$$

so that

$$c_*(c^*(\xi) \cap x) = \xi \cap x$$

as required. Now suppose $r < k$. Then the map

$$c^* : \widetilde{H}^r(\hat{X}) = H^r(M, \partial M) \to \widetilde{H}^r(I^{\bar{m}}X)$$

is an isomorphism. Define

$$\widetilde{H}^r(I^{\bar{m}}X) \otimes \widetilde{H}_i(\hat{X}) \xrightarrow{\cap} \widetilde{H}_{i-r}(I^{\bar{m}}X)$$

by

$$\xi \cap x = b_*((c^*)^{-1}(\xi) \cap x),$$

where b_* is the map $b_* : \widetilde{H}_{i-r}(M) \to \widetilde{H}_{i-r}(I^{\bar{m}}X)$ and the cap product used on the right-hand side is

$$\cap : H^r(M, \partial M) \otimes H_i(M, \partial M) \longrightarrow H_{i-r}(M) \to H_{i-r}(M, *) = \widetilde{H}_{i-r}(M).$$

(If $i - r > k$, then b_* is an isomorphism.) Using the commutativity of the diagram

$$
\begin{array}{ccc}
H^r(M, \partial M) \otimes H_i(M, \partial M) & \xrightarrow{\cap} & \widetilde{H}_{i-r}(M) \\
\| & & \Big\downarrow{a_*} \\
\widetilde{H}^r(\hat{X}) \otimes \widetilde{H}_i(\hat{X}) & \xrightarrow{\cap} & \widetilde{H}_{i-r}(\hat{X})
\end{array}
$$

we compute for $\xi \in \widetilde{H}^r(\hat{X}), x \in \widetilde{H}_i(\hat{X})$:

$$
\begin{aligned}
c_*(c^*(\xi) \cap x) &= c_*(b_*(\xi \cap x)) \quad \text{(by definition)} \\
&= a_*(\xi \cap x) \\
&= \xi \cap x.
\end{aligned}
$$

$\square$

Proposition 2.32. *Suppose* $n = \dim X \equiv 0 \mod 4$. *Then there exists a cap product*

$$\widetilde{H}^{2l}(I^{\bar{m}}X) \otimes \widetilde{H}_i(\hat{X}) \xrightarrow{\cap} \widetilde{H}_{i-2l}(I^{\bar{m}}X)$$

for $2l \neq n/2$ *such that*

$$
\begin{array}{ccc}
\widetilde{H}^{2l}(I^{\bar{m}}X) \otimes \widetilde{H}_i(\hat{X}) & \xrightarrow{\cap} & \widetilde{H}_{i-2l}(I^{\bar{m}}X) \\
\Big\uparrow{c^* \otimes \mathrm{id}} & & \Big\downarrow{c_*} \\
\widetilde{H}^{2l}(\hat{X}) \otimes \widetilde{H}_i(\hat{X}) & \xrightarrow{\cap} & \widetilde{H}_{i-2l}(\hat{X})
\end{array}
$$

commutes.

Proof. Write $n = 4m$. We may assume $i \leq n$. From $\bar{m}(2p) = p - 1$ it follows that

$$k = n - 1 - \bar{m}(n) = 4m - 1 - (2m - 1) = 2m = n/2.$$

Thus for $r = 2l \neq n/2$, we have either $r > k$ or $r < k$. Suppose $r > k$. In this case, the map $b^* : H^r(I^{\bar{m}}X) \to H^r(M)$ is an isomorphism. As in the case 2 mod 4,

$$i - r \leq n - r < n - k = n/2 = k,$$

so the construction can proceed precisely as in the proof of Proposition 2.31. When $r < k$, the cap product can be defined by the formula $\xi \cap x = b_*((c^*)^{-1}(\xi) \cap x)$, just as in the proof of Proposition 2.31. $\square$

Proposition 2.33. *Suppose $n = \dim X \equiv 1$ mod 4. Then there exists a cap product*

$$\widetilde{H}^{2l}(I^{\bar{m}}X) \otimes \widetilde{H}_i(\hat{X}) \xrightarrow{\cap} \widetilde{H}_{i-2l}(I^{\bar{n}}X)$$

such that

$$\widetilde{H}^{2l}(I^{\bar{m}}X) \otimes \widetilde{H}_i(\hat{X}) \xrightarrow{\cap} \widetilde{H}_{i-2l}(I^{\bar{n}}X)$$

$$c^* \otimes \mathrm{id} \uparrow \qquad\qquad\qquad \downarrow c_*$$

$$\widetilde{H}^{2l}(\hat{X}) \otimes \widetilde{H}_i(\hat{X}) \xrightarrow{\cap} \widetilde{H}_{i-2l}(\hat{X})$$

commutes.

Proof. Write $n = 4m + 1$. As usual, we may assume $i \leq n$. From $\bar{m}(n) = (n-3)/2$, $\bar{n}(n) = (n-1)/2$ it follows that for $I^{\bar{m}}X$,

$$k_{\bar{m}} = n - 1 - \bar{m}(n) = 4m - \frac{4m-2}{2} = 2m + 1$$

is odd and for $I^{\bar{n}}X$,

$$k_{\bar{n}} = n - 1 - \bar{n}(n) = 4m - \frac{4m}{2} = 2m$$

is even. Thus for $r = 2l$, we have either $r > k_{\bar{m}}$ or $r < k_{\bar{m}}$. Suppose $r > k_{\bar{m}}$. In this case, the map $b^*_{\bar{m}} : \widetilde{H}^r(I^{\bar{m}}X) \to \widetilde{H}^r(M)$ is an isomorphism. Since

$$i - r \leq n - r < n - k_{\bar{m}} = 4m + 1 - (2m+1) = 2m = k_{\bar{n}},$$

the map

$$c^{\bar{n}}_* : \widetilde{H}_{i-r}(I^{\bar{n}}X) \to H_{i-r}(M, \partial M) = \widetilde{H}_{i-r}(\hat{X})$$

is an isomorphism. Define

$$\widetilde{H}^r(I^{\bar{m}}X) \otimes \widetilde{H}_i(\hat{X}) \xrightarrow{\cap} \widetilde{H}_{i-r}(I^{\bar{n}}X)$$

through the diagram

$$\begin{array}{ccc}
\widetilde{H}^r(I^{\bar{m}}X) \otimes \widetilde{H}_i(\hat{X}) & \overset{\cap}{\dashrightarrow} & \widetilde{H}_{i-r}(I^{\bar{n}}X) \\
b_{\bar{m}}^* \otimes \mathrm{id} \Big\downarrow \cong & & \cong \Big\downarrow c_*^{\bar{n}} \\
\widetilde{H}^r(M) \otimes H_i(M, \partial M) & \overset{\cap}{\longrightarrow} & H_{i-r}(M, \partial M).
\end{array}$$

Since the diagrams

$$\begin{array}{ccc}
\widetilde{H}^r(\hat{X}) \otimes \widetilde{H}_i(\hat{X}) & \overset{\cap}{\longrightarrow} & \widetilde{H}_{i-r}(\hat{X}) \\
a^* \otimes \mathrm{id} \Big\downarrow & & \Big\| \\
\widetilde{H}^r(M) \otimes H_i(M, \partial M) & \overset{\cap}{\longrightarrow} & H_{i-r}(M, \partial M)
\end{array}$$
(2.26)

and

$$\begin{array}{ccc}
\widetilde{H}^r(\hat{X}) & \overset{c_{\bar{m}}^*}{\longrightarrow} & \widetilde{H}^r(I^{\bar{m}}X) \\
 & \varrho_* \searrow & \cong \Big\downarrow b_{\bar{m}}^* \\
 & & \widetilde{H}^r(M)
\end{array}$$

commute, we have for $\xi \in \widetilde{H}^r(\hat{X})$ and $x \in \widetilde{H}_i(\hat{X})$:

$$\begin{aligned}
c_{\bar{m}}^*(\xi) \cap x &= (b_{\bar{m}}^*)^{-1} a^*(\xi) \cap x \\
&= (c_*^{\bar{n}})^{-1}(a^*(\xi) \cap x) \text{ (by definition)} \\
&= (c_*^{\bar{n}})^{-1}(\xi \cap x) \quad \text{(by the commutativity of (2.26))}
\end{aligned}$$

so that

$$c_*^{\bar{n}}(c_{\bar{m}}^*(\xi) \cap x) = \xi \cap x$$

as required. Now suppose $r < k_{\bar{m}}$. Then the map $c_{\bar{m}}^* : \widetilde{H}^r(\hat{X}) = H^r(M, \partial M) \rightarrow \widetilde{H}^r(I^{\bar{m}}X)$ is an isomorphism. Define

$$\widetilde{H}^r(I^{\bar{m}}X) \otimes \widetilde{H}_i(\hat{X}) \overset{\cap}{\longrightarrow} \widetilde{H}_{i-r}(I^{\bar{n}}X)$$

by

$$\xi \cap x = b_*^{\bar{n}}((c_{\bar{m}}^*)^{-1}(\xi) \cap x),$$

where $b_*^{\bar{n}}$ is the map $b_*^{\bar{n}} : \tilde{H}_{i-r}(M) \to \tilde{H}_{i-r}(I^{\bar{n}}X)$ and the cap product used on the right-hand side is

$$\cap : H^r(M, \partial M) \otimes H_i(M, \partial M) \longrightarrow H_{i-r}(M) \to H_{i-r}(M, *) = \tilde{H}_{i-r}(M).$$

(If $i - r > k_{\bar{n}}$, then $b_*^{\bar{n}}$ is an isomorphism.) Using the commutativity of the diagram

$$
\begin{array}{ccc}
H^r(M, \partial M) \otimes H_i(M, \partial M) & \overset{\cap}{\longrightarrow} & \tilde{H}_{i-r}(M) \\
\| & & \downarrow{a_*} \\
\tilde{H}^r(\hat{X}) \otimes \tilde{H}_i(\hat{X}) & \overset{\cap}{\longrightarrow} & \tilde{H}_{i-r}(\hat{X})
\end{array}
$$

we compute for $\xi \in \tilde{H}^r(\hat{X})$, $x \in \tilde{H}_i(\hat{X})$:

$$
\begin{aligned}
c_*^{\bar{n}}(c_{\bar{m}}^*(\xi) \cap x) &= c_*^{\bar{n}}(b_*^{\bar{n}}(\xi \cap x)) \text{ (by definition)} \\
&= a_*(\xi \cap x) \\
&= \xi \cap x.
\end{aligned}
$$

$\square$

Proposition 2.34. *Suppose* $n = \dim X \equiv 3 \mod 4$. *Then there exists a cap product*

$$\tilde{H}^{2l}(I^{\bar{n}}X) \otimes \tilde{H}_i(\hat{X}) \overset{\cap}{\longrightarrow} \tilde{H}_{i-2l}(I^{\bar{m}}X)$$

such that

$$
\begin{array}{ccc}
\tilde{H}^{2l}(I^{\bar{n}}X) \otimes \tilde{H}_i(\hat{X}) & \overset{\cap}{\to} & \tilde{H}_{i-2l}(I^{\bar{m}}X) \\
\uparrow{c^* \otimes \mathrm{id}} & & \downarrow{c_*} \\
\tilde{H}^{2l}(\hat{X}) \otimes \tilde{H}_i(\hat{X}) & \overset{\cap}{\longrightarrow} & \tilde{H}_{i-2l}(\hat{X})
\end{array}
$$

commutes.

Proof. Write $n = 4m + 3$. We may assume $i \le n$. From $\bar{m}(n) = (n-3)/2$, $\bar{n}(n) = (n-1)/2$ it follows that for $I^{\bar{m}}X$,

$$k_{\bar{m}} = n - 1 - \bar{m}(n) = 4m + 2 - \frac{4m}{2} = 2m + 2$$

is even and for $I^{\bar{n}}X$,

$$k_{\bar{n}} = n - 1 - \bar{n}(n) = 4m + 2 - \frac{4m+2}{2} = 2m + 1$$

is odd. Thus for $r = 2l$, we have either $r > k_{\bar{n}}$ or $r < k_{\bar{n}}$. Suppose $r > k_{\bar{n}}$. In this case, the map $b_{\bar{n}}^* : \widetilde{H}^r(I^{\bar{n}}X) \to \widetilde{H}^r(M)$ is an isomorphism. Since

$$i - r \le n - r < n - k_{\bar{n}} = 4m + 3 - (2m + 1) = 2m + 2 = k_{\bar{m}},$$

the map

$$c_*^{\bar{m}} : \widetilde{H}_{i-r}(I^{\bar{m}}X) \to H_{i-r}(M, \partial M) = \widetilde{H}_{i-r}(\hat{X})$$

is an isomorphism. Define

$$\widetilde{H}^r(I^{\bar{n}}X) \otimes \widetilde{H}_i(\hat{X}) \xrightarrow{\cap} \widetilde{H}_{i-r}(I^{\bar{m}}X)$$

through the diagram

$$
\begin{array}{ccc}
\widetilde{H}^r(I^{\bar{n}}X) \otimes \widetilde{H}_i(\hat{X}) & \dashrightarrow^{\cap} & \widetilde{H}_{i-r}(I^{\bar{m}}X) \\
b_{\bar{n}}^* \otimes \mathrm{id} \downarrow \cong & & \cong \downarrow c_*^{\bar{m}} \\
\widetilde{H}^r(M) \otimes H_i(M, \partial M) & \xrightarrow{\cap} & H_{i-r}(M, \partial M).
\end{array}
$$

Since the diagrams

$$
\begin{array}{ccc}
\widetilde{H}^r(\hat{X}) \otimes \widetilde{H}_i(\hat{X}) & \xrightarrow{\cap} & \widetilde{H}_{i-r}(\hat{X}) \\
a^* \otimes \mathrm{id} \downarrow & & \| \\
\widetilde{H}^r(M) \otimes H_i(M, \partial M) & \xrightarrow{\cap} & H_{i-r}(M, \partial M)
\end{array}
$$

(2.27)

and

$$
\begin{array}{ccc}
\widetilde{H}^r(\hat{X}) & \xrightarrow{c_{\bar{n}}^*} & \widetilde{H}^r(I^{\bar{n}}X) \\
& {}_{a^*}\searrow & \cong \downarrow b_{\bar{n}}^* \\
& & \widetilde{H}^r(M)
\end{array}
$$

commute, we have for $\xi \in \widetilde{H}^r(\hat{X})$ and $x \in \widetilde{H}_i(\hat{X})$:

$$
\begin{aligned}
c_{\bar{n}}^*(\xi) \cap x &= (b_{\bar{n}}^*)^{-1} a^*(\xi) \cap x \\
&= (c_*^{\bar{m}})^{-1}(a^*(\xi) \cap x) \quad \text{(by definition)} \\
&= (c_*^{\bar{m}})^{-1}(\xi \cap x) \quad \text{(by the commutativity of (2.27))}
\end{aligned}
$$

so that

$$c_*^{\tilde{m}}(c_{\tilde{n}}^*(\xi) \cap x) = \xi \cap x$$

as required. Now suppose $r < k_{\tilde{n}}$. Then the map $c_{\tilde{n}}^* : \tilde{H}^r(\hat{X}) = H^r(M, \partial M) \to \tilde{H}^r(I^{\tilde{n}}X)$ is an isomorphism. Define

$$\tilde{H}^r(I^{\tilde{n}}X) \otimes \tilde{H}_i(\hat{X}) \xrightarrow{\cap} \tilde{H}_{i-r}(I^{\tilde{m}}X)$$

by

$$\xi \cap x = b_*^{\tilde{m}}((c_{\tilde{n}}^*)^{-1}(\xi) \cap x),$$

where $b_*^{\tilde{m}}$ is the map $b_*^{\tilde{m}} : \tilde{H}_{i-r}(M) \to \tilde{H}_{i-r}(I^{\tilde{m}}X)$ and the cap product used on the right-hand side is

$$\cap : H^r(M, \partial M) \otimes H_i(M, \partial M) \longrightarrow H_{i-r}(M) \to H_{i-r}(M, *) = \tilde{H}_{i-r}(M).$$

(If $i - r > k_{\tilde{m}}$, then $b_*^{\tilde{m}}$ is an isomorphism.) Using the commutativity of the diagram

$$
\begin{array}{ccc}
H^r(M, \partial M) \otimes H_i(M, \partial M) & \xrightarrow{\cap} & \tilde{H}_{i-r}(M) \\
\| & & \downarrow a_* \\
\tilde{H}^r(\hat{X}) \otimes \tilde{H}_i(\hat{X}) & \xrightarrow{\cap} & \tilde{H}_{i-r}(\hat{X})
\end{array}
$$

we compute for $\xi \in \tilde{H}^r(\hat{X})$, $x \in \tilde{H}_i(\hat{X})$:

$$
\begin{aligned}
c_*^{\tilde{m}}(c_{\tilde{n}}^*(\xi) \cap x) &= c_*^{\tilde{m}}(b_*^{\tilde{m}}(\xi \cap x)) \quad \text{(by definition)} \\
&= a_*(\xi \cap x) \\
&= \xi \cap x.
\end{aligned}
$$

$\square$

2.7 L-Theory

Let $\mathbb{L}^\bullet$ be the 0-connective symmetric L-spectrum, as in [Ran92, §16, page 173], with homotopy groups

$$\pi_i(\mathbb{L}^\bullet) = L^i(\mathbb{Z}) = \begin{cases} \mathbb{Z}, & i \equiv 0 \mod 4 \text{ (signature)} \\ \mathbb{Z}/2, & i \equiv 1 \mod 4 \text{ (de Rham invariant)} \\ 0, & i \equiv 2, 3 \mod 4 \end{cases}$$

for $i \geq 0$, and $\pi_i(\mathbb{L}^{\bullet}) = 0$ for negative i. Rationally, $\mathbb{L}^{\bullet}$ has the homotopy type of a product of Eilenberg–MacLane spectra

$$\mathbb{L}^{\bullet} \otimes \mathbb{Q} \simeq \prod_{i \geq 0} K(\mathbb{Q}, 4i).$$

A compact oriented smooth n-manifold-with-boundary $(M, \partial M)$ possesses a canonical $\mathbb{L}^{\bullet}$-orientation $[M, \partial M]_{\mathbb{L}} \in H_n(M, \partial M; \mathbb{L}^{\bullet})$ which is given rationally by the homology L-class of M:

$$[M, \partial M]_{\mathbb{L}} \otimes 1 = \mathcal{L}_*(M, \partial M) = \mathcal{L}^*(M) \cap [M, \partial M]$$

$$\in H_n(M, \partial M; \mathbb{L}^{\bullet}) \otimes \mathbb{Q} = \bigoplus_{i \geq 0} H_{n-4i}(M, \partial M; \mathbb{Q}),$$

where $\mathcal{L}^*(M) \in H^{4*}(M; \mathbb{Q})$ denotes the Hirzebruch L-class of (the tangent bundle of) M and $[M, \partial M] \in H_n(M, \partial M; \mathbb{Q})$ denotes the fundamental class in ordinary homology. There is defined a cap product

$$\cap : H^i(M; \mathbb{L}^{\bullet}) \otimes H_n(M, \partial M; \mathbb{L}^{\bullet}) \longrightarrow H_{n-i}(M, \partial M; \mathbb{L}^{\bullet})$$

such that

$$- \cap [M, \partial M]_{\mathbb{L}} : H^i(M; \mathbb{L}^{\bullet}) \longrightarrow H_{n-i}(M, \partial M; \mathbb{L}^{\bullet})$$

is an isomorphism. This product induces a cap product on the rationalized groups,

$$\cap : H^i(M; \mathbb{L}^{\bullet}) \otimes \mathbb{Q} \otimes H_n(M, \partial M; \mathbb{L}^{\bullet}) \otimes \mathbb{Q} \longrightarrow H_{n-i}(M, \partial M; \mathbb{L}^{\bullet}) \otimes \mathbb{Q}$$

such that the diagram

$$
\begin{array}{ccc}
H^i(M; \mathbb{L}^{\bullet}) \otimes \mathbb{Q} \otimes H_n(M, \partial M; \mathbb{L}^{\bullet}) \otimes \mathbb{Q} & \xrightarrow{\cap} & H_{n-i}(M, \partial M; \mathbb{L}^{\bullet}) \otimes \mathbb{Q} \\
\Big\downarrow{\text{proj}} & & \Big\downarrow{\text{proj}} \\
H^{i+4l}(M; \mathbb{Q}) \otimes H_{n-4r}(M, \partial M; \mathbb{Q}) & \xrightarrow{\cap} & H_{n-i-4(l+r)}(M, \partial M; \mathbb{Q})
\end{array}
$$

commutes, where the lower product is the usual cap product in ordinary homology.

Let X^n be an oriented, compact pseudomanifold of positive dimension n with only isolated singularities and let $(M, \partial M)$ be the exterior, assumed to be smooth, of the singular set. We define the *reduced $\mathbb{L}^{\bullet}$-orientation* $[\hat{X}]_{\mathbb{L}}$ of $\hat{X}$ to be

$$[\hat{X}]_{\mathbb{L}} = [M, \partial M]_{\mathbb{L}} \in H_n(M, \partial M; \mathbb{L}^{\bullet}) = \widetilde{H}_n(\hat{X}; \mathbb{L}^{\bullet}).$$

(The "denormalization" $\hat{X}$ of X was defined in Section 2.6.2). We define the *reduced L-class* $\mathcal{L}_*(\hat{X})$ of $\hat{X}$ to be

$$\mathcal{L}_*(\hat{X}) = \mathcal{L}_*(M, \partial M) \in H_{n-4*}(M, \partial M; \mathbb{Q}) = \widetilde{H}_{n-4*}(\hat{X}; \mathbb{Q}).$$

Thus $[\hat{X}]_{\mathbb{L}} \otimes 1 = \mathcal{L}_*(\hat{X})$.

Definition 2.35. A homology class

$$u = u_n + u_{n-4} + u_{n-8} + \cdots \in \widetilde{H}_{n-4*}(\hat{X}; \mathbb{Q})$$

is called *unipotent* if

$$- \cap u_n : H^r(M, \partial M; \mathbb{Q}) \longrightarrow H_{n-r}(M; \mathbb{Q})$$

(and

$$- \cap u_n : H^r(M; \mathbb{Q}) \longrightarrow H_{n-r}(M, \partial M; \mathbb{Q}))$$

are isomorphisms for all r. An $\mathbb{L}^\bullet$-homology class $u \in \widetilde{H}_n(\hat{X}; \mathbb{L}^\bullet)$ is called *rationally unipotent* if $u \otimes 1 \in \widetilde{H}_n(\hat{X}; \mathbb{L}^\bullet) \otimes \mathbb{Q}$ is unipotent.

Examples 2.36. If X is an oriented compact pseudomanifold, then the fundamental class $u = [X] \in H_n(X; \mathbb{Q})$ is unipotent. Thus any class u with $u_n = [X]$ is unipotent. In particular, the L-class $u = \mathcal{L}_*(\hat{X})$ is unipotent, since the top-component of the homology L-class of a pseudomanifold is the fundamental class. The $\mathbb{L}^\bullet$-homology fundamental class $[\hat{X}]_{\mathbb{L}} \in \widetilde{H}_n(\hat{X}; \mathbb{L}^\bullet)$ is rationally unipotent as $[\hat{X}]_{\mathbb{L}} \otimes 1 = \mathcal{L}_*(\hat{X})$.

The following duality theorem covers all dimensions n, except $n \equiv 0(8)$.

Theorem 2.37. *Let X be an n-dimensional compact pseudomanifold, $n > 0$, with only isolated singularities. Capping with a rationally unipotent class $u \in \widetilde{H}_n(\hat{X}; \mathbb{L}^\bullet)$ induces an isomorphism*

$$- \cap u \otimes 1 : \widetilde{H}^0(I^{\bar{m}}X; \mathbb{L}^\bullet) \otimes \mathbb{Q} \overset{\cong}{\longrightarrow} \widetilde{H}_n(I^{\bar{m}}X; \mathbb{L}^\bullet) \otimes \mathbb{Q}$$

for $n \equiv 2 \mod 4$ and $n \equiv 4 \mod 8$ such that

$$
\begin{array}{ccc}
\widetilde{H}^0(I^{\bar{m}}X; \mathbb{L}^\bullet) \otimes \mathbb{Q} & \xrightarrow[-\cap u \otimes 1]{\cong} & \widetilde{H}_n(I^{\bar{m}}X; \mathbb{L}^\bullet) \otimes \mathbb{Q} \\
\big\uparrow c^* & & \big\downarrow c_* \\
\widetilde{H}^0(\hat{X}; \mathbb{L}^\bullet) \otimes \mathbb{Q} & \xrightarrow[-\cap u \otimes 1]{} & \widetilde{H}_n(\hat{X}; \mathbb{L}^\bullet) \otimes \mathbb{Q}
\end{array}
$$

commutes, an isomorphism

$$- \cap u \otimes 1 : \widetilde{H}^0(I^{\bar{m}}X; \mathbb{L}^\bullet) \otimes \mathbb{Q} \overset{\cong}{\longrightarrow} \widetilde{H}_n(I^{\bar{n}}X; \mathbb{L}^\bullet) \otimes \mathbb{Q}$$

for $n \equiv 1$ mod 4 such that

$$\widetilde{H}^0(I^{\bar{m}}X;\mathbf{L}^{\bullet}) \otimes \mathbb{Q} \xrightarrow[-\cap u \otimes 1]{\cong} \widetilde{H}_n(I^{\bar{n}}X;\mathbf{L}^{\bullet}) \otimes \mathbb{Q}$$

$$\left\uparrow\, c^* \qquad\qquad\qquad\qquad\qquad \downarrow c_*\right.$$

$$\widetilde{H}^0(\hat{X};\mathbf{L}^{\bullet}) \otimes \mathbb{Q} \xrightarrow[-\cap u \otimes 1]{} \widetilde{H}_n(\hat{X};\mathbf{L}^{\bullet}) \otimes \mathbb{Q}$$

commutes, and an isomorphism

$$-\cap u \otimes 1 : \widetilde{H}^0(I^{\bar{n}}X;\mathbf{L}^{\bullet}) \otimes \mathbb{Q} \xrightarrow{\cong} \widetilde{H}_n(I^{\bar{m}}X;\mathbf{L}^{\bullet}) \otimes \mathbb{Q}$$

for $n \equiv 3$ mod 4 such that

$$\widetilde{H}^0(I^{\bar{n}}X;\mathbf{L}^{\bullet}) \otimes \mathbb{Q} \xrightarrow[-\cap u \otimes 1]{\cong} \widetilde{H}_n(I^{\bar{m}}X;\mathbf{L}^{\bullet}) \otimes \mathbb{Q}$$

$$\left\uparrow\, c^* \qquad\qquad\qquad\qquad\qquad \downarrow c_*\right.$$

$$\widetilde{H}^0(\hat{X};\mathbf{L}^{\bullet}) \otimes \mathbb{Q} \xrightarrow[-\cap u \otimes 1]{} \widetilde{H}_n(\hat{X};\mathbf{L}^{\bullet}) \otimes \mathbb{Q}$$

commutes.

Proof. Let us provide the details for the case $n \equiv 2$ mod 4 first. We have

$$\widetilde{H}^0(I^{\bar{m}}X;\mathbf{L}^{\bullet}) \otimes \mathbb{Q} = \bigoplus_{l \geq 0} \widetilde{H}^{4l}(I^{\bar{m}}X;\mathbb{Q})$$
$$\cong \bigoplus_{4l<k} H^{4l}(M,\partial M;\mathbb{Q}) \oplus \bigoplus_{4l>k} H^{4l}(M;\mathbb{Q}),$$

where $k = n/2$ (an odd number). Let $\{\varepsilon_1^l, \ldots, \varepsilon_{j_l}^l\}$ be a basis for $H^{4l}(M,\partial M;\mathbb{Q})$ when $4l < k$ and for $H^{4l}(M;\mathbb{Q})$ when $4l > k$. The homology groups are rationally given by

$$\widetilde{H}_n(I^{\bar{m}}X;\mathbf{L}^{\bullet}) \otimes \mathbb{Q} = \bigoplus_{l \geq 0} \widetilde{H}_{n-4l}(I^{\bar{m}}X;\mathbb{Q})$$
$$\cong \bigoplus_{n-4l>k} H_{n-4l}(M;\mathbb{Q}) \oplus \bigoplus_{n-4l<k} H_{n-4l}(M,\partial M;\mathbb{Q}).$$

Since u is rationally unipotent, capping with the top component u_n of

$$u \otimes 1 = u_n + u_{n-4} + \ldots \in \widetilde{H}_{n-4*}(\hat{X};\mathbb{Q})$$

yields isomorphisms

$$-\cap u_n : H^{4l}(M,\partial M;\mathbb{Q}) \xrightarrow{\cong} H_{n-4l}(M;\mathbb{Q})$$

and

$$- \cap u_n : H^{4l}(M; \mathbb{Q}) \xrightarrow{\cong} H_{n-4l}(M, \partial M; \mathbb{Q}).$$

Thus, setting $e^l_j = \varepsilon^l_j \cap u_n$ yields bases $\{e^l_1, \ldots, e^l_{j_l}\}$ for $H_{n-4l}(M; \mathbb{Q})$ when $4l < k$ and for $H_{n-4l}(M, \partial M; \mathbb{Q})$ when $4l > k$. With respect to the basis

$$\{\varepsilon^0_1, \ldots, \varepsilon^0_{j_0}, \varepsilon^1_1, \ldots, \varepsilon^1_{j_1}, \ldots\}$$

of $\widetilde{H}^0(I^{\bar{m}}X; \mathbb{L}^\bullet) \otimes \mathbb{Q}$ and the basis

$$\{e^0_1, \ldots, e^0_{j_0}, e^1_1, \ldots, e^1_{j_1}, \ldots\}$$

of $\widetilde{H}_n(I^{\bar{m}}X; \mathbb{L}^\bullet) \otimes \mathbb{Q}$, the linear map $- \cap u \otimes 1$ can be expressed as a matrix U. The image of a basis vector ε^l_p, $1 \le p \le j_l$, is

$$\varepsilon^l_p \cap (u \otimes 1) = \varepsilon^l_p \cap u_n + \varepsilon^l_p \cap u_{n-4} + \varepsilon^l_p \cap u_{n-8} + \ldots \in H_{n-4l-4*}$$

$$= e^l_p + \sum_{j=1}^{j_{l+1}} \lambda^{l+1}_j e^{l+1}_j + \sum_{j=1}^{j_{l+2}} \lambda^{l+2}_j e^{l+2}_j + \ldots.$$

(The cap product used here is of course the one provided by Proposition 2.31. For $\varepsilon^l_p \cap u_{n-4i}$ with $4l < k < 4(i+l)$, this involves the map $b_* : \widetilde{H}_*(M) \to \widetilde{H}_*(I^{\bar{m}}X)$.) Hence, the ^l_p-column of U is

$$(\underbrace{0, \ldots, 0}_{j_0}, \ldots, \underbrace{0, \ldots, 0}_{j_{l-1}}, \underbrace{0, \ldots, \overset{p}{1}, \ldots, 0}_{j_l}, \lambda^{l+1}_1, \ldots, \lambda^{l+1}_{j_{l+1}}, \ldots)^T.$$

In terms of $(j_l \times j_r)$-block matrices, U has thus the form

$$U = \begin{pmatrix} I_{j_0} & 0 & 0 & \cdots \\ * & I_{j_1} & 0 & \cdots \\ * & * & I_{j_2} & \\ \vdots & \vdots & & \ddots \end{pmatrix},$$

where I_q denotes the $q \times q$ identity matrix. We see that U is a lower triangular matrix with entries 1 on the diagonal, i.e. a unipotent matrix. In particular, it is invertible and so $- \cap u \otimes 1$ is an isomorphism. The commutativity of the diagram follows from the commutative diagram of Proposition 2.31.

Let us explain why the other cases concerning the dimension n can be treated analogously and why the argument breaks down when $n \equiv 0 \mod 8$. Let $k = n - 1 - \bar{p}(n)$ be the cut-off value for the cohomology perversity ($\bar{p} = \bar{m}$ or $\bar{n}$) and $k' = n - 1 - \bar{q}(n)$ be the cut-off value for the homology perversity ($\bar{q} = \bar{n}$ or $\bar{m}$). In order for the above argument to work, the rational $\mathbb{L}^\bullet$-cohomology has to be decomposed into degrees $4l < k$ and $4l > k$, so we need $k \not\equiv 0(4)$. In addition, the rational $\mathbb{L}^\bullet$-homology has to be decomposed into degrees $n - 4l > k'$ and $n - 4l < k'$, so we also need $n - k' \not\equiv 0(4)$. Since for complementary middle perversities we have $k + k' = n$, the two conditions are equivalent. The following table shows that this condition is satisfied for all n (using appropriate complementary middle perversities), except when $n \equiv 0(8)$.

n	$4q+1$	$4q+2$	$4q+3$	$8q+4$	$8q$
$\bar{p}$	$\bar{m}$	$\bar{m}$	$\bar{n}$	$\bar{m}$	$\bar{m}$
k	$2q+1$	$2q+1$	$2q+1$	$4q+2$	$4q$
$\bar{q}$	$\bar{n}$	$\bar{m}$	$\bar{m}$	$\bar{m}$	$\bar{m}$
k'	$2q$	$2q+1$	$2q+2$	$4q+2$	$4q$

Once the decomposition has been carried out, using for the homology decomposition k' instead of k, the rest of the argument is the same. $\qquad\qquad \square$

Corollary 2.38. *Let X be an n-dimensional, compact, oriented pseudomanifold with only isolated singularities. Capping with the $\mathbb{L}^\bullet$-homology fundamental class $[\hat{X}]_{\mathbb{L}} \in \tilde{H}_n(\hat{X}; \mathbb{L}^\bullet)$ induces rationally an isomorphism*

$$- \cap [\hat{X}]_{\mathbb{L}} \otimes 1 : \tilde{H}^0(I^{\bar{m}}X; \mathbb{L}^\bullet) \otimes \mathbb{Q} \xrightarrow{\cong} \tilde{H}_n(I^{\bar{m}}X; \mathbb{L}^\bullet) \otimes \mathbb{Q}$$

for $n \equiv 2 \mod 4$ and $n \equiv 4 \mod 8$ such that

$$
\begin{array}{ccc}
\tilde{H}^0(I^{\bar{m}}X; \mathbb{L}^\bullet) \otimes \mathbb{Q} & \xrightarrow[-\cap[\hat{X}]_{\mathbb{L}}\otimes 1]{\cong} & \tilde{H}_n(I^{\bar{m}}X; \mathbb{L}^\bullet) \otimes \mathbb{Q} \\
\uparrow{\scriptstyle c^*} & & \downarrow{\scriptstyle c_*} \\
\tilde{H}^0(\hat{X}; \mathbb{L}^\bullet) \otimes \mathbb{Q} & \xrightarrow[-\cap[\hat{X}]_{\mathbb{L}}\otimes 1]{} & \tilde{H}_n(\hat{X}; \mathbb{L}^\bullet) \otimes \mathbb{Q}
\end{array}
$$

commutes, an isomorphism

$$- \cap [\hat{X}]_{\mathbb{L}} \otimes 1 : \tilde{H}^0(I^{\bar{m}}X; \mathbb{L}^\bullet) \otimes \mathbb{Q} \xrightarrow{\cong} \tilde{H}_n(I^{\bar{n}}X; \mathbb{L}^\bullet) \otimes \mathbb{Q}$$

for $n \equiv 1 \mod 4$ such that

$$
\begin{array}{ccc}
\tilde{H}^0(I^{\bar{m}}X; \mathbb{L}^\bullet) \otimes \mathbb{Q} & \xrightarrow[-\cap[\hat{X}]_{\mathbb{L}}\otimes 1]{\cong} & \tilde{H}_n(I^{\bar{n}}X; \mathbb{L}^\bullet) \otimes \mathbb{Q} \\
\uparrow{\scriptstyle c^*} & & \downarrow{\scriptstyle c_*} \\
\tilde{H}^0(\hat{X}; \mathbb{L}^\bullet) \otimes \mathbb{Q} & \xrightarrow[-\cap[\hat{X}]_{\mathbb{L}}\otimes 1]{} & \tilde{H}_n(\hat{X}; \mathbb{L}^\bullet) \otimes \mathbb{Q}
\end{array}
$$

commutes, and an isomorphism

$$-\cap[\hat{X}]_{\mathrm{L}}\otimes 1 : \tilde{H}^0(I^{\bar{n}}X;\mathbf{L}^\bullet)\otimes\mathbb{Q} \xrightarrow{\cong} \tilde{H}_n(I^{\bar{m}}X;\mathbf{L}^\bullet)\otimes\mathbb{Q}$$

for $n \equiv 3 \mod 4$ *such that*

$$
\begin{array}{ccc}
\tilde{H}^0(I^{\bar{n}}X;\mathbf{L}^\bullet)\otimes\mathbb{Q} & \xrightarrow[-\cap[\hat{X}]_{\mathrm{L}}\otimes 1]{\cong} & \tilde{H}_n(I^{\bar{m}}X;\mathbf{L}^\bullet)\otimes\mathbb{Q} \\
\Big\uparrow{\scriptstyle c^*} & & \Big\downarrow{\scriptstyle c_*} \\
\tilde{H}^0(\hat{X};\mathbf{L}^\bullet)\otimes\mathbb{Q} & \xrightarrow[-\cap[\hat{X}]_{\mathrm{L}}\otimes 1]{} & \tilde{H}_n(\hat{X};\mathbf{L}^\bullet)\otimes\mathbb{Q}
\end{array}
$$

commutes.

Proof. The class $u = [\hat{X}]_{\mathrm{L}}$ is rationally unipotent. $\qquad\qquad\square$

Example 2.39. Let us work out the duality for the 12-dimensional pseudomanifold

$$X^{12} = D^4 \times \mathbb{P}^4 \cup_{S^3 \times \mathbb{P}^4} \mathrm{cone}(S^3 \times \mathbb{P}^4),$$

where $\mathbb{P}^4$ denotes complex projective space. Let $g \in H^2(\mathbb{P}^4)$ be the negative of the first Chern class of the tautological line bundle over $\mathbb{P}^4$ so that $\langle g^4, [\mathbb{P}^4]\rangle = 1$ and the Pontrjagin class is $p(\mathbb{P}^4) = (1+g^2)^5 = 1 + 5g^2 + 10g^4$. As $\mathcal{L}^1(p_1) = \frac{1}{3}p_1$, we have $\mathcal{L}^1(\mathbb{P}^4) = \frac{5}{3}g^2$. Since the signature of $\mathbb{P}^4$ is 1, we have $\mathcal{L}^2(\mathbb{P}^4) = g^4$ by the Hirzebruch signature theorem, so that

$$\mathcal{L}^*(\mathbb{P}^4) = 1 + \frac{5}{3}g^2 + g^4.$$

In this example $(M,\partial M) = (D^4 \times \mathbb{P}^4, S^3 \times \mathbb{P}^4)$ and

$$\mathcal{L}^*(M) = 1 \times \mathcal{L}^*(\mathbb{P}^4) = 1 \times 1 + \frac{5}{3}1 \times g^2 + 1 \times g^4.$$

Let $\mu = [D^4, S^3] \in H_4(D^4, S^3)$ and $[\mathbb{P}^4] \in H_8(\mathbb{P}^4)$ be the fundamental classes. Then the homology L-class of $(M, \partial M)$ is given by

$$\begin{aligned}
\mathcal{L}_*(M,\partial M) &= (1 \times 1) \cap \mu \times [\mathbb{P}^4] + \tfrac{5}{3}(1 \times g^2) \cap \mu \times [\mathbb{P}^4] + (1 \times g^4) \cap \mu \times [\mathbb{P}^4] \\
&= \mu \times [\mathbb{P}^4] + \tfrac{5}{3}\mu \times [\mathbb{P}^2] + \mu \times [\mathbb{P}^0].
\end{aligned}$$

The link L of the singularity of X is $L = S^3 \times \mathbb{P}^4$. The cut-off-value k for the middle-perversity intersection space $I^{\bar{m}}X$ is $k = n-1-\bar{m}(n) = 11 - \bar{m}(12) = 6$. Since all boundary operators in the cellular chain complexes $C_*(S^3)$ and $C_*(\mathbb{P}^4)$ vanish, the boundary operators in the complex $C_*(S^3 \times \mathbb{P}^4)$ vanish also because they are given by the Leibniz formula. Thus $L_{<k} = L_{<6}$ is the 5-skeleton of $S^3 \times \mathbb{P}^4$ and $I^{\bar{m}}X$ is the cofiber of the composite cofibration

$$(S^3 \times \mathbb{P}^4)^5 \hookrightarrow L = \partial M \hookrightarrow D^4 \times \mathbb{P}^4.$$

Let $d \in H^4(D^4, S^3)$ be the unique generator such that $d \cap \mu = [\text{pt}] \in H_0(D^4)$. For the $\mathbb{L}^\bullet$-homology we have

$$\begin{aligned}
\tilde{H}_{12}(I^{\bar{m}}X; \mathbb{L}^\bullet) \otimes \mathbb{Q} &= \tilde{H}_{12}(I^{\bar{m}}X) \oplus \tilde{H}_8(I^{\bar{m}}X) \oplus \tilde{H}_4(I^{\bar{m}}X) \oplus \tilde{H}_0(I^{\bar{m}}X) \\
&= H_{12}(M) \oplus H_8(M) \oplus H_4(M, \partial M) \oplus H_0(M, \partial M) \\
&= \quad 0 \quad \oplus \mathbb{Q}[\text{pt}] \times [\mathbb{P}^4] \oplus \mathbb{Q}\mu \times [\mathbb{P}^0] \oplus \quad 0,
\end{aligned}$$

and for the $\mathbb{L}^\bullet$-cohomology

$$\begin{aligned}
\tilde{H}^0(I^{\bar{m}}X; \mathbb{L}^\bullet) \otimes \mathbb{Q} &= \tilde{H}^0(I^{\bar{m}}X) \oplus \tilde{H}^4(I^{\bar{m}}X) \oplus \tilde{H}^8(I^{\bar{m}}X) \oplus \tilde{H}^{12}(I^{\bar{m}}X) \\
&= H^0(M, \partial M) \oplus H^4(M, \partial M) \oplus H^8(M) \oplus H^{12}(M) \\
&= \quad 0 \quad \oplus \quad \mathbb{Q}d \times 1 \quad \oplus \mathbb{Q}1 \times g^4 \oplus \quad 0.
\end{aligned}$$

Setting $\varepsilon^1 = d \times 1$ and $\varepsilon^2 = 1 \times g^4$, we obtain a basis $\{\varepsilon^1, \varepsilon^2\}$ for $\tilde{H}^0(I^{\bar{m}}X; \mathbb{L}^\bullet) \otimes \mathbb{Q}$. The dual basis for $\tilde{H}_{12}(I^{\bar{m}}X; \mathbb{L}^\bullet) \otimes \mathbb{Q}$ is $\{e^1, e^2\}$, with

$$\begin{aligned}
e^1 &= \varepsilon^1 \cap u_{12} = d \times 1 \cap \mu \times [\mathbb{P}^4] = [\text{pt}] \times [\mathbb{P}^4], \\
e^2 &= \varepsilon^2 \cap u_{12} = 1 \times g^4 \cap \mu \times [\mathbb{P}^4] = \mu \times [\mathbb{P}^0].
\end{aligned}$$

The images of the basis elements under cap product with the reduced L-class of $\hat{X}$ are

$$\begin{aligned}
\varepsilon^1 \cap \mathcal{L}_*(\hat{X}) &= (d \times 1) \cap (\mu \times [\mathbb{P}^4] + \tfrac{5}{3}\mu \times [\mathbb{P}^2] + \mu \times [\mathbb{P}^0]) \\
&= e^1 + \tfrac{5}{3}b_*((d \times 1) \cap (\mu \times [\mathbb{P}^2])) \\
&= e^1,
\end{aligned}$$

since the map $b_* : \tilde{H}_4(M) \to H_4(M, \partial M)$ is zero (its neighboring maps in the sequence of the pair are isomorphisms), and

$$\begin{aligned}
\varepsilon^2 \cap \mathcal{L}_*(\hat{X}) &= (1 \times g^4) \cap (\mu \times [\mathbb{P}^4] + \tfrac{5}{3}\mu \times [\mathbb{P}^2] + \mu \times [\mathbb{P}^0]) \\
&= e^2.
\end{aligned}$$

Thus in the bases $\{\varepsilon^1, \varepsilon^2\}$ and $\{e^1, e^2\}$, the map

$$-\cap [\hat{X}]_{\mathbb{L}} \otimes 1 : \tilde{H}^0(I^{\bar{m}}X; \mathbb{L}^\bullet) \otimes \mathbb{Q} \xrightarrow{\cong} \tilde{H}_{12}(I^{\bar{m}}X; \mathbb{L}^\bullet) \otimes \mathbb{Q}$$

is given by the identity matrix

$$U = \begin{pmatrix} 1 & 0 \\ 0 & 1 \end{pmatrix}.$$

The pseudomanifold X itself does not possess Poincaré duality. The cohomology group $H^4(X;\mathbb{Q})$ is generated by $d \times 1$, which is dual to $\mathrm{pt} \times [\mathbb{P}^4]$. However, the cycle $\mathrm{pt} \times [\mathbb{P}^4]$ is zero in $H_8(X;\mathbb{Q}) = \mathbb{Q}\langle \mu \times [\mathbb{P}^2] \rangle$.

2.8 Intersection Vector Bundles and K-Theory

Given a pseudomanifold X with fixed intersection space $I^{\bar{p}}X$, we may define a $\bar{p}$-intersection vector bundle ξ on X to be an actual vector bundle ξ on $I^{\bar{p}}X$. That is, the isomorphism classes $I^{\bar{p}}VB_{\mathbb{R}}(X)$ of real n-plane $\bar{p}$-intersection vector bundles on X may be defined by

$$I^{\bar{p}}VB_{\mathbb{R}}(X) = [I^{\bar{p}}X, BO_n]$$

and similarly for complex intersection vector bundles using BU_n in place of BO_n. More generally, given any structure group G, one may describe *principal intersection G-bundles* over X as

$$I^{\bar{p}}\mathrm{Princ}_G(X) = [I^{\bar{p}}X, BG].$$

The variation of these notions over different choices of $I^{\bar{p}}X$ for fixed X remains to be investigated. Any vector bundle over $\hat{X}$ determines a $\bar{p}$-intersection vector bundle on X by pulling back under the canonical map $c : I^{\bar{p}}X \to \hat{X}$. Naturally, a complex intersection vector bundle on X has Chern classes in the intersection space cohomology of X.

As in the previous section, there are Poincaré duality statements between the reduced rational K-theory of the intersection space, $\widetilde{K}^*(I^{\bar{m}}X) \otimes \mathbb{Q}$, and reduced rational K-homology $\widetilde{K}_*(I^{\bar{n}}X) \otimes \mathbb{Q}$. These can be worked out in analogy with the previous section, observing that the rational type of the K-spectrum and KO-spectrum can be easily understood using the Chern and the Pontrjagin character, respectively.

Let M, as usual, denote the exterior of the singular set of X. If M is smooth, for example X Whitney stratified, then it has a tangent bundle TM, which defines an element in $\widetilde{KO}^0(M)$. Even in the isolated singularity situation, X itself will not have a tangent bundle in the classical sense of vector bundle theory, restricting to TM, unless the link of the singularity is parallelizable. Let $a : M \to X$ and $b : M \to I^{\bar{p}}X$ be the canonical maps, see Section 2.6.2. It may very well happen (see Example 2.40 below) that the tangent bundle element does not lift under

$$\widetilde{KO}^0(X) \xrightarrow{a^*} \widetilde{KO}^0(M),$$

but *does* lift back to the KO-theory of the intersection space $I^{\bar{p}}X$ under

$$\widetilde{KO}^0(I^{\bar{p}}X) \xrightarrow{b^*} \widetilde{KO}^0(M).$$

Indeed, the higher the perversity $\bar{p}$, the closer $I^{\bar{p}}X$ is to M, and the easier it becomes to lift. The intersection space $I^{\bar{p}}X$ (in the isolated singularity case) is the mapping cone cone(g) of a map $g : L_{<k} \to M$. The cofibration sequence

$$L_{<k} \xrightarrow{g} M \xrightarrow{b} I^{\bar{p}}X = \text{cone}(g) \longrightarrow S(L_{<k}),$$

where $S(-)$ denotes reduced suspension, induces an exact sequence

$$\widetilde{KO}^{-1}(L_{<k}) \longrightarrow \widetilde{KO}^{0}(I^{\bar{p}}X) \xrightarrow{b^*} \widetilde{KO}^{0}(M) \xrightarrow{g^*} \widetilde{KO}^{0}(t_{<k}L),$$

which can be used to investigate existence and uniqueness of such lifts. Thus singular pseudomanifolds may have (stable classes of) $\bar{p}$-*intersection tangent bundles*, even when they do not have actual tangent bundles. Such a $\bar{p}$-intersection tangent bundle will have characteristic classes, for example Chern classes $c_i \in H^{2i}(I^{\bar{p}}X)$ in the complex case, or Pontrjagin classes $p_i \in H^{4i}(I^{\bar{p}}X)$ in the real case. Using the cap products of Section 2.6, one can multiply these characteristic classes with any homology class in X and will get a class in the homology of an intersection space of X, not merely an ordinary homology class of X. (If $\bar{p}$ is a middle perversity, then the resulting class will again lie in the homology of a middle perversity intersection space.)

Example 2.40. By surgery theory, there exist infinitely many smooth manifolds L_i, $i = 1, 2, \ldots$, in the homotopy type of $S^2 \times S^4$, distinguished by the first Pontrjagin class of their tangent bundle, $p_1(TL_i) \in H^4(S^2 \times S^4) \cong \mathbb{Z}$, namely, $p_1(TL_i) = Pi$, P a fixed integer $\neq 0$. Let L^6 be any such manifold, $p_1(L) \neq 0$. A smooth triangulation, for example, gives L a CW-structure. Since the bordism group Ω_6^{SO} is trivial, there exists a smooth compact oriented manifold M^7 with $\partial M = L$. Set

$$X = M \cup_L \text{cone}(L).$$

We will show that the tangent bundle element $t = [TM] - [\theta_M^7] \in \widetilde{KO}^{0}(M)$, where θ_X^r is the (isomorphism class of the) trivial r-plane bundle over a space X, has no lift under

$$\widetilde{KO}^{0}(X) \xrightarrow{a^*} \widetilde{KO}^{0}(M),$$

but *does* have a lift under

$$\widetilde{KO}^{0}(I^{\bar{n}}X) \xrightarrow{b^*} \widetilde{KO}^{0}(M),$$

$a : M \to X$, $b : M \to I^{\bar{n}}X$. Since $X \cong M/L$ and $a : M \to X \cong M/L$ is homotopic to the quotient map, the cofibration sequence $L \xhookrightarrow{j} M \xrightarrow{a} X$ induces an exact sequence

$$\widetilde{KO}^{0}(X) \xrightarrow{a^*} \widetilde{KO}^{0}(M) \xrightarrow{j^*} \widetilde{KO}^{0}(L).$$

Thus t lifts back to $\widetilde{KO}^0(X)$ if and only if $j^*(t) = 0$. To show that in fact $j^*(t) \neq 0$, we use the Pontrjagin character ph as a detector,

$$ph : KO^0(-) \longrightarrow \bigoplus_{i \geq 0} H^{4i}(-; \mathbb{Q}),$$

$$ph = \mathrm{rank} + p_1 + \frac{1}{12}(p_1^2 - 2p_2) + \cdots.$$

Using the naturality of the Pontrjagin classes and observing that classes of degree 8 and higher vanish in the cohomology of M as M is seven-dimensional, we calculate

$$\begin{aligned} ph(j^*t) &= j^* ph([TM] - [\theta_M^7]) = j^*(\mathrm{rk}(TM) + p_1(TM) - \mathrm{rk}(\theta_M^7) - p_1(\theta_M^7)) \\ &= j^* p_1(TM) = p_1(TM|_{\partial M}) = p_1(TL \oplus \theta_L^1) = p_1(L) \neq 0. \end{aligned}$$

Thus $j^*t \neq 0$ and t cannot be lifted back to $\widetilde{KO}^0(X)$.

The manifold L, being homotopy equivalent to $S^2 \times S^4$, is an object of the interleaf category **ICW**. (See also Example 1.65(2).) Thus to construct the spatial homology truncation $t_{<k}L$, where $k = n - 1 - \bar{n}(n) = 3$, we may use the functor $t_{<k} : \mathbf{ICW} \to \mathbf{HoCW}$ of Section 1.9. The natural transformation $emb_3 : t_{<3} \to t_{<\infty}$ gives a homotopy class $[f] = emb_3(L) : t_{<3}L \to L$, whose canonical representative f is $t_{<3}L = E(L)^2 \overset{incl}{\hookrightarrow} E(L) \overset{h'_L}{\to} L$, with h'_L a cellular homotopy equivalence. Since h'_L is cellular, its restriction $(h'_L)^2$ to the 2-skeleton maps into the 2-skeleton L^2 of L and we have a factorization

$$\begin{CD} E(L)^2 @>f>> L \\ @V(h'_L)^2VV @AAi_2A \\ & L^2. \end{CD}$$

The intersection space $I^{\bar{n}}X$ is the mapping cone of $g : t_{<3}L \to M$, where g is the composition

$$t_{<3}L \overset{f}{\to} L \overset{j}{\hookrightarrow} M.$$

The cofibration sequence $t_{<3}L \overset{g}{\to} M \overset{b}{\to} I^{\bar{n}}X = \mathrm{cone}(g)$ induces an exact sequence

$$\widetilde{KO}^0(I^{\bar{n}}X) \overset{b^*}{\to} \widetilde{KO}^0(M) \overset{g^*}{\to} \widetilde{KO}^0(t_{<3}L),$$

which shows that t lifts back to $\widetilde{KO}^0(I^{\bar{n}}X)$ if and only if $g^*(t) = 0$. Let us prove first that L is spinnable, i.e. the restriction of its tangent bundle TL to the 2-skeleton is trivial. While the Pontrjagin classes of closed manifolds are of course not homotopy invariant, Wu's formula implies that the Stiefel-Whitney classes w_i of closed manifolds are homotopy invariants. Thus $w_1(L) = w_1(S^2 \times S^4)$, $w_2(L) = w_2(S^2 \times S^4)$. As $H^1(S^2 \times S^4; \mathbb{Z}/2) = 0$, we have $w_1(L) = 0$. The second Wu class

$v_2 = v_2(S^2 \times S^4) \in H^2(S^2 \times S^4; \mathbb{Z}/2) \cong \mathbb{Z}/2$ is determined by $v_2 \cup x = Sq^2(x)$ for all x. Since $Sq^2 : H^*(S^2 \times S^4; \mathbb{Z}/2) \to H^{*+2}(S^2 \times S^4; \mathbb{Z}/2)$ is zero, as follows, for instance, from the Cartan formula , we have $v_2 = 0$. By Wu's formula, $w_2(L) = w_2(S^2 \times S^4) = v_2(S^2 \times S^4) = 0$. Let $V_5(TL)$ denote the 5-frame Stiefel manifold bundle associated to TL. There exists a cross-section of $V_5(TL)$ over the 1-skeleton L^1 of L. There exists a cross-section over the 2-skeleton L^2 if and only if a primary obstruction class in $H^2(L; \mathbb{Z}/2)$ vanishes, and that class is $w_2(L)$, indeed zero. Thus $TL|_{L^2} \cong \theta_{L^2}^5 \oplus \lambda^1$, where λ^1 is some line bundle over L^2. Now

$$w_1(\lambda^1) = w_1(\lambda^1 \oplus \theta_{L^2}^5) = w_1(TL|_{L^2}) = i_2^* w_1(L) = 0,$$

whence λ^1 is trivial also. Hence $TL|_{L^2} \cong \theta_{L^2}^6$ and the element $g^*(t)$ is

$$\begin{aligned}
g^*(t) &= f^* j^* ([TM] - [\theta_M^7]) \\
&= f^* ([TM|_L] - [\theta_M^7|_L]) \\
&= f^* ([TL \oplus \theta_L^1] - [\theta_L^7]) \\
&= f^* ([TL] - [\theta_L^6]) \\
&= (h_L')^{2*} i_2^* ([TL] - [\theta_L^6]) \\
&= (h_L')^{2*} ([TL|_{L^2}] - [\theta_L^6|_{L^2}]) \\
&= (h_L')^{2*} ([\theta_{L^2}^6] - [\theta_{L^2}^6]) \\
&= 0.
\end{aligned}$$

Therefore, t lifts back to $\widetilde{KO}^0(I^{\bar{n}}X)$.

2.9 Beyond Isolated Singularities

Let X be an n-dimensional, compact, stratified pseudomanifold with two strata

$$X = X_n \supset X_{n-c}.$$

The singular set $\Sigma = X_{n-c}$ is thus an $(n - c)$-dimensional closed manifold and the singularities are not isolated, unless $c = n$. Assume that X has a trivial link bundle, that is, a neighborhood of Σ in X looks like $\Sigma \times \overset{\circ}{\text{cone}} L$, where L is a $(c - 1)$-dimensional closed manifold, the link of Σ. We assume furthermore that L is a simply connected CW-complex in order to be able to apply the spatial homology truncation machine of Section 1.1. For such a pseudomanifold X, we shall construct associated *perversity $\bar{p}$ intersection spaces* $I^{\bar{p}}X$ by performing truncation fiberwise. If $k = c - 1 - \bar{p}(c) \geq 3$, we can and do fix a completion (L, Y) of L so that (L, Y) is an object in $\mathbf{CW}_{k \supset \partial}$. If $k \leq 2$, no group Y has to be chosen and we simply apply the low-degree truncation of Section 1.1.5. Applying the truncation $t_{<k} : \mathbf{CW}_{k \supset \partial} \to \mathbf{HoCW}_{k-1}$ as defined on page 50, we obtain a

CW-complex $t_{<k}(L,Y) \in Ob\, \mathbf{HoCW}_{k-1}$. The natural transformation $\mathrm{emb}_k : t_{<k} \to t_{<\infty}$ of Theorem 1.41 gives a homotopy class

$$f = \mathrm{emb}_k(L,Y) : t_{<k}(L,Y) \longrightarrow L$$

such that for $r < k$,

$$f_* : H_r(t_{<k}(L,Y)) \cong H_r(L),$$

while $H_r(t_{<k}(L,Y)) = 0$ for $r \geq k$. Let M^n be the compact manifold-with-boundary obtained by removing from X an open neighborhood $\Sigma \times \overset{\circ}{\mathrm{cone}} L$ of Σ. Thus the boundary of M is $\partial M = \Sigma \times L$. Let

$$g : \Sigma \times t_{<k}(L,Y) \longrightarrow M$$

be the composition

$$\Sigma \times t_{<k}(L,Y) \xrightarrow{\mathrm{id}_\Sigma \times f} \Sigma \times L = \partial M \xrightarrow{j} M.$$

The intersection space will be the homotopy cofiber of g:

Definition 2.41. The *perversity $\bar{p}$ intersection space* $I^{\bar{p}} X$ of X is defined to be

$$\boxed{I^{\bar{p}} X = \mathrm{cone}(g) = M \cup_g \mathrm{cone}(\Sigma \times t_{<k}(L,Y)).}$$

(More precisely, $I^{\bar{p}} X$ is a homotopy type of a space.) As pointed out in Section 2.2, the construction simplifies if the link happens to lie in the interleaf category **ICW**, for then we apply $t_{<k} : \mathbf{ICW} \to \mathbf{HoCW}$ instead of $t_{<k} : \mathbf{CW}_{k \supset \partial} \to \mathbf{HoCW}_{k-1}$.

Rational coefficients for homology and cohomology will be understood for the rest of this section. If N is a simply connected CW-complex, k an integer, and $N_{<k}$ a homological k-truncation of N with structure map $f : N_{<k} \to N$ (so that f_* on homology is an isomorphism in degrees less than k), then we shall often think of f up to homotopy as an inclusion, by replacing N with the mapping cylinder of f. We shall thus also use the notation $H_*(N,N_{<k})$ for the reduced homology of the mapping cone of f. A statement similar to Lemma 2.42 below was already discussed in Proposition 1.75; nevertheless we shall provide details.

Lemma 2.42. *Let N be a simply connected CW-complex. Then the map*

$$\pi_* : H_r(N) \longrightarrow H_r(N,N_{<k})$$

induced on homology by the inclusion map is an isomorphism when $r \geq k$, while $H_r(N,N_{<k}) = 0$ when $r < k$.

Proof. If $r < k$, then the long exact homology sequence of f has the form

$$H_r(N_{<k}) \xrightarrow{\cong} H_r(N) \xrightarrow{0} H_r(N,N_{<k}) \xrightarrow{\partial_*=0} H_{r-1}(N_{<k}) \xrightarrow{\cong} H_{r-1}(N),$$

whence $H_r(N, N_{<k}) = 0$. For $r = k$, it has the form

$$0 = H_k(N_{<k}) \longrightarrow H_k(N) \longrightarrow H_k(N, N_{<k}) \xrightarrow{\partial_* = 0} H_{k-1}(N_{<k}) \xrightarrow{\cong} H_{k-1}(N),$$

so that $H_k(N) \longrightarrow H_k(N, N_{<k})$ is an isomorphism. Finally, if $r > k$, then the exact sequence

$$0 = H_r(N_{<k}) \longrightarrow H_r(N) \longrightarrow H_r(N, N_{<k}) \xrightarrow{\partial_*} H_{r-1}(N_{<k}) = 0$$

again exhibits $H_r(N) \longrightarrow H_r(N, N_{<k})$ as an isomorphism.   □

Proposition 2.43. *Let N^n be a closed, oriented, simply connected manifold equipped with a CW-structure. Let k be an integer. Let $N_{<k}$ be any homological k-truncation and $N_{<n-k+1}$ be any homological $(n-k+1)$-truncation of N.*
(1) There exists a cap product

$$H^{n-r}(N_{<k}) \otimes H_n(N) \xrightarrow{\cap} H_r(N, N_{<n-k+1})$$

such that

$$
\begin{array}{ccc}
H^{n-r}(N_{<k}) \otimes H_n(N) & \xrightarrow{\cap} & H_r(N, N_{<n-k+1}) \\
\Big\uparrow{\scriptstyle f^* \otimes \mathrm{id}} & & \Big\uparrow{\scriptstyle \pi_*} \\
H^{n-r}(N) \otimes H_n(N) & \xrightarrow{\ \cap\ } & H_r(N)
\end{array}
\tag{2.28}
$$

commutes.

(2) Capping with the fundamental class $[N] \in H_n(N)$ is an isomorphism

$$- \cap [N] : H^{n-r}(N_{<k}) \xrightarrow{\cong} H_r(N, N_{<n-k+1}).$$

Proof. (1) We consider the two cases $n - r < k$ and $n - r \geq k$ separately. Suppose $n - r < k$. Then f^* is an isomorphism and we define the cap product of $\xi \in H^{n-r}(N_{<k})$ and $x \in H_n(N)$ by

$$\xi \cap x = \pi_*((f^*)^{-1}(\xi) \cap x).$$

By definition, diagram (2.28) commutes. If $n - r \geq k$ then $H^{n-r}(N_{<k}) = 0$ and we set $\xi \cap x = 0 \in H_r(N, N_{<n-k+1})$. This is in fact the only available value, since $n - r \geq k$ implies $r < n - k + 1$, and by Lemma 2.42, $H_r(N, N_{<n-k+1}) = 0$. In particular, the diagram commutes in this case as well.

(2) Suppose $n - r < k$. As this implies $r \geq n - k + 1$, Lemma 2.42 asserts that

$$\pi_* : H_r(N) \longrightarrow H_r(N, N_{<n-k+1})$$

is an isomorphism. The map f^* is an isomorphism, too, and the claim follows from Poincaré duality for the manifold N and the commutativity of the diagram

$$
\begin{array}{ccc}
H^{n-r}(N_{<k}) & \xrightarrow{\;-\cap[N]\;} & H_r(N,N_{<n-k+1}) \\
\uparrow{\scriptstyle f^*}\ {\scriptstyle\cong} & & {\scriptstyle\cong}\ \uparrow{\scriptstyle \pi_*} \\
H^{n-r}(N) & \xrightarrow[\;\cong\;]{\;-\cap[N]\;} & H_r(N).
\end{array}
$$

If $n - r \geq k$, then both $H^{n-r}(N_{<k})$ and $H_r(N,N_{<n-k+1})$ are zero, using Lemma 2.42.
□

Proposition 2.44. *Let Σ^s, N^n be closed, oriented manifolds with N simply connected and equipped with a CW-structure. Let k be an integer.*
(1) There exists a cap product

$$
H^{s+n-r}(\Sigma \times N_{<k}) \otimes H_{s+n}(\Sigma \times N) \xrightarrow{\;\cap\;} H_r(\Sigma \times (N,N_{<n-k+1}))
$$

such that

$$
\begin{array}{ccc}
H^{s+n-r}(\Sigma \times N_{<k}) \otimes H_{s+n}(\Sigma \times N) & \xrightarrow{\;\cap\;} & H_r(\Sigma \times (N,N_{<n-k+1})) \\
\uparrow{\scriptstyle (\mathrm{id}_\Sigma \times f)^* \otimes \mathrm{id}} & & \uparrow{\scriptstyle \mathrm{incl}_*} \\
H^{s+n-r}(\Sigma \times N) \otimes H_{s+n}(\Sigma \times N) & \xrightarrow{\;\cap\;} & H_r(\Sigma \times N)
\end{array}
\tag{2.29}
$$

commutes.

(2) Capping with the fundamental class $[\Sigma \times N] \in H_{s+n}(\Sigma \times N)$ is an isomorphism

$$
-\cap[\Sigma \times N] : H^{s+n-r}(\Sigma \times N_{<k}) \xrightarrow{\;\cong\;} H_r(\Sigma \times (N,N_{<n-k+1})).
$$

Proof. In the interest of better readability, we shall denote the product to be constructed by $\cap'$ and the product of Proposition 2.43 by $\tilde{\cap}$. (1) Let $\xi \in H^{s+n-r}(\Sigma \times N_{<k})$ and $x \in H_{s+n}(\Sigma \times N)$. By the Künneth theorem, these elements can be uniquely written as

$$
\xi = \sum_{p+q=s+n-r} \sum_i \sigma_p^{(i)} \times v_q^{(i)}, \quad \sigma_p^{(i)} \in H^p(\Sigma),\ v_q^{(i)} \in H^q(N_{<k}),
$$
$$
x = u \times v,\ u \in H_s(\Sigma),\ v \in H_n(N).
$$

(For the latter equation, observe that Σ need not be connected, but N is connected by assumption.) We define

$$\xi \cap' x = \sum_{p+q=s+n-r} (-1)^{p(n-q)} \sum_i (\sigma_p^{(i)} \cap u) \times (v_q^{(i)} \widetilde{\cap} v),$$

with $\sigma_p^{(i)} \cap u \in H_{s-p}(\Sigma)$ and $v_q^{(i)} \widetilde{\cap} v \in H_{n-q}(N, N_{<n-k+1})$. (Recall that we are using the sign conventions of [Spa66].) Let us verify that diagram (2.29) commutes. Given $\eta \in H^{s+n-r}(\Sigma \times N)$ and $x \in H_{s+n}(\Sigma \times N)$, write

$$\eta = \sum_{p+q=s+n-r} \sum_i \sigma_p^{(i)} \times \mu_q^{(i)}, \quad \sigma_p^{(i)} \in H^p(\Sigma), \ \mu_q^{(i)} \in H^q(N),$$

$$x = u \times v, \ u \in H_s(\Sigma), \ v \in H_n(N).$$

Then

$$
\begin{aligned}
\mathrm{incl}_*(\eta \cap x) &= (\mathrm{id}_\Sigma \times \pi)_* ((\textstyle\sum \sigma_p^{(i)} \times \mu_q^{(i)}) \cap (u \times v)) \\
&= \textstyle\sum (-1)^{p(n-q)} (\mathrm{id}_\Sigma \times \pi)_* (\sigma_p^{(i)} \cap u) \times (\mu_q^{(i)} \cap v) \\
&= \textstyle\sum (-1)^{p(n-q)} (\sigma_p^{(i)} \cap u) \times \pi_*(\mu_q^{(i)} \cap v) \\
&= \textstyle\sum (-1)^{p(n-q)} (\sigma_p^{(i)} \cap u) \times (f^*\mu_q^{(i)} \widetilde{\cap} v) \qquad \text{(by Proposition 2.43)} \\
&= \textstyle\sum (\sigma_p^{(i)} \times f^*\mu_q^{(i)}) \cap' (u \times v) \\
&= \textstyle\sum ((\mathrm{id}_\Sigma \times f)^*(\sigma_p^{(i)} \times \mu_q^{(i)})) \cap' x \\
&= ((\mathrm{id}_\Sigma \times f)^* \eta) \cap' x.
\end{aligned}
$$

(2) Let $\{\sigma_p^{(i)}\}_i$ be a basis for $H^p(\Sigma)$ and let $\{v_q^{(j)}\}_j$ be a basis for $H^q(N_{<k})$. Then $\{\sigma_p^{(i)} \otimes v_q^{(j)}\}_{i,j}$ is a basis for $H^p(\Sigma) \otimes H^q(N_{<k})$ and thus, by the Künneth theorem,

$$\{\sigma_p^{(i)} \times v_q^{(j)}\}_{\substack{i,j,p,q \\ p+q=s+n-r}}$$

is a basis for $H^{s+n-r}(\Sigma \times N_{<k})$. We shall show that

$$\{(\sigma_p^{(i)} \times v_q^{(j)}) \cap' [\Sigma \times N]\}_{\substack{i,j,p,q \\ p+q=s+n-r}}$$

is a basis for $H_r(\Sigma \times (N, N_{<n-k+1}))$. Set

$$
\begin{aligned}
a_p^{(i)} &= (-1)^{p(p+r-s)} \sigma_p^{(i)} \cap [\Sigma] \in H_{s-p}(\Sigma), \\
b_q^{(j)} &= v_q^{(j)} \widetilde{\cap} [N] \in H_{n-q}(N, N_{<n-k+1}).
\end{aligned}
$$

Since

$$H^p(\Sigma) \xrightarrow{-\cap[\Sigma]} H_{s-p}(\Sigma)$$

is an isomorphism by Poincaré duality, $\{a_p^{(i)}\}_i$ is a basis for $H_{s-p}(\Sigma)$. By Proposition 2.43(2),

$$H^q(N_{<k}) \xrightarrow{-\widetilde{\cap}[N]} H_{n-q}(N,N_{<n-k+1})$$

is an isomorphism, so that $\{b_q^{(j)}\}_j$ is a basis for $H_{n-q}(N,N_{<n-k+1})$. Thus $\{a_p^{(i)} \otimes b_q^{(j)}\}_{i,j}$ is a basis for $H_{s-p}(\Sigma) \otimes H_{n-q}(N,N_{<n-k+1})$ and

$$\{a_p^{(i)} \otimes b_q^{(j)}\}_{\substack{i,j,p,q \\ p+q=s+n-r}}$$

is a basis for

$$\bigoplus_{(s-p)+(n-q)=r} H_{s-p}(\Sigma) \otimes H_{n-q}(N,N_{<n-k+1}).$$

By the Künneth theorem,

$$\{a_p^{(i)} \times b_q^{(j)}\}_{\substack{i,j,p,q \\ p+q=s+n-r}}$$

is a basis for $H_r(\Sigma \times (N,N_{<n-k+1}))$. Since

$$
\begin{aligned}
(\sigma_p^{(i)} \times v_q^{(j)}) \cap' [\Sigma \times N] &= (\sigma_p^{(i)} \times v_q^{(j)}) \cap' ([\Sigma] \times [N]) \\
&= (-1)^{p(n-q)} (\sigma_p^{(i)} \cap [\Sigma]) \times (v_q^{(j)} \widetilde{\cap} [N]) \\
&= (-1)^{p(p+r-s)} (\sigma_p^{(i)} \cap [\Sigma]) \times (v_q^{(j)} \widetilde{\cap} [N]) \\
&= a_p^{(i)} \times b_q^{(j)}
\end{aligned}
$$

for $p+q = s+n-r$, the claim is established. $\square$

We return to the notation present in the definition of $I^{\bar{p}}X$. The manifold Σ thus has dimension $n-c$ and the link L has dimension $c-1$. Assume that X^n is oriented and that the singular stratum Σ and the link L are oriented in a compatible way, that is, for the fundamental classes we have $[\partial M] = [\Sigma \times L] = [\Sigma] \times [L]$, where M, and hence ∂M, receive their orientation from the orientation of X. Put $L_{<k} = t_{<k}(L,Y)$. Choose a Y' such that (L,Y') is an object of $\mathbf{CW}_{(c-k)\supset\partial}$ and set $L_{<c-k} = t_{<c-k}(L,Y')$. (If $c-k \leq 2$, no Y' has to be chosen and we apply low-degree truncations, as usual.)

Lemma 2.45. *The diagram*

$$
\begin{CD}
H^{n-r}(M) @>{g^*}>> H^{n-r}(\Sigma \times L_{<k}) \\
@V{-\cap[M,\partial M]}V{\cong}V @V{\cong}V{-\cap[\Sigma \times L]}V \\
H_r(M,\partial M) @>{\partial_*}>> H_{r-1}(\Sigma \times (L,L_{<c-k}))
\end{CD}
\qquad (2.30)
$$

commutes (there is no sign here), where ∂_* is the connecting homomorphism for the triple $(M, \partial M = \Sigma \times L, \Sigma \times L_{<c-k})$.

Proof. The connecting homomorphism $\partial_* : H_n(M, \partial M) \to H_{n-1}(\partial M)$ sends the fundamental class $[M, \partial M]$ to $\partial_*[M, \partial M] = [\partial M] = [\Sigma \times L]$. Since for $j^* : H^{n-r}(M) \to H^{n-r}(\Sigma \times L)$ and $\xi \in H^{n-r}(M)$ we have

$$\partial_*(\xi \cap [M, \partial M]) = j^*\xi \cap \partial_*[M, \partial M] = j^*\xi \cap [\Sigma \times L]$$

(see [Spa66], Chapter 5, Section 6, 20, page 255), the square

$$
\begin{array}{ccc}
H^{n-r}(M) & \xrightarrow{\ j^*\ } & H^{n-r}(\Sigma \times L) \\
{\scriptstyle -\cap[M,\partial M]}\downarrow{\scriptstyle \cong} & & {\scriptstyle \cong}\downarrow{\scriptstyle -\cap[\Sigma\times L]} \\
H_r(M, \partial M) & \xrightarrow{\ \partial_*\ } & H_{r-1}(\Sigma \times L)
\end{array}
\tag{2.31}
$$

commutes. By Proposition 2.44, the square

$$
\begin{array}{ccc}
H^{n-r}(\Sigma \times L) & \xrightarrow{\ (\mathrm{id}_\Sigma \times f)^*\ } & H^{n-r}(\Sigma \times L_{<k}) \\
{\scriptstyle -\cap[\Sigma\times L]}\downarrow{\scriptstyle \cong} & & {\scriptstyle \cong}\downarrow{\scriptstyle -\cap[\Sigma\times L]} \\
H_{r-1}(\Sigma \times L) & \xrightarrow{\ \mathrm{incl}_*\ } & H_{r-1}(\Sigma \times (L, L_{<c-k}))
\end{array}
\tag{2.32}
$$

commutes as well. Since $g^* = (\mathrm{id}_\Sigma \times f)^* \circ j^*$ and the connecting homomorphism

$$\partial_* : H_r(M, \partial M) \longrightarrow H_{r-1}(\Sigma \times (L, L_{<c-k}))$$

of the triple factors as

$$H_r(M, \partial M) \xrightarrow{\ \partial_*\ } H_{r-1}(\Sigma \times L) \xrightarrow{\ \mathrm{incl}_*\ } H_{r-1}(\Sigma \times (L, L_{<c-k})),$$

diagram (2.30) is the composition of diagram (2.31) and diagram (2.32) and therefore commutes as well. $\square$

Lemma 2.46. *Let*

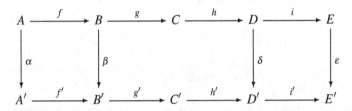

be a commutative diagram of rational vector spaces with exact rows. Then there exists a map $\gamma : C \to C'$ completing the diagram commutatively.

Proof. Let $s : \operatorname{im} h \to C$ be a splitting for $h| : C \twoheadrightarrow \operatorname{im} h$ and let $s' : \operatorname{im} h' \to C'$ be a splitting for $h'| : C' \twoheadrightarrow \operatorname{im} h'$. Then $C = \operatorname{im} g \oplus \operatorname{im} s$ and an element $c \in C$ can be uniquely written as $c = c_g + c_s$, with $c_g \in \operatorname{im} g$ and $c_s \in \operatorname{im} s$. We set

$$\gamma(c) = g'\beta(b) + s'\delta h(c_s),$$

where $b \in B$ is any element such that $g(b) = c_g$. Note that indeed $\delta h(c_s) \in \operatorname{im} h' = \ker i'$, since $i'\delta h = \varepsilon i h = 0$. To show that γ is well-defined, consider $b' \in B$ with $g(b') = c_g$. Then $b - b' = f(a)$ for some $a \in A$ and thus

$$g'\beta(b) - g'\beta(b') = g'\beta f(a) = g'f'\alpha(a) = 0.$$

Furthermore,

$$h'\gamma(c) = h'g'\beta(b) + h's'\delta h(c_s) = \delta h(c_s) = \delta h g(b) + \delta h(c_s) = \delta h(c)$$

and for any $b \in B$,

$$\gamma g(b) = g'\beta(b)$$

by definition. $\qquad\square$

Theorem 2.47. *Let X be an n-dimensional, compact, oriented, stratified pseudo-manifold with one singular stratum Σ of dimension $n - c$ and trivial link bundle. The link L is assumed to be simply connected and X, Σ and L are oriented compatibly. Let $I^{\bar{p}}X$ and $I^{\bar{q}}X$ be $\bar{p}$- and $\bar{q}$-intersection spaces of X with $\bar{p}$ and $\bar{q}$ complementary perversities. Then there exists a generalized Poincaré duality isomorphism*

$$D : \widetilde{H}^{n-r}(I^{\bar{p}}X) \xrightarrow{\cong} \widetilde{H}_r(I^{\bar{q}}X)$$

such that

$$
\begin{array}{ccc}
\widetilde{H}^{n-r}(I^{\bar{p}}X) & \longrightarrow & H^{n-r}(M) \\
D \Big\downarrow \cong & & \cong \Big\downarrow {-\cap[M,\partial M]} \\
\widetilde{H}_r(I^{\bar{q}}X) & \longrightarrow & H_r(M,\partial M)
\end{array}
$$

commutes, where $(M, \partial M)$ is the complement of an open tube neighborhood of Σ, and

$$
\begin{array}{ccc}
H^{n-r-1}(\Sigma \times L_{<k}) & \xrightarrow{\delta^*} & \widetilde{H}^{n-r}(I^{\bar{p}}X) \\
{-\cap[\partial M]} \Big\downarrow \cong & & \cong \Big\downarrow D \\
H_r(\partial M, \Sigma \times L_{<c-k}) & \longrightarrow & \widetilde{H}_r(I^{\bar{q}}X)
\end{array}
$$

commutes, where $k = c - 1 - \bar{p}(c)$.

Proof. We have
$$I^{\bar{p}}X = \mathrm{cone}(g_{\bar{p}}) = M \cup_{g_{\bar{p}}} \mathrm{cone}(\Sigma \times L_{<k})$$

for $g_{\bar{p}} : \Sigma \times L_{<k} \to M$ the composition

$$\Sigma \times L_{<k} \xrightarrow{\mathrm{id}_{\Sigma} \times f_{\bar{p}}} \Sigma \times L = \partial M \xhookrightarrow{j} M,$$

and, since $c - k = c - 1 - \bar{q}(c)$,

$$I^{\bar{q}}X = \mathrm{cone}(g_{\bar{q}}) = M \cup_{g_{\bar{q}}} \mathrm{cone}(\Sigma \times L_{<c-k})$$

for $g_{\bar{q}} : \Sigma \times L_{<c-k} \to M$ the composition

$$\Sigma \times L_{<c-k} \xrightarrow{\mathrm{id}_{\Sigma} \times f_{\bar{q}}} \Sigma \times L = \partial M \xhookrightarrow{j} M.$$

Hence

$$\widetilde{H}^*(I^{\bar{p}}X) = H^*(g_{\bar{p}}) = H^*(M, \Sigma \times L_{<k}),$$

$$\widetilde{H}^*(I^{\bar{q}}X) = H^*(g_{\bar{q}}) = H^*(M, \Sigma \times L_{<c-k}),$$

and similarly for homology. Consider the diagram

$$
\begin{array}{ccc}
H^{n-r-1}(M) & \xrightarrow[\cong]{-\cap[M,\partial M]} & H_{r+1}(M, \partial M) \\
\downarrow{\scriptstyle g_{\bar{p}}^*} & & \downarrow{\scriptstyle \partial_*} \\
H^{n-r-1}(\Sigma \times L_{<k}) & \xrightarrow[\cong]{-\cap[\Sigma \times L]} & H_r(\Sigma \times (L, L_{<c-k})) \\
\downarrow{\scriptstyle \delta^*} & & \downarrow \\
H^{n-r}(M, \Sigma \times L_{<k}) & & H_r(M, \Sigma \times L_{<c-k}) \\
\downarrow & & \downarrow \\
H^{n-r}(M) & \xrightarrow[\cong]{-\cap[M,\partial M]} & H_r(M, \partial M) \\
\downarrow{\scriptstyle g_{\bar{p}}^*} & & \downarrow{\scriptstyle \partial_*} \\
H^{n-r}(\Sigma \times L_{<k}) & \xrightarrow[\cong]{-\cap[\Sigma \times L]} & H_{r-1}(\Sigma \times (L, L_{<c-k})),
\end{array}
$$

whose left hand column is the long exact sequence of the pair $(M, \Sigma \times L_{<k})$ and whose right hand column is the long exact sequence of the triple $(M, \Sigma \times L, \Sigma \times L_{<c-k})$. By Lemma 2.45, the top and bottom squares commute. By Lemma 2.46, there exists a map

$$D : H^{n-r}(M, \Sigma \times L_{<k}) \longrightarrow H_r(M, \Sigma \times L_{<c-k})$$

filling in the diagram commutatively. By the 5-lemma, D is an isomorphism. $\qquad \square$

Example 2.48. Set $L = S^3 \times S^4$ and $M^{14} = D^3 \times S^2 \times S^2 \times L$. We will compute the duality in the homology of the intersection space $I^{\bar{m}}X$ for the pseudomanifold

$$X^{14} = M \cup_{\partial M} S^2 \times S^2 \times S^2 \times \text{cone} L.$$

This pseudomanifold is to be stratified in the intrinsic manner, with singular set $\Sigma = S^2 \times S^2 \times S^2 \times \{\sigma\}$, where σ is the cone point of $\text{cone}(L)$, and link L. Since the codimension c of Σ is 8, the cut-off value k is $k = c - 1 - \bar{m}(c) = 4$. Hence $L_{<k} = L_{<4} = S^3 \times \text{pt}$ and

$$f : L_{<4} = S^3 \times \text{pt} \hookrightarrow S^3 \times S^4$$

is the inclusion. The intersection space $I^{\bar{m}}X$ is the mapping cone of

$$g : S^2 \times S^2 \times S^2 \times S^3 \times \text{pt} \hookrightarrow D^3 \times S^2 \times S^2 \times S^3 \times S^4,$$

that is,

$$I^{\bar{m}}X \simeq \frac{D^3 \times S^2 \times S^2 \times S^3 \times S^4}{S^2 \times S^2 \times S^2 \times S^3 \times \text{pt}}.$$

If A, B are cycles in a 2-sphere and C is a cycle in the 3-sphere, then

$$D^3 \times A \times B \times C \times \text{pt} \cup_{S^2 \times A \times B \times C \times \text{pt}} \text{cone}(S^2 \times A \times B \times C \times \text{pt})$$

is a cycle in the space $I^{\bar{m}}X$. We shall denote the homology class of such a cycle briefly by $[D^3 \times A \times B \times C \times \text{pt}]^\wedge$. The following table lists all generating cycles of the homology of the intersection space. Dual cycles are next to each other in the same row.

	$\widetilde{H}_*(I^{\bar{m}}X)$	$\widetilde{H}_{14-*}(I^{\bar{m}}X)$
$* = 0$	0	0
$* = 1$	0	0
$* = 2$	0	0
$* = 3$	$[D^3 \times \text{pt} \times \text{pt} \times \text{pt} \times \text{pt}]^\wedge$	$[\text{pt} \times S^2 \times S^2 \times S^3 \times S^4]$
$* = 4$	$[\text{pt} \times \text{pt} \times \text{pt} \times \text{pt} \times S^4]$	$[D^3 \times S^2 \times S^2 \times S^3 \times \text{pt}]^\wedge$
$* = 5$	$[D^3 \times S^2 \times \text{pt} \times \text{pt} \times \text{pt}]^\wedge$ $[D^3 \times \text{pt} \times S^2 \times \text{pt} \times \text{pt}]^\wedge$	$[\text{pt} \times \text{pt} \times S^2 \times S^3 \times S^4]$ $[\text{pt} \times S^2 \times \text{pt} \times S^3 \times S^4]$
$* = 6$	$[\text{pt} \times S^2 \times \text{pt} \times \text{pt} \times S^4]$ $[\text{pt} \times \text{pt} \times S^2 \times \text{pt} \times S^4]$ $[D^3 \times \text{pt} \times \text{pt} \times S^3 \times \text{pt}]^\wedge$	$[D^3 \times \text{pt} \times S^2 \times S^3 \times \text{pt}]^\wedge$ $[D^3 \times S^2 \times \text{pt} \times S^3 \times \text{pt}]^\wedge$ $[\text{pt} \times S^2 \times S^2 \times \text{pt} \times S^4]$
$* = 7$	$[\text{pt} \times \text{pt} \times \text{pt} \times S^3 \times S^4]$ $[D^3 \times S^2 \times S^2 \times \text{pt} \times \text{pt}]^\wedge$	$[D^3 \times S^2 \times S^2 \times \text{pt} \times \text{pt}]^\wedge$ $[\text{pt} \times \text{pt} \times \text{pt} \times S^3 \times S^4]$

Let us indicate how one may form candidates for intersection spaces $I^{\bar{p}}X$ for pseudomanifolds X having more than two strata and whose link bundle may be nontrivial. Up to now, we have used only a small fraction of the spatial homology truncation machine as developed in Chapter 1, namely, we have only invoked it on the object level. For general stratifications, the full range of capabilities of the machine will have to be employed. Let us start out with some remarks on gluing constructions and homotopy pushouts. A *3-diagram* Γ *of spaces* is a diagram of the form

$$X \xleftarrow{f} A \xrightarrow{g} Y,$$

where A, X, Y are topological spaces and f, g are continuous maps. The *realization* $|\Gamma|$ of Γ is the pushout of f and g. A *morphism* $\Gamma \to \Gamma'$ of 3-diagrams is a commutative diagram

$$
\begin{array}{ccccc}
X & \xleftarrow{\ f\ } & A & \xrightarrow{\ g\ } & Y \\
\downarrow & & \downarrow & & \downarrow \\
X' & \xleftarrow{\ f'\ } & A' & \xrightarrow{\ g'\ } & Y'
\end{array}
\tag{2.33}
$$

in the category of topological spaces. The universal property of the pushout implies that a morphism $\Gamma \to \Gamma'$ induces a map $|\Gamma| \to |\Gamma'|$ between realizations. A homotopy theoretic weakening of a morphism is the notion of an *h-morphism* $\Gamma \to_h \Gamma'$. This is again a diagram of the above form (2.33), but the two squares are required to commute only up to homotopy. An h-morphism does not induce a map between realizations. The remedy is to use the *homotopy pushout*, or double mapping cylinder. This is a special case of the notion of a *homotopy colimit*. To a 3-diagram Γ we associate another 3-diagram $H(\Gamma)$ given by

$$X \cup_f A \times I = \mathrm{cyl}(f) \xleftarrow{\text{at } 0} A \xrightarrow{\text{at } 0} \mathrm{cyl}(g) = Y \cup_g A \times I.$$

We define the homotopy pushout, or homotopy colimit, of Γ to be

$$\mathrm{hocolim}(\Gamma) = |H(\Gamma)|.$$

The morphism $H(\Gamma) \to \Gamma$ given by

$$
\begin{array}{ccccc}
X \cup_f A \times I & \longleftarrow & A & \longrightarrow & Y \cup_g A \times I \\
\downarrow{\scriptstyle r} & & \downarrow{\scriptstyle \mathrm{id}_A} & & \downarrow{\scriptstyle r} \\
X & \xleftarrow{\ f\ } & A & \xrightarrow{\ g\ } & Y,
\end{array}
$$

where the maps r are the canonical mapping cylinder retractions, induces a canonical map

$$\mathrm{hocolim}(\Gamma) \longrightarrow |\Gamma|.$$

An h-morphism $\Gamma \to_h \Gamma'$ together with a choice of homotopies between clockwise and counterclockwise compositions will induce a map on the homotopy pushout,

$$\text{hocolim}(\Gamma) \longrightarrow |\Gamma'|.$$

Indeed, let

$$
\begin{array}{ccccc}
X & \xleftarrow{\ f\ } & A & \xrightarrow{\ g\ } & Y \\
\downarrow{\scriptstyle \xi} & & \downarrow{\scriptstyle \alpha} & & \downarrow{\scriptstyle \eta} \\
X' & \xleftarrow{\ f'\ } & A' & \xrightarrow{\ g'\ } & Y'
\end{array}
$$

be the given h-morphism. Let $F : A \times I \to X'$ be a homotopy between $F_0 = f'\alpha$ and $F_1 = \xi f$. Let $G : A \times I \to Y'$ be a homotopy between $G_0 = g'\alpha$ and $G_1 = \eta g$. Then

$$
\begin{array}{ccccc}
X \cup_f A \times I & \xleftarrow{\ \text{at } 0\ } A \hookrightarrow & A & \xrightarrow{\ \text{at } 0\ } & Y \cup_g A \times I \\
\downarrow{\scriptstyle \xi \cup_f F} & & \downarrow{\scriptstyle \alpha} & & \downarrow{\scriptstyle \eta \cup_g G} \\
X' & \xleftarrow{\ f'\ } & A' & \xrightarrow{\ g'\ } & Y',
\end{array}
$$

commutes (on the nose) and thus defines a morphism $H(\Gamma) \to \Gamma'$. This morphism induces a continuous map on realizations $\text{hocolim}(\Gamma) = |H(\Gamma)| \to |\Gamma'|$.

Let X^n be a PL stratified pseudomanifold with a stratification of the form $X_n = X^n \supset X_1 \supset X_0, X_1 \cong S^1, X_0 = \{x_0\}$. There are thus three strata. If the link-type at x_0 is the same as for points in $X_1 - X_0$, then X can be restratified as $\hat{X}_n = X^n \supset \hat{X}_1 \cong S^1$, $\hat{X}_0 = \varnothing$, and the link bundle around the circle $\hat{X}_1$ may be a twisted mapping torus. Let N_0 be a regular neighborhood of x_0 in X. Then $N_0 = \text{cone}(L_0)$, where L_0 is a compact PL stratified pseudomanifold of dimension $n-1$, the link of x_0. Set $X' = X - \text{int}(N_0)$, a compact pseudomanifold with boundary. This X' has one singular stratum, $X_1' = X_1 \cap X' \cong \Delta^1$, where Δ^1 is a 1-simplex (closed interval). Let L_1 be the link of X_1', a closed manifold of dimension $n-2$. The link L_0 may be singular with singular stratum $L_0 \cap X_1 = L_0 \cap X_1' = \partial \Delta^1 = \{\Delta_0^0, \Delta_1^0\}$ (two points). A regular neighborhood of Δ_i^0, $i = 0,1$, in L_0 is isomorphic to $\text{cone}(L_1)$. If we remove the interiors of these two cones from L_0, we obtain a compact $(n-1)$-manifold W, which is a bordism between L_1 at Δ_0^0 and L_1 at Δ_1^0. A normal regular neighborhood of X_1' in X' is isomorphic to a product $\Delta^1 \times \text{cone}(L_1)$ since Δ^1 is contractible. Removing the interior of this neighborhood from X', we get a compact n-manifold M with boundary ∂M. The boundary is $\Delta^1 \times L_1$ glued to W along the boundary $\partial W = \partial \Delta^1 \times L_1 = \{\Delta_0^0, \Delta_1^0\} \times L_1$. Thus ∂M has the form

$$\partial M = |\Gamma|,$$

where Γ is a 3-diagram of spaces

$$W \xleftarrow{\ f_0 \sqcup f_1\ } \partial \Delta^1 \times L_1 \xhookrightarrow{\ \text{incl} \times \text{id}\ } \Delta^1 \times L_1,$$

for suitable maps $f_i : \Delta_i^0 \times L_1 \to W$, $i = 0, 1$. For example, if the link-type does not change running along $X_1 - X_0$ into x_0, then L_0 is the suspension of L_1 and W is the cylinder $W = I \times L_1$. The boundary of M is a mapping torus with fiber L_1. We may take f_0 to be the identity and f_1 the monodromy of the mapping torus.

Given a perversity $\bar{p}$, set cut-off degrees

$$k_L = n - 2 - \bar{p}(n-1), \quad k_W = n - 1 - \bar{p}(n).$$

We observe that the inequality $k_W \geq k_L$ holds because $\bar{p}(n) \leq \bar{p}(n-1) + 1$. Two cases arise. If $\bar{p}(n) = \bar{p}(n-1) + 1$, then $k_L = k_W$; if $\bar{p}(n) = \bar{p}(n-1)$, then $k_W = k_L + 1$. Suppose the perversity value actually increases and we are thus in the case $k_L = k_W$ (denote this value simply by k). Next, and this is the only point where an obstruction could occur, you have to be able to choose Y_L and Y_W such that $f_0, f_1 : (L_1, Y_L) \to (W, Y_W)$ become morphisms in $\mathbf{CW}_{k \supset \partial}$. If f_0 and f_1 are inclusions, then Proposition 1.57 is frequently helpful to settle this. If L_1 or W lie in the interleaf category $\mathbf{ICW}$, then no Y_L or no Y_W has to be chosen and dealing with the obstructions simplifies considerably. Once f_0 and f_1 are known to be morphisms in $\mathbf{CW}_{k \supset \partial}$, we can apply spatial homology truncation and receive diagrams

$$
\begin{array}{ccc}
t_{<k}(L_1, Y_L) & \xrightarrow{\mathrm{emb}_k(L_1,Y_L)} & L_1 \\
{\scriptstyle t_{<k}(f_i)}\downarrow & & \downarrow{\scriptstyle [f_i]} \\
t_{<k}(W, Y_W) & \xrightarrow{\mathrm{emb}_k(W,Y_W)} & W,
\end{array}
$$

$i = 0, 1$, which commute in $\mathbf{HoCW}_{k-1}$. Let $(f_i)_{<k}$ be a representative of the homotopy class $t_{<k}(f_i)$, $i = 0, 1$, and let $t_{<k}\Gamma$ be the 3-diagram of spaces

$$t_{<k}(W, Y_W) \xleftarrow{(f_0)_{<k} \sqcup (f_1)_{<k}} \partial\Delta^1 \times t_{<k}(L_1, Y_L) \xhookrightarrow{\mathrm{incl} \times \mathrm{id}} \Delta^1 \times t_{<k}(L_1, Y_L).$$

Let e_L be a representative of the homotopy class $\mathrm{emb}_k(L_1, Y_L)$ and let e_W be a representative of the homotopy class $\mathrm{emb}_k(W, Y_W)$. An h-morphism $t_{<k}\Gamma \to_h \Gamma$ is given by

$$
\begin{array}{ccc}
t_{<k}(W, Y_W) \xleftarrow{(f_0)_{<k} \sqcup (f_1)_{<k}} \partial\Delta^1 \times t_{<k}(L_1, Y_L) \xhookrightarrow{\mathrm{incl} \times \mathrm{id}} \Delta^1 \times t_{<k}(L_1, Y_L) \\
{\scriptstyle e_W}\downarrow \qquad\qquad\qquad {\scriptstyle \mathrm{id} \times e_L}\downarrow \qquad\qquad\qquad {\scriptstyle \mathrm{id} \times e_L}\downarrow \\
W \xleftarrow{\quad f_0 \sqcup f_1 \quad} \partial\Delta^1 \times L_1 \xhookrightarrow{\mathrm{incl} \times \mathrm{id}} \Delta^1 \times L_1.
\end{array}
$$

(The right-hand square commutes on the nose.) Once the requisite homotopy has been chosen, this h-morphism induces a map

$$\mathrm{hocolim}(t_{<k}\Gamma) \xrightarrow{f} |\Gamma| = \partial M.$$

Let g be the composition

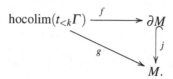

It is consistent with our earlier constructions to consider $\mathrm{cone}(g)$ as a candidate for $I^{\bar{p}}X$.

If the perversity value does not increase, so that $k_W = k_L + 1$, then one must use iterated truncation techniques to form a 3-diagram $t_{<k}\Gamma$. If $f_0, f_1 : L_1 \to W$ can be promoted to morphisms $f_0, f_1 : (L_1, Y_1) \to (W, Y_W)$ in $\mathbf{CW}_{k_W \supset \partial}$ by choosing suitable Y_1, Y_W, then there are truncations

$$t_{<k_W}(f_i) : t_{<k_W}(L_1, Y_1) \longrightarrow t_{<k_W}(W, Y_W).$$

By Proposition 1.58, there is a homotopy equivalence

$$t_{<k_L}(t_{<k_W}(L_1, Y_1), Y_L) \simeq t_{<k_L}(L_1, Y_L),$$

where $(L_1, Y_L) \in Ob\mathbf{CW}_{k_L \supset \partial}$. Choosing a representative for the result of applying the natural transformation emb_{k_L} to the pair $(t_{<k_W}(L_1, Y_1), Y_L)$ gives a map

$$e : t_{<k_L}(t_{<k_W}(L_1, Y_1), Y_L) \to t_{<k_W}(L_1, Y_1).$$

Let

$$a : \partial\Delta^1 \times t_{<k_L}(L_1, Y_L) \longrightarrow t_{<k_W}(W, Y_W)$$

be the composition

$$\partial\Delta^1 \times t_{<k_L}(L_1, Y_L) \xrightarrow{\mathrm{id} \times \simeq} \partial\Delta^1 \times t_{<k_L}(t_{<k_W}(L_1, Y_1), Y_L)$$

$$\Big\downarrow \mathrm{id} \times e$$

$$\partial\Delta^1 \times t_{<k_W}(L_1, Y_1)$$

$$\Big\downarrow (f_0)_{<k_W} \sqcup (f_1)_{<k_W}$$

$$t_{<k_W}(W, Y_W),$$

with a the diagonal arrow from $\partial\Delta^1 \times t_{<k_L}(L_1, Y_L)$ to $t_{<k_W}(W, Y_W)$,

where $(f_i)_{<k_W}$ is a representative of $t_{<k_W}(f_i)$, $i = 0, 1$. Let $t_{<k}\Gamma$ be the 3-diagram

$$t_{<k_W}(W, Y_W) \xleftarrow{\ a\ } \partial\Delta^1 \times t_{<k_L}(L_1, Y_L) \xhookrightarrow{\mathrm{incl} \times \mathrm{id}} \Delta^1 \times t_{<k_L}(L_1, Y_L).$$

For an appropriate $t_{<k}\Gamma \to_h \Gamma$, one will get f, g and a candidate for $I^{\bar{p}}X$ as above.

Chapter 3
String Theory

3.1 Introduction

String theory models physical phenomena by closed vibrating loops ("strings") moving in space. As the string moves, it forms a surface, its world sheet Σ. The movement in space is described by a map $\Sigma \rightarrow T$ to some target space T. (This is the starting point for the data of a nonlinear sigma model.) This space is usually required to be $10 = 4 + 6$-dimensional and is often assumed to be of the form $T = M^4 \times X^6$, where M^4 is a 4-manifold which, at least locally, may be thought of as the space–time of special relativity. The additional 6 dimensions are necessary because a string needs a sufficient number of directions in which it can vibrate. If this number is smaller than 6, then problems such as negative probabilities occur. The space X carries a Riemannian metric and is very small compared to M. Among other constraints, supersymmetry imposes conditions on the metric of X that imply that it has to be a Calabi–Yau space. A Calabi–Yau manifold has a complex structure such that the first Chern class vanishes, and the metric is Kähler for this complex structure. (A large class of examples of Kähler manifolds are complex submanifolds of complex projective spaces.) Calabi conjectured that all Kähler manifolds with vanishing first Chern class admit a Ricci-flat metric, which was later proven by S. T. Yau. Many examples of Calabi–Yau manifolds are obtained as complete intersections in products of projective spaces. Consider for instance the quintic

$$P_\varepsilon(z) = z_0^5 + z_1^5 + z_2^5 + z_3^5 + z_4^5 - 5(1 + \varepsilon)z_0 z_1 z_2 z_3 z_4,$$

depending on a complex structure parameter ε. The variety

$$X_\varepsilon = \{z \in \mathbb{C}P^4 \mid P_\varepsilon(z) = 0\}$$

is Calabi–Yau. It is smooth for small $\varepsilon \neq 0$ and becomes singular for $\varepsilon = 0$. (For X_ε to be singular, $1 + \varepsilon$ must be fifth root of unity, so X_ε is smooth for $0 < |\varepsilon| < |e^{2\pi i/5} - 1|$.) It is at present not known which Calabi–Yau space is the physically correct choice. Thus it is very important to analyze the moduli space of all Calabi–Yau threefolds and to find ways to navigate in it. One such way is the

M. Banagl, *Intersection Spaces, Spatial Homology Truncation, and String Theory*, Lecture Notes in Mathematics 1997, DOI 10.1007/978-3-642-12589-8_3, © Springer-Verlag Berlin Heidelberg 2010

conifold transition. The term "conifold" arose in physics and we shall here adopt the following definition:

Definition 3.1. A *topological conifold* is a six-dimensional topological stratified pseudomanifold S, whose singular set consists of isolated points, each of which has link $S^2 \times S^3$. That is, S possesses a subset Σ, the singular set, such that $S - \Sigma$ is a 6-manifold, every point s of Σ is isolated and has an open neighborhood homeomorphic to the open cone on $S^2 \times S^3$.

An example is the above space X_0. The singularities are those points where the gradient of P_0 vanishes. If one of the five homogeneous coordinates $z_0, \ldots, z_4$ vanishes, then the gradient equations imply that all the others must vanish, too. This is not a point on $\mathbb{C}P^4$, and so all coordinates of a singularity must be nonzero. We may then normalize the first one to be $z_0 = 1$. From the gradient equation $z_0^4 = z_1 z_2 z_3 z_4$ it follows that z_1 is determined by the last three coordinates, $z_1 = (z_2 z_3 z_4)^{-1}$. The gradient equations also imply that

$$1 = z_0^5 = z_0 z_1 z_2 z_3 z_4 = z_1^5 = z_2^5 = z_3^5 = z_4^5,$$

so that all coordinates of a singularity are fifth roots of unity. Let (ω, ξ, η) be any triple of fifth roots of unity. (There are 125 distinct such triples.) The 125 points

$$(1 : (\omega \xi \eta)^{-1} : \omega : \xi : \eta)$$

lie on X_0 and the gradient vanishes there. These are thus the 125 singularities of X_0. Each one of them is a node, whose neighborhood therefore looks topologically like the cone on the 5-manifold $S^2 \times S^3$.

3.2 The Topology of 3-Cycles in 6-Manifolds

Middle dimensional homology classes in a Calabi–Yau 3-manifold have particularly nice representative cycles, namely embedded 3-spheres, as we shall now prove.

Proposition 3.2. *Every three-dimensional homology class in a simply connected smooth 6-manifold X, in particular in a (simply connected) complex three-dimensional Calabi–Yau manifold, can be represented by a smoothly embedded 3-sphere $S^3 \subset X$ with trivial normal bundle.*

Proof. As X is simply connected, the Hurewicz theorem implies that the Hurewicz map $\pi_2(X) \to H_2(X)$ is an isomorphism and the Hurewicz map $\pi_3(X) \to H_3(X)$ is onto. Thus, given a homology class $x \in H_3(X)$, there exists a continuous map $f : S^3 \to X$ such that $f_*[S^3] = x$, where $[S^3] \in H_3(S^3)$ is the fundamental class. Let us recall part of the Whitney embedding theorem [Whi36, Whi44]: Let N^n, M^{2n} be smooth manifolds, $n \geq 3$. If M is simply connected, then every map $f : N^n \to M^{2n}$ is homotopic to a smooth embedding $N \hookrightarrow M$. Hence, with $n = 3$, our f is homotopic to a smooth embedding $f' : S^3 \hookrightarrow X$, $f'_*[S^3] = f_*[S^3] = x$. So x is represented by an embedded S^3. The transition function for the normal bundle of f' lies in $\pi_2(GL(3,\mathbb{R})) = \pi_2(O(3)) = \pi_2(SO(3)) = 0$. Thus the normal bundle is trivial. □

This result implies in particular that one can do (smooth) surgery on any three-dimensional homology class in a Calabi–Yau 3-manifold. One represents the class by a smoothly embedded 3-sphere. Since the normal bundle is trivial, this cycle has an open tubular neighborhood diffeomorphic to $S^3 \times \text{int}(D^3)$. Removing this neighborhood, one gets a manifold with boundary $S^3 \times S^2$. The surgery is completed by gluing in $D^4 \times S^2$ along the boundary $\partial(D^4 \times S^2) = S^3 \times S^2$.

3.3 The Conifold Transition

The conifold transition takes as its input a Calabi–Yau manifold and produces another (topologically different) Calabi–Yau manifold as an output by passing through a Calabi–Yau conifold. Let X_ε be a Calabi Yau threefold whose complex structure depends on a complex parameter ε. The dependence is such that for small $\varepsilon \neq 0$, X_ε is smooth and the homotopy type of X_ε is independent of ε, while in the limit $\varepsilon \to 0$, one obtains a singular space S which is a conifold in the above sense. We will refer to this process

$$X_\varepsilon \rightsquigarrow S$$

as a *deformation* of complex structures. Let us assume that the singularities are all nodes. This implies that the link of every singularity is a product of spheres $S^2 \times S^3$ and the neighborhood of every singularity thus is topologically a cone on $S^2 \times S^3$. Topologically, the deformation $X_\varepsilon \rightsquigarrow S$ collapses S^3-shaped cycles in X_ε to the singular points and there is a collapse map $X_\varepsilon \to S$. The singular space S admits a *small resolution* $Y \to S$, which replaces every node in S by a $\mathbb{C}P^1$. The resulting space Y is a smooth Calabi–Yau manifold. The transition

$$X_\varepsilon \rightsquigarrow S \rightsquigarrow Y$$

is an instance of a *conifold transition*. (Other instances may involve singularities worse than nodes.) Suitable generalizations of such transitions connect the parameter spaces of many large families of simply connected Calabi–Yau manifolds, see [GH88, GH89], and may indeed connect all of them.

3.4 Breakdown of the Low Energy Effective Field Theory Near a Singularity

Let X be a Calabi–Yau manifold of complex dimension 3. By Poincaré duality, there exists a symplectic basis $A_1, \ldots, A_r, B^1, \ldots, B^r$ for $H_3(X;\mathbb{Z})$, that is, a basis with the intersections

$$A_i \cap B^j = -B^j \cap A_i = \delta_{ij}, \; A_i \cap A_j = 0 = B^i \cap B^j.$$

By Proposition 3.2, we may think of the A_i and B^j as smoothly embedded 3-spheres with trivial normal bundle. Let Ω be the holomorphic 3-form on X, which is unique up to a nonzero complex rescaling ($b_{3,0} = 1$). Then a complex structure on X is characterized by the periods

$$F_i = \int_{A_i} \Omega, \; Z^j = \int_{B^j} \Omega.$$

The Z^j can serve as projective coordinates on the moduli space $\mathcal{M}$ of complex structures on X. Locally, the F_i may be regarded as functions of the Z^j. When one of the periods, say Z^1, goes to zero, the corresponding 3-cycle B^1 collapses to a singular point and X becomes a conifold. On $\mathcal{M}$ there is a natural metric $\mathcal{G}$, the Petersson-Weil metric [Tia87]. According to [Str95], see also [Pol00], near $Z^1 = 0$,

$$F_1(Z^1) \sim \text{const} + \frac{1}{2\pi i} Z^1 \log Z^1,$$

and one obtains

$$\mathcal{G}_{1\bar{1}} \sim \log(Z^1 \bar{Z}^1)$$

for the metric near $Z^1 = 0$. Thus, while the distance with respect to $\mathcal{G}$ to $Z^1 = 0$ is finite, the metric blows up at the conifold. The conifold is hence a singularity for $\mathcal{M}$ in this sense. This singularity is responsible for generic inconsistencies in low-energy effective field theories arising from the Calabi–Yau string compactification.

3.5 Massless D-Branes

The problem is rectified in type II string theories by (nonperturbative quantum effects due to) the presence of D-branes that become massless at the conifold, see [Str95, Hüb97]. In ten-dimensional type IIB theory, there is a charged threebrane that wraps around (a minimal representative of) the 3-cycle B^1, which collapses to a singularity for $Z^1 \to 0$. The mass of the threebrane is proportional to the volume of B^1. In the limit

$$\varepsilon = Z^1 \to 0, \; X_\varepsilon \rightsquigarrow S,$$

this volume goes to zero and the threebrane becomes massless. If the conifold S has n nodes arising from the collapse of n 3-cycles, and there are m homology relations between these n cycles in X_ε, then there will be $n - m$ massless threebranes present, since a D-brane is really an object associated to a homology class.

In type IIA theory, there are charged twobranes that wrap around (minimal representatives of) the 2-cycles $\mathbb{C}P^1$ of Y, where

$$S \rightsquigarrow Y$$

is the second part of the conifold transition (the small resolution) and the curves $\mathbb{C}P^1$ resolve the nodes. Again, the mass of the twobrane is proportional to the volume of the $\mathbb{C}P^1$. As the resolution map $Y \to S$ collapses the $\mathbb{C}P^1$, this volume goes to zero and the twobrane becomes massless. If n and m are as before, then there will be m massless twobranes present, as we will see in Section 3.7 below. For a nonsingular description of the physics, these extra massless particles arising from the D-branes must be explicitly kept present in the effective theory.

3.6 Cohomology and Massless States

Following [GSW87], we will explain that cohomology classes on X, that is, harmonic forms on X, are manifested in four dimensions as massless particles. Let ω be an antisymmetric tensor field, i.e. a differential form, on $T = M^4 \times X$. For such a form to be physically realistic, it must satisfy the field equation

$$d^* d\omega = 0$$

(if ω is a 1-form, this is the Maxwell equation) and the generalization

$$d^* \omega = 0$$

of the Lorentz gauge condition in electrodynamics, where d^* is the adjoint operator[1] $d^* : \Omega^k(T) \to \Omega^{k-1}(T)$ and $* : \Omega^k(T) \to \Omega^{10-k}(T)$ is the Hodge star-operator. If $\Delta_T = dd^* + d^*d$ denotes the Hodge-de Rham Laplacian on T, then the two equations imply

$$\Delta_T \omega = 0.$$

The Laplacian on the product manifold decomposes as

$$\Delta_T = \Delta_M + \Delta_X,$$

where Δ_M and Δ_X are the Hodge-de Rham Laplacians of M and X, respectively. Hence, ω satisfies the wave equation

$$(\Delta_M + \Delta_X)\omega = 0. \tag{3.1}$$

This equation suggests the interpretation of Δ_X as a kind of "mass" operator for four-dimensional fields, whose eigenvalues are masses as seen in four dimensions. (Compare this to the Klein-Gordon equation $(\Box_M + m^2)\omega = 0$ for a free particle, where m denotes mass and $\Box_M$ is the d'Alembert operator, i.e. the Laplace operator of Minkowski space.) In particular, for the zero modes of Δ_X (the harmonic forms

---

[1] On an even-dimensional manifold the mathematical literature usually uses $d^* = - * \circ d \circ *$, whereas physicists seem to prefer $d^* = + * \circ d \circ *$ in the present context.

on X), one sees in the four-dimensional reduction massless forms. For example if ξ is the unique harmonic representative of a cohomology class in X and $\omega = \mu \wedge \xi$, where μ is a differential form on M, then the wave equation (3.1) implies that

$$\Delta_M \mu = 0$$

so that μ is indeed massless. Therefore, a good cohomology theory for X should capture all physically present massless particles. This is the case for intersection cohomology in type IIA theory, but is not the case for ordinary cohomology, nor for intersection cohomology or L^2-cohomology, in type IIB theory, as we shall see in the next section.

3.7 The Homology of Intersection Spaces and Massless D-Branes

In the present section, homology will be understood with rational coefficients. Let

$$X_\varepsilon \rightsquigarrow S \rightsquigarrow Y$$

be a conifold transition as in Section 3.3, with some of the 3-cycles (3-spheres) B^j collapsing to points. Let $\Sigma \subset S$ be the singular set of S and let $n = \mathrm{card}(\Sigma)$ denote the number of nodes in S. Let $X_\varepsilon \to S$ denote the collapse map. Set

$$p = b_2(X_\varepsilon), \ q = \mathrm{rk}(H_3(S - \Sigma) \to H_3(S)) = \mathrm{rk}\, IH_3(S),$$

and

$$m = \mathrm{rk}\,\mathrm{coker}(H_4(X_\varepsilon) \to H_4(S)).$$

(Here, $b_i(\cdot)$ is the ith ordinary Betti number of a space and $IH_* = IH_*^{\bar{m}}$ denotes middle-perversity intersection homology.)

Lemma 3.3. *The conifold transition is accompanied by the following Betti numbers:*

(1) The map $H_3(X_\varepsilon) \to H_3(S)$ is surjective.
(2) The map $H_4(X_\varepsilon) \to H_4(S)$ is injective.
(3) $\mathrm{rk}\,H_4(S) = p + m$.
(4) $\mathrm{rk}\,\mathrm{ker}(H_3(X_\varepsilon) \to H_3(S)) = n - m$.
(5) $\mathrm{rk}\,H_2(Y) = \mathrm{rk}\,H_4(Y) = p + m$.
(6) $\mathrm{rk}\,H_3(Y) = q$.
(7) $\mathrm{rk}\,H_3(X_\varepsilon) = q + 2(n - m)$.
(8) $\mathrm{rk}\,H_3(S) = q + (n - m)$.
(9) $\mathrm{rk}\,H_2(S) = p$.

Proof. We shall briefly write X for X_ε. Let $C = \bigsqcup_{j=1}^n S_j^3 \subset X$ be the disjoint union of those 3-spheres S_j^3 that are collapsed to the n nodes in S. The collapse map $X \to X/C = S$ induces an isomorphism $H_*(X,C) \xrightarrow{\cong} \tilde{H}_*(S)$. Let $D = \bigsqcup_{j=1}^n \mathbb{CP}_j^1 \subset Y$

be the disjoint union of those 2-spheres $\mathbb{C}P_j^1$ that are collapsed to the n nodes in S by the small resolution $Y \to S$. The collapse map $Y \to Y/D = S$ induces an isomorphism $H_*(Y,D) \xrightarrow{\cong} \tilde{H}_*(S)$.

(1) The diagram

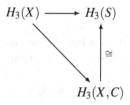

commutes. Consequently, it suffices to show that $H_3(X) \to H_3(X,C)$ is surjective. This follows from the exactness of the homology sequence of the pair (X,C),

$$H_3(X) \longrightarrow H_3(X,C) \xrightarrow{\partial_*} H_2(C) = \bigoplus_{j=1}^n H_2(S_j^3) = 0.$$

(2) As in (1), it suffices to show that $H_4(X) \to H_4(X,C)$ is injective. This follows from the exactness of the sequence

$$0 = H_4(C) \longrightarrow H_4(X) \longrightarrow H_4(X,C).$$

(3) Consider the exact sequence

$$H_4(X) \xrightarrow{\alpha} H_4(S) \xrightarrow{\partial_*} H_3(C) \xrightarrow{\beta} H_3(X) \xrightarrow{\gamma} H_3(S).$$

(The first map, α, is injective by (2).) By Poincaré duality in the manifold X, $\operatorname{rk} H_4(X) = \operatorname{rk} H_2(X) = p$ and by (2) and the definition of m, $\operatorname{rk} H_4(S) = p+m$.

(4) By exactness of the sequence in (3),

$$\begin{aligned}
\operatorname{rk}\ker\beta &= \operatorname{rk}\partial_* \\
&= \operatorname{rk} H_4(S) - \operatorname{rk}\ker\partial_* \\
&= p+m - \operatorname{rk}\alpha \\
&= p+m-p \\
&= m.
\end{aligned}$$

Since

$$\operatorname{rk} H_3(C) = \sum_{j=1}^n \operatorname{rk} H_3(S_j^3) = n,$$

we have

$$\operatorname{rk}\ker\gamma = \operatorname{rk}\beta = \operatorname{rk} H_3(C) - \operatorname{rk}\ker\beta = n - m.$$

(5) The exact homology sequence of the pair (Y,D),

$$0 = H_4(D) \longrightarrow H_4(Y) \longrightarrow H_4(S) \longrightarrow H_3(D) = 0$$

shows that the small resolution $Y \to S$ induces an isomorphism $H_4(Y) \cong H_4(S)$. In particular, $\mathrm{rk}\, H_4(Y) = \mathrm{rk}\, H_4(S) = p+m$, see (3). By Poincaré duality, $\mathrm{rk}\, H_2(Y) = \mathrm{rk}\, H_4(Y)$.

(6) The intersection homology does not change under a small resolution of singularities, and the intersection homology of a manifold equals the ordinary homology of the manifold. Thus

$$\mathrm{rk}\, H_3(Y) = \mathrm{rk}\, IH_3(S) = q.$$

(7) The Euler characteristic of X is given by

$$\chi(X) = 2 + 2p - b_3(X).$$

By (5) and (6), the Euler characteristic of Y is given by

$$\chi(Y) = 2 + 2(p+m) - q.$$

By the Mayer-Vietoris sequence,

$$\chi(Y) = \chi(Y - D) + \chi(D) - \chi(\bigsqcup S_j^3 \times S_j^2)$$

and

$$\chi(X) = \chi(X - C) + \chi(C) - \chi(\bigsqcup S_j^3 \times S_j^2).$$

Subtracting these two equations and observing that $X - C \cong S - \Sigma \cong Y - D$, we obtain

$$\chi(Y) - \chi(X) = \chi(D) - \chi(C) = 2n,$$

as noted also in [Hüb92]. Therefore,

$$2m - q + b_3(X) = 2n,$$

that is, $b_3(X) = q + 2(n - m)$.

(8) By (1), $H_3(X) \to H_3(S)$ is surjective. Thus

$$\mathrm{rk}\, H_3(S) = \mathrm{rk}\, H_3(X) - \mathrm{rk}\, \ker(H_3X \to H_3S) = q + 2(n-m) - (n-m) = q + (n-m),$$

using (7) and (4).

(9) This follows from the exactness of the sequence

$$0 = H_2(C) \to H_2(X) \to H_2(S) \xrightarrow{\partial_*} H_1(C) = 0.$$

$\square$

In general, the set of the n collapsed 3-spheres does not define a set of linearly independent homology classes. The number

$$m = \mathrm{rk}\,\mathrm{coker}(H_4(X_\varepsilon) \to H_4(S))$$

is precisely the number of homology relations between these 3-spheres. In type IIB theory, there will therefore, as we have already mentioned in Section 3.5, be $n-m$ massless threebranes present, since a D-brane is a homological object. Similarly, the set of the n two-spheres collapsed by the resolution map does not generally define a set of linearly independent homology classes. The number of homology relations between these two-spheres is

$$\mathrm{rk}\,\mathrm{coker}(H_3(Y) \to H_3(S)).$$

From the exact homology sequence of the pair (Y,D) (notation as in the proof of Lemma 3.3) we see that $H_3(Y) \to H_3(S)$ is injective. So the rank of the cokernel is $q + (n-m) - q = n - m$ using Lemma 3.3. Hence there are n two-spheres with $n-m$ relations between them. Consequently, in type IIA theory, the number of two-branes is $n - (n-m) = m$. By Lemma 3.3 and Section 3.6, we obtain the following summary of the topology and physics of the conifold transition.

Type	dim	X_ε	S	Y
	2	p	p	$p+m$
Elem. Massless	3	$q+2(n-m)$	$q+(n-m)$	q
	4	p	$p+m$	$p+m$
	2		m (massless)	m (IIA 2-Branes, massive)
D-Branes	3	$n-m$ (IIB 3-Branes, massive)	$n-m$ (massless)	
	2	p	$\mathbf{p+m}$	$p+m$
Total Massless IIA	3	$q+2(n-m)$	$\mathbf{q+(n-m)}$	q
	4	p	$\mathbf{p+m}$	$p+m$
	2	p	p	$p+m$
Total Massless IIB	3	$q+2(n-m)$	$\mathbf{q+2(n-m)}$	q
	4	p	$\mathbf{p+m}$	$p+m$
	2	p	p	$p+m$
$\mathrm{rk}\,H_*$	3	$q+2(n-m)$	$q+(n-m)$	q
	4	p	$p+m$	$p+m$
				$H_*(Y) = IH_*(S)$

In type IIB string theory, a good homology theory $\mathcal{H}_*^{IIB}$ for singular Calabi–Yau varieties should ideally satisfy Poincaré duality (actually the entire Kähler package would be desirable) and record all massless particles. But as we see from the above table, these two requirements are mutually inconsistent; the total IIB numbers of massless particles do not satisfy Poincaré duality. Thus one has a choice of modifying one of the two requirements. Either we do not insist on Poincaré duality or we omit some massless particles. In the present monograph we investigate theories that do possess Poincaré duality. Which massless particles, then, should be omitted? Clearly the ones that have no geometrically dual partner in the singular space. As the table suggests, in the IIB regime, these are m four-dimensional classes that are not dually paired to classes in dimension 2. But these classes correspond to elementary massless particles. Thus the $n - m$ threebrane classes that repair the physical inconsistencies discussed in Section 3.4 *are* recorded by such a theory, as required, and they will have geometrically Poincaré dual classes in the theory.

An analogous discussion applies to type IIA string theory. If we do insist on Poincaré duality for a good homology theory $\mathcal{H}_*^{IIA}$ for singular Calabi–Yau varieties, then, according to the above table, we must omit those $n - m$ three-dimensional classes that do not have dual partners. Again, these correspond to elementary massless particles and the m twobrane classes that repair the physical inconsistencies *are* recorded by $\mathcal{H}_*^{IIA}$. We thus adopt the following axiomatics.

Let $\mathcal{C}$ be a class of possibly singular Calabi–Yau threefolds such that the singular ones all sit in the middle of a conifold transition.

Definition 3.4. A homology theory $\mathcal{H}_*^{IIA}$ defined on $\mathcal{C}$ is called *IIA conifold calibrated*, if

(CCA1) for every space $S \in \mathcal{C}$, $\mathcal{H}_*^{IIA}(S)$ (or its reduced version) satisfies Poincaré duality; for singular $S \in \mathcal{C}$ one has
(CCA2) $\mathrm{rk}\,\mathcal{H}_2^{IIA}(S) = p + m$,
(CCA3) $\mathrm{rk}\,\mathcal{H}_3^{IIA}(S) = q$; and
(CCA4) it agrees with ordinary homology on nonsingular $S \in \mathcal{C}$.

A homology theory $\mathcal{H}_*^{IIB}$ defined on $\mathcal{C}$ is called *IIB conifold calibrated*, if

(CCB1) for every space $S \in \mathcal{C}$, $\mathcal{H}_*^{IIB}(S)$ (or its reduced version) satisfies Poincaré duality; for singular $S \in \mathcal{C}$ one has
(CCB2) $\mathrm{rk}\,\mathcal{H}_2^{IIB}(S) = p$,
(CCB3) $\mathrm{rk}\,\mathcal{H}_3^{IIB}(S) = q + 2(n - m)$; and
(CCB4) it agrees with ordinary homology on nonsingular $S \in \mathcal{C}$.

Examples 3.5. If S sits in the conifold transition $X \rightsquigarrow S \rightsquigarrow Y$, then setting

$$\mathcal{H}_*^{IIA}(S) = H_*(Y; \mathbb{Q})$$

and

$$\mathcal{H}_*^{\text{IIB}}(S) = H_*(X;\mathbb{Q})$$

yields conifold calibrated theories according to the above table. However, these theories are not intrinsic to the space S as they use extrinsic data associated to the surrounding conifold transition. A mathematically superior construction of such theories should have access only to S itself, not to its process of formation. (For example, one advantage is that such an intrinsic construction may then generalize to singular spaces that do not arise in the course of a conifold transition.) In type IIA theory, a solution is given by (middle perversity) intersection homology $IH_*(S)$. Since $IH_*(S) = H_*(Y)$, taking

$$\mathcal{H}_*^{\text{IIA}}(S) = IH_*(S)$$

gives us a IIA conifold calibrated theory which only uses the geometry of S. A solution for type IIB theory is given by taking the homology of the (middle perversity) intersection space IS of S.

Proposition 3.6. *The theory*

$$\mathcal{H}_*^{\text{IIB}}(S) = H_*(IS;\mathbb{Q})$$

is IIB conifold calibrated on $\mathcal{C}$.

Proof. Axiom (CCB4) follows from $IS = S$ for a one-stratum space S. Poincaré duality (CCB1) is established in Theorem 2.12. Let M denote the exterior manifold of the singular set with boundary ∂M and let $\hat{S} = M/\partial M$, see Section 2.6.2 for this "denormalization." Axiom (CCB2) is verified by

$$\text{rk}\, H_2(IS) = \text{rk}\, H_2(M,\partial M) = \text{rk}\, H_2(\hat{S}) = \text{rk}\, H_2(S) = p,$$

using Lemma 3.3. By Theorem 3.9 below, there is a short exact sequence

$$0 \to K \longrightarrow H_3(IS) \longrightarrow H_3(S) \to 0,$$

where $K = \ker(H_3(S - \Sigma) \to H_3(S))$. Each of the n singular points has a small open neighborhood of the form $\overset{\circ}{\text{cone}}(S^3 \times S^2)$. Thus the singular set Σ possesses an open neighborhood U of the form $U = \bigsqcup_{j=1}^{n} \overset{\circ}{\text{cone}}_j(S^3 \times S^2)$. Removing this neighborhood, one obtains a compact manifold M^6 with boundary ∂M consisting of n disjoint copies of $S^3 \times S^2$. From the exact sequence

$$0 = H_3(U) \longrightarrow H_3(S) \longrightarrow H_3(S,U) \longrightarrow H_2(U) = 0$$

we conclude that

$$H_3(S) \cong H_3(S, \bigsqcup_{j=1}^{n} \overset{\circ}{\text{cone}}_j(S^3 \times S^2)) \cong H_3(M, \partial M),$$

where the second isomorphism is given by excision and homotopy invariance. By Poincaré duality and the universal coefficient theorem,

$$\text{rk}\, H_3(M, \partial M) = \text{rk}\, H^3(M) = \text{rk}\, H_3(M).$$

Since M and $S - \Sigma$ are homotopy equivalent, we have $\text{rk}\, H_3(M) = \text{rk}\, H_3(S - \Sigma)$. Hence

$$\text{rk}\, H_3(S - \Sigma) = \text{rk}\, H_3(S) = q + (n - m),$$

and consequently,

$$\begin{aligned}
\text{rk}\, H_3(IS) &= \text{rk}\, H_3(S) + \text{rk}\, K \\
&= q + (n - m) + \text{rk}\, H_3(S - \Sigma) - \text{rk}(H_3(S - \Sigma) \to H_3(S)) \\
&= 2q + 2(n - m) - q \\
&= q + 2(n - m).
\end{aligned}$$

Thus (CCB3) holds. $\square$

How would one characterize theories that faithfully record the physically correct number of massless D-branes if one does not know that the singular space sits in a conifold transition? Let $\mathcal{C}$ be any class of six-dimensional compact oriented pseudomanifolds with only isolated singularities and simply connected links, not necessarily arising from conifold transitions.

Definition 3.7. A homology theory $\mathcal{H}_*^{\text{IIA}}$ defined on $\mathcal{C}$ is called *IIA-brane-complete*, if

(BCA1) for every space $S \in \mathcal{C}$, $\mathcal{H}_*^{\text{IIA}}(S)$ (or its reduced version) satisfies Poincaré duality,
(BCA2) $\mathcal{H}_2^{\text{IIA}}(S)$ is an extension of $H_2(S)$ by $\ker(H_2(S - \Sigma) \to H_2(S))$ for singular $S \in \mathcal{C}$, and
(BCA3) $\mathcal{H}_*^{\text{IIA}}$ agrees with ordinary homology on nonsingular $S \in \mathcal{C}$.

A homology theory $\mathcal{H}_*^{\text{IIB}}$ defined on $\mathcal{C}$ is called *IIB-brane-complete*, if

(BCB1) for every space $S \in \mathcal{C}$, $\mathcal{H}_*^{\text{IIB}}(S)$ (or its reduced version) satisfies Poincaré duality,
(BCB2) $\mathcal{H}_3^{\text{IIB}}(S)$ is an extension of $H_3(S)$ by $\ker(H_3(S - \Sigma) \to H_3(S))$ for singular $S \in \mathcal{C}$, and
(BCB3) $\mathcal{H}_*^{\text{IIB}}$ agrees with ordinary homology on nonsingular $S \in \mathcal{C}$.

In the IIA context, provided all links have vanishing first homology, there is actually an obvious candidate for the extension required by axiom (BCA2), namely

$$\mathcal{H}_2^{\mathrm{IIA}}(S) = H_2(S - \Sigma),$$

as follows from identifying the map $H_2(S - \Sigma) \to H_2(S)$ up to isomorphism with $H_2(M) \to H_2(M, \partial M)$ and observing that the latter is onto, since $H_1(\partial M)$ vanishes. Since intersection homology satisfies

$$IH_2(S) = H_2(S - \Sigma),$$

as well as (BCA1) and (BCA3), we obtain

Proposition 3.8. *Middle perversity intersection homology is IIA-brane-complete on the class $\mathcal{C}$ of six-dimensional compact oriented pseudomanifolds with only isolated singularities and simply connected links (or more generally, links with zero first homology).*

In the IIB situation, on the other hand, there is a priori no obvious space around, whose homology gives the sought extension.

Theorem 3.9. *The theory*
$$\mathcal{H}_*^{\mathrm{IIB}}(S) = H_*(IS; \mathbb{Q})$$

is IIB-brane-complete on the class $\mathcal{C}$ of six-dimensional compact oriented pseudomanifolds with only isolated singularities and simply connected links.

Proof. Axiom (BCB3) follows from $IS = S$ for a one-stratum space S. Poincaré duality (BCB1) is established in Theorem 2.12. To prove (BCB2), we observe that the diagram

$$
\begin{array}{c}
H_3(M) \\
{}^{j_*}\nearrow \quad \searrow{}^{\alpha} \\
H_3(L) \xrightarrow[\alpha_- j_*]{} \widetilde{H}_3(IS) \xrightarrow{\alpha_+} H_3(j) = H_3(M, \partial M) \cong H_3(S) \longrightarrow 0,
\end{array}
$$

with exact bottom row ($L = \partial M$), yields a short exact sequence

$$0 \to \operatorname{im}(\alpha_- j_*) \longrightarrow \widetilde{H}_3(IS) \xrightarrow{\alpha_+} H_3(S) \to 0.$$

Since α_- is injective, it induces an isomorphism $\operatorname{im} j_* \cong \operatorname{im}(\alpha_- j_*)$. By the exactness of the sequence

$$H_3(\partial M) \xrightarrow{j_*} H_3(M) \longrightarrow H_3(M, \partial M),$$

we have

$$\operatorname{im} j_* = \ker(H_3(M) \to H_3(M, \partial M)).$$

From the commutative diagram

$$
\begin{array}{ccc}
H_3(M) & \longrightarrow & H_3(M, \partial M) \\
\cong \Big\downarrow & & \cong \Big\downarrow \\
H_3(S - \Sigma) & \longrightarrow & H_3(S)
\end{array}
$$

we see that

$$
\ker(H_3(M) \to H_3(M, \partial M)) \cong \ker(H_3(S - \Sigma) \to H_3(S)).
$$

$\square$

3.8 Mirror Symmetry

Let us turn to the behavior of these theories with respect to mirror symmetry. We begin by reviewing this phenomenon briefly, following [CK99]. Supersymmetry interchanges bosons and fermions. The Lie algebra of the symmetry group of a supersymmetric string theory contains two generators $Q, \overline{Q}$ called *supersymmetric charges* that are only well-defined up to sign. Replacing Q by $-Q$ and leaving $\overline{Q}$ unchanged is a physically valid operation. Regarding $Q, \overline{Q}$ as operators on a Hilbert space of states, e.g. some complex of differential forms on a manifold, particles are assigned eigenvalues of $(Q, \overline{Q})$ that indicate their charge. A given Calabi–Yau three-fold M together with a complexified Kähler class ω determines such an algebra. In particular, it determines the pair $(Q, \overline{Q})$, and for $p, q \geq 0$, the (p,q)-eigenspace can be computed to be $H^q(M; \wedge^p TM)$, while the $(-p,q)$-eigenspace turns out to be $H^q(M; \Omega_M^p)$. Replacing Q by $-Q$ (leaving $\overline{Q}$ unchanged), the (p,q)- and $(-p,q)$-eigenspaces are interchanged. Roughly, a space M° together with a complexified Kähler class ω° is called a *mirror* of (M, ω) if the supersymmetry charges determined by (M°, ω°) are $(-Q, \overline{Q})$ and the field theories of (M, ω) and (M°, ω°) are isomorphic. This implies identifications

$$
H^q(M; \wedge^p TM) \cong H^q(M^\circ; \Omega_{M^\circ}^p),
$$
$$
H^q(M; \Omega_M^p) \cong H^q(M^\circ; \wedge^p TM^\circ).
$$

Using the nonvanishing holomorphic 3-form on M,

$$
H^q(M; \wedge^p TM) \cong H^q(M; \Omega_M^{3-p}).
$$

We obtain thus isomorphisms

$$
H^q(M; \Omega_M^{3-p}) \cong H^q(M^\circ; \Omega_{M^\circ}^p).
$$

The two interesting Hodge numbers $b_{p,q}(M) = \dim H^q(M; \Omega_M^p)$ of a simply connected smooth Calabi–Yau threefold M are $b_{1,1}(M)$ and $b_{2,1}(M)$. We have seen that mirror symmetry interchanges these:

$$b_{1,1}(M) = b_{2,1}(M^\circ),\ b_{2,1}(M) = b_{1,1}(M^\circ).$$

For the ordinary Betti numbers

$$b_2 = b_{1,1} = b_4,\ b_3 = 2 + 2b_{2,1},$$

this means

$$b_3(M) = b_2(M^\circ) + b_4(M^\circ) + 2,$$
$$b_3(M^\circ) = b_2(M) + b_4(M) + 2.$$

In the conifold transition context, we shall answer below the following question: What is the correct version of these formulae if M is allowed to be singular and in either the left or right hand side, the ordinary Betti numbers are replaced by intersection Betti numbers?

Definition 3.10. A class $\mathcal{C}$ of possibly singular Calabi–Yau threefolds is called *mirror-closed*, if it is closed under the formation of mirrors in the sense of mirror symmetry.

Definition 3.11. Let $\mathcal{C}$ be a mirror-closed class and $(\mathcal{H}_*^{\mathrm{IIA}}, \mathcal{H}_*^{\mathrm{IIB}})$ a pair of homology theories on $\mathcal{C}$. We call $(\mathcal{H}_*^{\mathrm{IIA}}, \mathcal{H}_*^{\mathrm{IIB}})$ a *mirror-pair*, if

$$\mathrm{rk}\,\mathcal{H}_3^{\mathrm{IIA}}(S) = \mathrm{rk}\,\mathcal{H}_2^{\mathrm{IIB}}(S^\circ) + \mathrm{rk}\,\mathcal{H}_4^{\mathrm{IIB}}(S^\circ) + 2,$$
$$\mathrm{rk}\,\mathcal{H}_3^{\mathrm{IIA}}(S^\circ) = \mathrm{rk}\,\mathcal{H}_2^{\mathrm{IIB}}(S) + \mathrm{rk}\,\mathcal{H}_4^{\mathrm{IIB}}(S) + 2,$$
$$\mathrm{rk}\,\mathcal{H}_3^{\mathrm{IIB}}(S) = \mathrm{rk}\,\mathcal{H}_2^{\mathrm{IIA}}(S^\circ) + \mathrm{rk}\,\mathcal{H}_4^{\mathrm{IIA}}(S^\circ) + 2,\ \text{and}$$
$$\mathrm{rk}\,\mathcal{H}_3^{\mathrm{IIB}}(S^\circ) = \mathrm{rk}\,\mathcal{H}_2^{\mathrm{IIA}}(S) + \mathrm{rk}\,\mathcal{H}_4^{\mathrm{IIA}}(S) + 2,$$

where S° denotes any mirror of S, for all $S \in \mathcal{C}$.

Example 3.12. Let $\mathcal{C}$ be any mirror-closed class of *smooth* Calabi–Yau threefolds. Then ordinary homology defines a mirror-pair $(\mathcal{H}_*^{\mathrm{IIA}} = H_*, \mathcal{H}_*^{\mathrm{IIB}} = H_*)$, as we have seen above.

It is conjectured in [Mor99] that the mirror of a conifold transition is again a conifold transition, performed in the reverse direction. Thus it is reasonable to consider mirror-closed classes $\mathcal{C}$ of Calabi–Yau threefolds all of whose singular members sit in a conifold transition.

Proposition 3.13. *Let $\mathcal{C}$ be a mirror-closed class of possibly singular Calabi–Yau threefolds such that all singular members of $\mathcal{C}$ arise in the course of a conifold transition. Then any pair of homology theories $(\mathcal{H}_*^{\mathrm{IIA}}, \mathcal{H}_*^{\mathrm{IIB}})$ with $\mathcal{H}_*^{\mathrm{IIA}}$ IIA conifold calibrated and $\mathcal{H}_*^{\mathrm{IIB}}$ IIB conifold calibrated is a mirror-pair.*

Proof. If $S \in \mathcal{C}$ is nonsingular, the statement follows from axioms (CCA4), (CCB4) and Example 3.12. Let $S \in \mathcal{C}$ be singular with conifold transition $X \rightsquigarrow S \rightsquigarrow Y$. If S° is a mirror of S, then by assumption it sits in a conifold transition $Y^\circ \rightsquigarrow S^\circ \rightsquigarrow X^\circ$, where X° is a mirror of X and Y° is a mirror of Y. According to the table on page 199, the ordinary homology ranks of these spaces are of the form

	X	S	Y	Y°	S°	X°
b_2	p	p	$p+m$	P	P	$P+M$
b_3	$q+2(n-m)$	$q+(n-m)$	q	$Q+2(N-M)$	$Q+(N-M)$	Q
b_4	p	$p+m$	$p+m$	P	$P+M$	$P+M$

Since X and X° are smooth,

$$Q = b_3(X^\circ) = b_2(X) + b_4(X) + 2 = 2p + 2$$

and

$$q + 2(n-m) = b_3(X) = b_2(X^\circ) + b_4(X^\circ) + 2 = 2(P+M) + 2.$$

Since Y and Y° smooth,

$$q = b_3(Y) = b_2(Y^\circ) + b_4(Y^\circ) + 2 = 2P + 2$$

and

$$Q + 2(N-M) = b_3(Y^\circ) = b_2(Y) + b_4(Y) + 2 = 2(p+m) + 2.$$

Thus, by axioms (CCA3), (CCB2) and (CCB1),

$$\mathrm{rk}\,\mathcal{H}_3^{\mathrm{IIA}}(S^\circ) = Q = 2p + 2 = \mathrm{rk}\,\mathcal{H}_2^{\mathrm{IIB}}(S) + \mathrm{rk}\,\mathcal{H}_4^{\mathrm{IIB}}(S) + 2,$$

and, by axioms (CCB3), (CCA2) and (CCA1),

$$\mathrm{rk}\,\mathcal{H}_3^{\mathrm{IIB}}(S) = q + 2(n-m) = 2(P+M) + 2 = \mathrm{rk}\,\mathcal{H}_2^{\mathrm{IIA}}(S^\circ) + \mathrm{rk}\,\mathcal{H}_4^{\mathrm{IIA}}(S^\circ) + 2.$$

Furthermore, by axioms (CCA3), (CCB2) and (CCB1),

$$\mathrm{rk}\,\mathcal{H}_3^{\mathrm{IIA}}(S) = q = 2P + 2 = \mathrm{rk}\,\mathcal{H}_2^{\mathrm{IIB}}(S^\circ) + \mathrm{rk}\,\mathcal{H}_4^{\mathrm{IIB}}(S^\circ) + 2,$$

and, by axioms (CCB3), (CCA2) and (CCA1),

$$\mathrm{rk}\,\mathcal{H}_3^{\mathrm{IIB}}(S^\circ) = Q + 2(N-M) = 2(p+m) + 2 = \mathrm{rk}\,\mathcal{H}_2^{\mathrm{IIA}}(S) + \mathrm{rk}\,\mathcal{H}_4^{\mathrm{IIA}}(S) + 2.$$

$\square$

Corollary 3.14. *Intersection homology and the homology of intersection spaces are a mirror-pair on any mirror-closed class of possibly singular Calabi–Yau threefolds all of whose singular members arise in the course of a conifold transition.*

Proof. We have observed above that intersection homology is IIA conifold cali-brated on such a class of spaces. By Proposition 3.6, the homology of intersection spaces is IIB conifold calibrated on such a class. The statement follows by applying Proposition 3.13.   □

In [Hüb97], T. Hübsch asks for a homology theory SH_* ("stringy homology") on threefolds with only isolated singularities such that

(SH1) SH_* satisfies Poincaré duality;
for singular S:
(SH2) $SH_r(S) \cong H_r(S - \Sigma)$ for $r < 3$,
(SH3) $SH_3(S)$ is an extension of $H_3(S)$ by $\ker(H_3(S - \Sigma) \to H_3(S))$,
(SH4) $SH_r(S) \cong H_r(S)$ for $r > 3$; and
(SH5) SH_* agrees with ordinary homology on nonsingular S.

(In fact, one may of course ask this more generally for n-folds.) Such a theory would record both the type IIA *and* the type IIB massless D-branes simultaneously. Intersection homology satisfies all of these axioms with the exception of axiom (SH3), and is thus not a solution. Regarding (SH3), Hübsch notes further that "the precise nature of this extension is to be determined from the as yet unspecified gen-eral cohomology theory." Using the homology of intersection spaces, $\widetilde{H}_*(IS)$, we have now provided an answer: By Theorem 3.9, the group $H_3(IS)$ satisfies axiom (SH3) for any threefold S with isolated singularities and simply connected links. The precise nature of the extension is given in the proof of that theorem. However, setting $SH_*(S) = \widetilde{H}_*(IS)$ does not satisfy axiom (SH2) (and thus, by Poincaré dual-ity, does not satisfy (SH4)), although is does satisfy (SH1), (SH3) and (SH5). The mirror-pair $(IH_*(S), H_*(IS))$ does contain all the information that a putative theory $SH_*(S)$ satisfying (SH1)–(SH5) would contain and so may be regarded as a solution to Hübsch' problem. In fact, one could set

$$SH_r(S) = \begin{cases} IH_r(S), & r \neq 3, \\ H_r(IS), & r = 3. \end{cases}$$

This SH_* then satisfies all axioms (SH1)–(SH5).

Since the intersection space IS has been constructed not just for singular spaces S with only isolated singularities, but for more general situations with nonisolated singular strata as well (see Section 2.9), one thus obtains an extension of the sought theories to these nonisolated scenarios.

An Ansatz for constructing SH_*, using the description of perverse sheaves due to MacPherson-Vilonen [MV86], has been given by A. Rahman in [Rah07] for iso-lated singularities.

3.9 An Example

Let us return to the quintic X_ε in $\mathbb{P}^4$ from the introduction (Section 3.1), and consider the conifold transition

$$X_\varepsilon \rightsquigarrow S \rightsquigarrow Y.$$

The conifold transition for this well-known quintic is described in [Hüb92], see also [Pol00]. We have seen that S has $n = 125$ nodes. Any smooth quintic hypersurface in $\mathbb{P}^4$ (is Calabi–Yau and) has Hodge numbers $b_{1,1} = 1$ and $b_{2,1} = 101$. Thus for $\varepsilon \neq 0$,

$$p = b_2(X_\varepsilon) = b_{1,1}(X_\varepsilon) = 1,$$

$$q + 2(n - m) = b_3(X_\varepsilon) = 2(1 + b_{2,1}) = 204.$$

By [Sch86], $b_{1,1}(Y) = 25$ for the small resolution Y. Hence

$$p + m = b_2(Y) = b_{1,1}(Y) = 25,$$

and so $m = 24$. From

$$204 = q + 2(n - m) = q + 2(125 - 24) = q + 202$$

we see that $q = 2$. So in this example the third homology of the intersection space IS sees

$$\operatorname{rk} H_3(IS) = q + 2(n - m) = 204$$

independent cycles, of which 202 remain invisible to intersection homology because the latter sees only

$$\operatorname{rk} IH_3(S) = q = 2$$

independent cycles. On the other hand, the second and fourth intersection homology of S sees

$$\operatorname{rk} IH_2(S) + \operatorname{rk} IH_4(S) = 50$$

independent cycles, of which 48 remain invisible to the homology of the intersection space because the latter sees only

$$\operatorname{rk} H_2(IS) + \operatorname{rk} H_4(IS) = 2$$

independent cycles. The above table for this example is:

Type	dim	X_ε	S	Y
	2	1	1	25
Elem. Massless	3	204	103	2
	4	1	25	25
D-Branes	2		24 (massless)	24 (IIA 2-Branes, massive)
	3	101 (IIB 3-Branes, massive)	101 (massless)	
Total Massless IIA	2	1	**25**	25
	3	204	**103**	2
	4	1	**25**	25
Total Massless IIB	2	1	**1**	25
	3	204	**204**	2
	4	1	**25**	25
$\mathrm{rk}\,H_*$	2	1	1	25
	3	204	103	2
	4	1	25	25
$\mathrm{rk}\,IH_*(S)$	2		25	
	3		2	
	4		25	
$\mathrm{rk}\,H_*(IS)$	2		1	
	3		204	
	4		1	

References

[Are46] R. Arens, *Topologies for homeomorphism groups*, Amer. J. Math. **68** (1946), 593 – 610.

[B⁺84] A. Borel et al., *Intersection cohomology*, Progr. Math., no. 50, Birkhäuser Verlag, Boston, 1984.

[Ban06a] M. Banagl, *Computing twisted signatures and L-classes of non-Witt spaces*, Proc. London Math. Soc. (3) **92** (2006), 428 – 470.

[Ban06b] _____, *The L-class of non-Witt spaces*, Annals of Math. **163** (2006), no. 3, 743 – 766.

[Ban07] _____, *Topological invariants of stratified spaces*, Springer Monographs in Mathematics, Springer-Verlag Berlin Heidelberg, 2007.

[Ban09] _____, *Singular spaces and generalized Poincaré complexes*, Electron. Res. Announc. Math. Sci. **16** (2009), 63 – 73.

[Bau88] H. J. Baues, *Algebraic homotopy*, Cambridge studies in advanced mathematics, vol. 15, Cambridge University Press, 1988.

[BCS03] M. Banagl, S. E. Cappell, and J. L. Shaneson, *Computing twisted signatures and L-classes of stratified spaces*, Math. Ann. **326** (2003), no. 3, 589 – 623.

[BG01] C. P. Boyer and K. Galicki, *New Einstein metrics in dimension five*, J. Diff. Geom. **57** (2001), 443 – 463.

[BGG73] I. N. Bernstein, I. M. Gelfand, and S. I. Gelfand, *Schubert cells and cohomology of the spaces G/P*, Russian Math. Surveys **28** (1973), no. 3, 1 – 26.

[BGN03] C. P. Boyer, K. Galicki, and M. Nakamaye, *On the geometry of Sasakian-Einstein 5-manifolds*, Math. Ann. **325** (2003), no. 3, 485 – 524.

[BJCJ59] E. H. Brown Jr. and A. H. Copeland Jr., *An homology analogue of Postnikov systems*, Michigan Math. J. **6** (1959), 313 – 330.

[Bre93] G. E. Bredon, *Topology and geometry*, Grad. Texts in Math., no. 139, Springer Verlag, 1993.

[Che79] J. Cheeger, *On the spectral geometry of spaces with cone-like singularities*, Proc. Natl. Acad. Sci. USA **76** (1979), 2103 – 2106.

[Che80] _____, *On the Hodge theory of Riemannian pseudomanifolds*, Proc. Sympos. Pure Math. **36** (1980), 91 – 146.

[Che83] _____, *Spectral geometry of singular Riemannian spaces*, J. Differential Geom. **18** (1983), 575 – 657.

[CK99] D. A. Cox and S. Katz, *Mirror symmetry and algebraic geometry*, Math. Surveys and Monographs, vol. 68, Amer. Math. Soc., 1999.

[Coo78] G. Cooke, *Replacing homotopy actions by topological actions*, Trans. Amer. Math. Soc. **237** (1978), 391 – 406.

[CW91] S. E. Cappell and S. Weinberger, *Classification de certaines espaces stratifiés*, C. R. Acad. Sci. Paris Sér. I Math. **313** (1991), 399 – 401.

[Dan78] V. I. Danilov, *The geometry of toric varieties*, Russian Math. Surveys **33** (1978), no. 2, 97 – 154.

[DT58] A. Dold and R. Thom, *Quasifaserungen und unendliche symmetrische Produkte*, Annals of Math. **67** (1958), no. 2, 239 – 281.

[Fad60] E. Fadell, *The equivalence of fiber spaces and bundles*, Bull. Amer. Math. Soc. **66** (1960), 50 – 53.

[FHT01] Y. Félix, S. Halperin, and J.-C. Thomas, *Rational homotopy theory*, Grad. Texts in Math., no. 205, Springer Verlag New York, 2001.

[FQ90] M. H. Freedman and F. Quinn, *Topology of 4-manifolds*, Princeton Mathematical Series, no. 39, Princeton Univ. Press, Princeton, New Jersey, 1990.

[Fuc71] M. Fuchs, *A modified Dold-Lashof construction that does classify H-principal fibrations*, Math. Ann. **192** (1971), 328 – 340.

[GH88] P. S. Green and T. Hübsch, *Possible phase transitions among Calabi-Yau compactifications*, Phys. Rev. Lett. **61** (1988), 1163 – 1166.

[GH89] _____, *Connecting moduli spaces of Calabi-Yau threefolds*, Commun. Math. Phys. **119** (1989), 431 – 441.

[GM80] M. Goresky and R. D. MacPherson, *Intersection homology theory*, Topology **19** (1980), 135 – 162.

[GM83] _____, *Intersection homology II*, Invent. Math. **71** (1983), 77 – 129.

[GM88] _____, *Stratified Morse theory*, Ergebnisse der Mathematik und ihrer Grenzgebiete, 3. Folge, no. 14, Springer Verlag, 1988.

[Gor84] M. Goresky, *Intersection homology operations*, Comm. Math. Helv. **59** (1984), 485 – 505.

[GR02] V. Gasharov and V. Reiner, *Cohomology of smooth Schubert varieties in partial flag manifolds*, J. London Math. Soc. (2) **66** (2002), 550 – 562.

[GSW87] M. B. Green, J. H. Schwarz, and E. Witten, *Superstring theory Vol. 2 Loop amplitudes, anomalies and phenomenology*, Cambridge Monographs on Mathematical Physics, Cambridge University Press, 1987.

[Hat02] A. Hatcher, *Algebraic topology*, Cambridge Univ. Press, 2002.

[Hil53] P. Hilton, *An introduction to homotopy theory*, Cambridge Tracts in Mathematics and Physics, no. 43, Cambridge University Press, 1953.

[Hil65] _____, *Homotopy theory and duality*, Notes on Mathematics and its Applications, Gordon and Breach Science Publishers, 1965.

[Hüb92] T. Hübsch, *Calabi-Yau manifolds: A bestiary for physicists*, World Scientific, 1992.

[Hüb97] _____, *On a stringy singular cohomology*, Mod. Phys. Lett. A **12** (1997), 521 – 533.

[Ill78] S. Illman, *Smooth equivariant triangulations of G-manifolds for G a finite group*, Math. Ann. **233** (1978), 199 – 220.

[KS77] R. C. Kirby and L. C. Siebenmann, *Foundational essays on topological manifolds, smoothings, and triangulations*, Ann. of Math. Studies, vol. 88, Princeton Univ. Press, Princeton, 1977.

[KW06] F. Kirwan and J. Woolf, *An introduction to intersection homology theory*, second ed., Chapman & Hall/CRC, 2006.

[MH73] J. Milnor and D. Husemoller, *Symmetric bilinear forms*, Springer, 1973.

[Mil58] J. Milnor, *On the existence of a connection with curvature zero*, Comment. Math. Helv. **32** (1958), 215 – 223.

[Mil68] _____, *Singular points of complex hypersurfaces*, Ann. of Math. Studies, vol. 61, Princeton Univ. Press, Princeton, 1968.

[Moi52] E. Moise, *Affine structures in 3-manifolds. V. The triangulation theorem and Hauptvermutung*, Ann. Math. (2) **56** (1952), 96 – 114.

[Moo] J. C. Moore, *Le théorème de Freudenthal, la suite exacte de James et l'invariant de Hopf généralisé*, Séminaire H. Cartan (1954–55).

[Mor99] D. Morrison, *Through the looking glass*, Mirror Symmetry III (D. H. Phong, L. Vinet, and S.-T. Yau, eds.), AMS/IP Studies in Advanced Mathematics, vol. 10, American Mathematical Society and International Press, 1999, pp. 263 – 277.

[Mun00] J. R. Munkres, *Topology*, second ed., Prentice Hall, 2000.

[MV86] R. D. MacPherson and K. Vilonen, *Elementary construction of perverse sheaves*, Invent. Math. **84** (1986), 403 – 435.

[Pfl01] M. J. Pflaum, *Analytic and geometric study of stratified spaces*, Lecture Notes in Math., no. 1768, Springer Verlag, 2001.

[Pol00] J. Polchinski, *String theory Vol. II Superstring theory and beyond*, Cambridge Monographs on Mathematical Physics, Cambridge University Press, 2000.

[Rah07] A. Rahman, *A perverse sheaf approach toward a cohomology theory for string theory*, math.AT/0704.3298, 2007.

[Ran92] A. A. Ranicki, *Algebraic L-theory and topological manifolds*, Cambridge Tracts in Math., no. 102, Cambridge University Press, 1992.

[Sch86] C. Schoen, *On the geometry of a special determinantal hypersurface associated to the Mumford-Horrocks vector bundle*, J. für Math. **364** (1986), 85 – 111.

[Sch03] J. Schürmann, *Topology of singular spaces and constructible sheaves*, Monografie Matematyczne, vol. 63, Birkhäuser Verlag, 2003.

[Sma62] S. Smale, *On the structure of 5-manifolds*, Annals of Math. **75** (1962), no. 1, 38 – 46.

[Spa66] E. H. Spanier, *Algebraic topology*, Springer Verlag, 1966.

[Sta63] J. Stasheff, *A classification theorem for fibre spaces*, Topology **2** (1963), 239 – 246.

[Str95] A. Strominger, *Massless black holes and conifolds in string theory*, Nucl. Phys. B **451** (1995), 96 – 108.

[Tia87] G. Tian, *Smoothness of the universal deformation space of compact Calabi-Yau manifolds and its Petersson Weil metric*, Mathematical Aspects of String Theory (S.-T. Yau, ed.), World Scientific, Singapore, 1987, pp. 629 – 646.

[Wei94] S. Weinberger, *The topological classification of stratified spaces*, Chicago Lectures in Math., Univ. of Chicago Press, Chicago, 1994.

[Whi36] H. Whitney, *Differentiable manifolds*, Ann. of Math. **37** (1936), 647 – 680.

[Whi44] _____, *The self-intersections of a smooth n-manifold in 2n-space*, Ann. of Math. **45** (1944), 220 – 246.

[Whi49] J. H. C. Whitehead, *Combinatorial homotopy I*, Bull. Amer. Math. Soc. **55** (1949), no. 3, 213 – 245.

[Whi78] G. W. Whitehead, *Elements of homotopy theory*, Graduate Texts in Math., no. 61, Springer Verlag, 1978.

[Zab76] Alexander Zabrodsky, *Hopf spaces*, North-Holland Mathematics Studies, no. 22, North Holland, 1976.

Index

Lecture Notes in Mathematics

For information about earlier volumes
please contact your bookseller or Springer
LNM Online archive: springerlink.com

Vol. 1851: O. Catoni, Statistical Learning Theory and Stochastic Optimization (2004)

Vol. 1852: A.S. Kechris, B.D. Miller, Topics in Orbit Equivalence (2004)

Vol. 1853: Ch. Favre, M. Jonsson, The Valuative Tree (2004)

Vol. 1854: O. Saeki, Topology of Singular Fibers of Differential Maps (2004)

Vol. 1855: G. Da Prato, P.C. Kunstmann, I. Lasiecka, A. Lunardi, R. Schnaubelt, L. Weis, Functional Analytic Methods for Evolution Equations. Editors: M. Iannelli, R. Nagel, S. Piazzera (2004)

Vol. 1856: K. Back, T.R. Bielecki, C. Hipp, S. Peng, W. Schachermayer, Stochastic Methods in Finance, Bressanone/Brixen, Italy, 2003. Editors: M. Fritelli, W. Runggaldier (2004)

Vol. 1857: M. Émery, M. Ledoux, M. Yor (Eds.), Séminaire de Probabilités XXXVIII (2005)

Vol. 1858: A.S. Cherny, H.-J. Engelbert, Singular Stochastic Differential Equations (2005)

Vol. 1859: E. Letellier, Fourier Transforms of Invariant Functions on Finite Reductive Lie Algebras (2005)

Vol. 1860: A. Borisyuk, G.B. Ermentrout, A. Friedman, D. Terman, Tutorials in Mathematical Biosciences I. Mathematical Neurosciences (2005)

Vol. 1861: G. Benettin, J. Henrard, S. Kuksin, Hamiltonian Dynamics – Theory and Applications, Cetraro, Italy, 1999. Editor: A. Giorgilli (2005)

Vol. 1862: B. Helffer, F. Nier, Hypoelliptic Estimates and Spectral Theory for Fokker-Planck Operators and Witten Laplacians (2005)

Vol. 1863: H. Führ, Abstract Harmonic Analysis of Continuous Wavelet Transforms (2005)

Vol. 1864: K. Efstathiou, Metamorphoses of Hamiltonian Systems with Symmetries (2005)

Vol. 1865: D. Applebaum, B.V. R. Bhat, J. Kustermans, J. M. Lindsay, Quantum Independent Increment Processes I. From Classical Probability to Quantum Stochastic Calculus. Editors: M. Schürmann, U. Franz (2005)

Vol. 1866: O.E. Barndorff-Nielsen, U. Franz, R. Gohm, B. Kümmerer, S. Thorbjønsen, Quantum Independent Increment Processes II. Structure of Quantum Lévy Processes, Classical Probability, and Physics. Editors: M. Schürmann, U. Franz, (2005)

Vol. 1867: J. Sneyd (Ed.), Tutorials in Mathematical Biosciences II. Mathematical Modeling of Calcium Dynamics and Signal Transduction. (2005)

Vol. 1868: J. Jorgenson, S. Lang, $Pos_n(R)$ and Eisenstein Series. (2005)

Vol. 1869: A. Dembo, T. Funaki, Lectures on Probability Theory and Statistics. Ecole d'Eté de Probabilités de Saint-Flour XXXIII-2003. Editor: J. Picard (2005)

Vol. 1870: V.I. Gurariy, W. Lusky, Geometry of Mntz Spaces and Related Questions. (2005)

Vol. 1871: P. Constantin, G. Gallavotti, A.V. Kazhikhov, Y. Meyer, S. Ukai, Mathematical Foundation of Turbulent Viscous Flows, Martina Franca, Italy, 2003. Editors: M. Cannone, T. Miyakawa (2006)

Vol. 1872: A. Friedman (Ed.), Tutorials in Mathematical Biosciences III. Cell Cycle, Proliferation, and Cancer (2006)

Vol. 1873: R. Mansuy, M. Yor, Random Times and Enlargements of Filtrations in a Brownian Setting (2006)

Vol. 1874: M. Yor, M. Émery (Eds.), In Memoriam Paul-Andr Meyer - Sminaire de Probabilits XXXIX (2006)

Vol. 1875: J. Pitman, Combinatorial Stochastic Processes. Ecole d'Et de Probabilits de Saint-Flour XXXII-2002. Editor: J. Picard (2006)

Vol. 1876: H. Herrlich, Axiom of Choice (2006)

Vol. 1877: J. Steuding, Value Distributions of L-Functions (2007)

Vol. 1878: R. Cerf, The Wulff Crystal in Ising and Percolation Models, Ecole d'Et de Probabilités de Saint-Flour XXXIV-2004. Editor: Jean Picard (2006)

Vol. 1879: G. Slade, The Lace Expansion and its Applications, Ecole d'Et de Probabilits de Saint-Flour XXXIV-2004. Editor: Jean Picard (2006)

Vol. 1880: S. Attal, A. Joye, C.-A. Pillet, Open Quantum Systems I, The Hamiltonian Approach (2006)

Vol. 1881: S. Attal, A. Joye, C.-A. Pillet, Open Quantum Systems II, The Markovian Approach (2006)

Vol. 1882: S. Attal, A. Joye, C.-A. Pillet, Open Quantum Systems III, Recent Developments (2006)

Vol. 1883: W. Van Assche, F. Marcellàn (Eds.), Orthogonal Polynomials and Special Functions, Computation and Application (2006)

Vol. 1884: N. Hayashi, E.I. Kaikina, P.I. Naumkin, I.A. Shishmarev, Asymptotics for Dissipative Nonlinear Equations (2006)

Vol. 1885: A. Telcs, The Art of Random Walks (2006)

Vol. 1886: S. Takamura, Splitting Deformations of Degenerations of Complex Curves (2006)

Vol. 1887: K. Habermann, L. Habermann, Introduction to Symplectic Dirac Operators (2006)

Vol. 1888: J. van der Hoeven, Transseries and Real Differential Algebra (2006)

Vol. 1889: G. Osipenko, Dynamical Systems, Graphs, and Algorithms (2006)

Vol. 1890: M. Bunge, J. Funk, Singular Coverings of Toposes (2006)

Vol. 1891: J.B. Friedlander, D.R. Heath-Brown, H. Iwaniec, J. Kaczorowski, Analytic Number Theory, Cetraro, Italy, 2002. Editors: A. Perelli, C. Viola (2006)

Vol. 1892: A. Baddeley, I. Bárány, R. Schneider, W. Weil, Stochastic Geometry, Martina Franca, Italy, 2004. Editor: W. Weil (2007)

Vol. 1893: H. Hanßmann, Local and Semi-Local Bifurcations in Hamiltonian Dynamical Systems, Results and Examples (2007)

Vol. 1894: C.W. Groetsch, Stable Approximate Evaluation of Unbounded Operators (2007)

Vol. 1895: L. Molnár, Selected Preserver Problems on Algebraic Structures of Linear Operators and on Function Spaces (2007)

Vol. 1896: P. Massart, Concentration Inequalities and Model Selection, Ecole d'Été de Probabilités de Saint-Flour XXXIII-2003. Editor: J. Picard (2007)

Vol. 1897: R. Doney, Fluctuation Theory for Lévy Processes, Ecole d'Été de Probabilités de Saint-Flour XXXV-2005. Editor: J. Picard (2007)

Vol. 1898: H.R. Beyer, Beyond Partial Differential Equations, On linear and Quasi-Linear Abstract Hyperbolic Evolution Equations (2007)

Vol. 1899: Séminaire de Probabilités XL. Editors: C. Donati-Martin, M. Émery, A. Rouault, C. Stricker (2007)

Vol. 1900: E. Bolthausen, A. Bovier (Eds.), Spin Glasses (2007)

Vol. 1901: O. Wittenberg, Intersections de deux quadriques et pinceaux de courbes de genre 1, Intersections of Two Quadrics and Pencils of Curves of Genus 1 (2007)

Vol. 1951: A. Moltó, J. Orihuela, S. Troyanski, M. Valdivia, A Non Linear Transfer Technique for Renorming (2009)

Vol. 1952: R. Mikhailov, I.B.S. Passi, Lower Central and Dimension Series of Groups (2009)

Vol. 1953: K. Arwini, C.T.J. Dodson, Information Geometry (2008)

Vol. 1954: P. Biane, L. Bouten, F. Cipriani, N. Konno, N. Privault, Q. Xu, Quantum Potential Theory. Editors: U. Franz, M. Schuermann (2008)

Vol. 1955: M. Bernot, V. Caselles, J.-M. Morel, Optimal Transportation Networks (2008)

Vol. 1956: C.H. Chu, Matrix Convolution Operators on Groups (2008)

Vol. 1957: A. Guionnet, On Random Matrices: Macroscopic Asymptotics, Ecole d'Eté de Probabilits de Saint-Flour XXXVI-2006 (2009)

Vol. 1958: M.C. Olsson, Compactifying Moduli Spaces for Abelian Varieties (2008)

Vol. 1959: Y. Nakkajima, A. Shiho, Weight Filtrations on Log Crystalline Cohomologies of Families of Open Smooth Varieties (2008)

Vol. 1960: J. Lipman, M. Hashimoto, Foundations of Grothendieck Duality for Diagrams of Schemes (2009)

Vol. 1961: G. Buttazzo, A. Pratelli, S. Solimini, E. Stepanov, Optimal Urban Networks via Mass Transportation (2009)

Vol. 1962: R. Dalang, D. Khoshnevisan, C. Mueller, D. Nualart, Y. Xiao, A Minicourse on Stochastic Partial Differential Equations (2009)

Vol. 1963: W. Siegert, Local Lyapunov Exponents (2009)

Vol. 1964: W. Roth, Operator-valued Measures and Integrals for Cone-valued Functions and Integrals for Cone-valued Functions (2009)

Vol. 1965: C. Chidume, Geometric Properties of Banach Spaces and Nonlinear Iterations (2009)

Vol. 1966: D. Deng, Y. Han, Harmonic Analysis on Spaces of Homogeneous Type (2009)

Vol. 1967: B. Fresse, Modules over Operads and Functors (2009)

Vol. 1968: R. Weissauer, Endoscopy for GSP(4) and the Cohomology of Siegel Modular Threefolds (2009)

Vol. 1969: B. Roynette, M. Yor, Penalising Brownian Paths (2009)

Vol. 1970: M. Biskup, A. Bovier, F. den Hollander, D. Ioffe, F. Martinelli, K. Netočný, F. Toninelli, Methods of Contemporary Mathematical Statistical Physics. Editor: R. Kotecký (2009)

Vol. 1971: L. Saint-Raymond, Hydrodynamic Limits of the Boltzmann Equation (2009)

Vol. 1972: T. Mochizuki, Donaldson Type Invariants for Algebraic Surfaces (2009)

Vol. 1973: M.A. Berger, L.H. Kauffmann, B. Khesin, H.K. Moffatt, R.L. Ricca, De W. Sumners, Lectures on Topological Fluid Mechanics. Cetraro, Italy 2001. Editor: R.L. Ricca (2009)

Vol. 1974: F. den Hollander, Random Polymers: École d'Été de Probabilités de Saint-Flour XXXVII – 2007 (2009)

Vol. 1975: J.C. Rohde, Cyclic Coverings, Calabi-Yau Manifolds and Complex Multiplication (2009)

Vol. 1976: N. Ginoux, The Dirac Spectrum (2009)

Vol. 1977: M.J. Gursky, E. Lanconelli, A. Malchiodi, G. Tarantello, X.-J. Wang, P.C. Yang, Geometric Analysis and PDEs. Cetraro, Italy 2001. Editors: A. Ambrosetti, S.-Y.A. Chang, A. Malchiodi (2009)

Vol. 1978: M. Qian, J.-S. Xie, S. Zhu, Smooth Ergodic Theory for Endomorphisms (2009)

Vol. 1979: C. Donati-Martin, M. Émery, A. Rouault, C. Stricker (Eds.), Séminaire de Probablitiés XLII (2009)

Vol. 1980: P. Graczyk, A. Stos (Eds.), Potential Analysis of Stable Processes and its Extensions (2009)

Vol. 1981: M. Chlouveraki, Blocks and Families for Cyclotomic Hecke Algebras (2009)

Vol. 1982: N. Privault, Stochastic Analysis in Discrete and Continuous Settings. With Normal Martingales (2009)

Vol. 1983: H. Ammari (Ed.), Mathematical Modeling in Biomedical Imaging I. Electrical and Ultrasound Tomographies, Anomaly Detection, and Brain Imaging (2009)

Vol. 1984: V. Caselles, P. Monasse, Geometric Description of Images as Topographic Maps (2010)

Vol. 1985: T. Linß, Layer-Adapted Meshes for Reaction-Convection-Diffusion Problems (2010)

Vol. 1986: J.-P. Antoine, C. Trapani, Partial Inner Product Spaces. Theory and Applications (2009)

Vol. 1987: J.-P. Brasselet, J. Seade, T. Suwa, Vector Fields on Singular Varieties (2010)

Vol. 1988: M. Broué, Introduction to Complex Reflection Groups and Their Braid Groups (2010)

Vol. 1989: I.M. Bomze, V. Demyanov, Nonlinear Optimization. Cetraro, Italy 2007. Editors: G. di Pillo, F. Schoen (2010)

Vol. 1990: S. Bouc, Biset Functors for Finite Groups (2010)

Vol. 1991: F. Gazzola, H.-C. Grunau, G. Sweers, Polyharmonic Boundary Value Problems (2010)

Vol. 1992: A. Parmeggiani, Spectral Theory of Non-Commutative Harmonic Oscillators: An Introduction (2010)

Vol. 1993: P. Dodos, Banach Spaces and Descriptive Set Theory: Selected Topics (2010)

Vol. 1994: A. Baricz, Generalized Bessel Functions of the First Kind (2010)

Vol. 1995: A.Y. Khapalov, Controllability of Partial Differential Equations Governed by Multiplicative Controls (2010)

Vol. 1996: T. Lorenz, Mutational Analysis. A Joint Framework for Cauchy Problems in and Beyond Vector Spaces (2010)

Vol. 1997: M. Banagl, Intersection Spaces, Spatial Homology Truncation, and String Theory (2010)

Recent Reprints and New Editions

Vol. 1702: J. Ma, J. Yong, Forward-Backward Stochastic Differential Equations and their Applications. 1999 – Corr. 3rd printing (2007)

Vol. 830: J.A. Green, Polynomial Representations of GL_n, with an Appendix on Schensted Correspondence and Littelmann Paths by K. Erdmann, J.A. Green and M. Schoker 1980 – 2nd corr. and augmented edition (2007)

Vol. 1693: S. Simons, From Hahn-Banach to Monotonicity (Minimax and Monotonicity 1998) – 2nd exp. edition (2008)

Vol. 470: R.E. Bowen, Equilibrium States and the Ergodic Theory of Anosov Diffeomorphisms. With a preface by D. Ruelle. Edited by J.-R. Chazottes. 1975 – 2nd rev. edition (2008)

Vol. 523: S.A. Albeverio, R.J. Høegh-Krohn, S. Mazzucchi, Mathematical Theory of Feynman Path Integral. 1976 – 2nd corr. and enlarged edition (2008)

Vol. 1764: A. Cannas da Silva, Lectures on Symplectic Geometry 2001 – Corr. 2nd printing (2008)

LECTURE NOTES IN MATHEMATICS  Springer

Edited by J.-M. Morel, F. Takens, B. Teissier, P.K. Maini

Editorial Policy (for the publication of monographs)

1. Lecture Notes aim to report new developments in all areas of mathematics and their applications - quickly, informally and at a high level. Mathematical texts analysing new developments in modelling and numerical simulation are welcome.

   Monograph manuscripts should be reasonably self-contained and rounded off. Thus they may, and often will, present not only results of the author but also related work by other people. They may be based on specialised lecture courses. Furthermore, the manuscripts should provide sufficient motivation, examples and applications. This clearly distinguishes Lecture Notes from journal articles or technical reports which normally are very concise. Articles intended for a journal but too long to be accepted by most journals, usually do not have this "lecture notes" character. For similar reasons it is unusual for doctoral theses to be accepted for the Lecture Notes series, though habilitation theses may be appropriate.

2. Manuscripts should be submitted either to Springer's mathematics editorial in Heidelberg, or to one of the series editors. In general, manuscripts will be sent out to 2 external referees for evaluation. If a decision cannot yet be reached on the basis of the first 2 reports, further referees may be contacted: The author will be informed of this. A final decision to publish can be made only on the basis of the complete manuscript, however a refereeing process leading to a preliminary decision can be based on a pre-final or incomplete manuscript. The strict minimum amount of material that will be considered should include a detailed outline describing the planned contents of each chapter, a bibliography and several sample chapters.

   Authors should be aware that incomplete or insufficiently close to final manuscripts almost always result in longer refereeing times and nevertheless unclear referees' recommendations, making further refereeing of a final draft necessary.

   Authors should also be aware that parallel submission of their manuscript to another publisher while under consideration for LNM will in general lead to immediate rejection.

3. Manuscripts should in general be submitted in English. Final manuscripts should contain at least 100 pages of mathematical text and should always include

   – a table of contents;
   – an informative introduction, with adequate motivation and perhaps some historical remarks: it should be accessible to a reader not intimately familiar with the topic treated;
   – a subject index: as a rule this is genuinely helpful for the reader.

   For evaluation purposes, manuscripts may be submitted in print or electronic form, in the latter case preferably as pdf- or zipped ps-files. Lecture Notes volumes are, as a rule, printed digitally from the authors' files. To ensure best results, authors are asked to use the LaTeX2e style files available from Springer's web-server at:

   ftp://ftp.springer.de/pub/tex/latex/svmonot1/ (for monographs).

Additional technical instructions, if necessary, are available on request from: lnm@springer.com.

4. Careful preparation of the manuscripts will help keep production time short besides ensuring satisfactory appearance of the finished book in print and online. After acceptance of the manuscript authors will be asked to prepare the final LaTeX source files (and also the corresponding dvi-, pdf- or zipped ps-file) together with the final printout made from these files. The LaTeX source files are essential for producing the full-text online version of the book (see www.springerlink.com/content/110312 for the existing online volumes of LNM).

   The actual production of a Lecture Notes volume takes approximately 12 weeks.

5. Authors receive a total of 50 free copies of their volume, but no royalties. They are entitled to a discount of 33.3% on the price of Springer books purchased for their personal use, if ordering directly from Springer.

6. Commitment to publish is made by letter of intent rather than by signing a formal contract. Springer-Verlag secures the copyright for each volume. Authors are free to reuse material contained in their LNM volumes in later publications: a brief written (or e-mail) request for formal permission is sufficient.

Addresses:
Professor J.-M. Morel, CMLA,
École Normale Supérieure de Cachan,
61 Avenue du Président Wilson, 94235 Cachan Cedex, France
E-mail: Jean-Michel.Morel@cmla.ens-cachan.fr

Professor F. Takens, Mathematisch Instituut,
Rijksuniversiteit Groningen, Postbus 800,
9700 AV Groningen, The Netherlands
E-mail: F.Takens@math.rug.nl

Professor B. Teissier, Institut Mathématique de Jussieu,
UMR 7586 du CNRS, Équipe "Géométrie et Dynamique",
175 rue du Chevaleret
75013 Paris, France
E-mail: teissier@math.jussieu.fr

For the "Mathematical Biosciences Subseries" of LNM:

Professor P.K. Maini, Center for Mathematical Biology,
Mathematical Institute, 24-29 St Giles,
Oxford OX1 3LP, UK
E-mail: maini@maths.ox.ac.uk

Springer, Mathematics Editorial I, Tiergartenstr. 17
69121 Heidelberg, Germany,
Tel.: +49 (6221) 487-8259
Fax: +49 (6221) 4876-8259
E-mail: lnm@springer.com